AF352766

Remote Sensing in Vessel Detection and Navigation

Remote Sensing in Vessel Detection and Navigation

Editors

Henning Heiselberg
Andrzej Stateczny

MDPI • Basel • Beijing • Wuhan • Barcelona • Belgrade • Manchester • Tokyo • Cluj • Tianjin

Editors
Henning Heiselberg
Technical University of Denmark
Denmark

Andrzej Stateczny
Gdansk Technical University
Poland

Editorial Office
MDPI
St. Alban-Anlage 66
4052 Basel, Switzerland

This is a reprint of articles from the Special Issue published online in the open access journal *Sensors* (ISSN 1424-8220) (available at: https://www.mdpi.com/journal/sensors/special_issues/RS_VDN).

For citation purposes, cite each article independently as indicated on the article page online and as indicated below:

LastName, A.A.; LastName, B.B.; LastName, C.C. Article Title. *Journal Name* **Year**, *Article Number*, Page Range.

ISBN 978-3-03943-609-5 (Hbk)
ISBN 978-3-03943-610-1 (PDF)

Contents

About the Editors

Henning Heiselberg holds a PhD in Physics (1987) and a BSc in Mathematics from the University of Aarhus, Denmark. Postdoc at the Niels Bohr Institute and the University of Illinois at Urbana-Champaign. Staff Scientist at Lawrence Berkeley Lab in California. Nordic associate prof. at the Nordic Institute of Theoretical Physics. Adjoint prof. at the University of South Denmark. Senior Scientist at Danish Defense Research for 15 years, contributing toward the MoD's New Fighter Program. Presently, head of the Center for Security at DTU (Technical University Denmark) Space. Heiselberg's research interests cover number of topics within physics, from nuclear astrophysics to quantum manybody systems, and sensor technology in general, remote sensing and Earth observation, and Arctic surveillance (see 120 publications in Orchid). The Center for Security DTU functions as a contact point between all departments at the Technical University Denmark (DTU) and the defense and security industry and public authorities. The center facilitates the research and development of technology at DTU as well as consultancy, education, and collaborations with industry and authorities.

Andrzej Stateczny is a Professor at Gdansk Technical University Poland and President of Marine Technology Ltd. His research interests are mainly centered on navigation, hydrography and geoinformatics. His current research activities include radar navigation, comparative navigation, hydrography, artificial intelligence methods focused on image processing and multisensory data fusion. He has been the Principal Investigator or Co-Investigator on a wide range of research projects in both civil and defense fields. He has published or presented over 200 journal and conference papers in the above areas, as well as several books, such as "Radar Navigation", "Comparative Navigation", "Methods of Comparative Navigation" and "Artificial Neural Networks for Marine Target Recognition". He was the head of numerous research projects. He was a promoter in 16 completed doctoral theses.

Editorial

Remote Sensing in Vessel Detection and Navigation

Henning Heiselberg [1,*] and Andrzej Stateczny [2]

[1] National Space Institute, Technical University of Denmark, 2800 Kgs. Lyngby, Denmark
[2] Department of Geodesy, Gdansk University of Technology, 80-233 Gdańsk, Poland;
     andrzej.stateczny@pg.edu.pl
*   Correspondence: hh@dtu.dk

Received: 10 October 2020; Accepted: 12 October 2020; Published: 15 October 2020

Abstract: The Special Issue (SI) "Remote Sensing in Vessel Detection and Navigation" highlighted a variety of topics related to remote sensing with navigational sensors. The sequence of articles included in this Special Issue is in line with the latest scientific trends. The latest developments in science, including artificial intelligence, were used. The 15 papers (from 23 submitted) were published.

Keywords: vessel detection; navigation; remote sensing

---

1. Introduction

Earth observation by multispectral, SAR (Synthetic Aperture Radar), and other sensors provides unique global as well as detailed local surveillance. Resolutions allow for vessel detection, classification, and discrimination from, e.g., icebergs and other objects. Important applications include vessel detection and navigation; trafficking and safety; and monitoring the oceans for fishing, oil spills, territorial violations, piracy, refugee boats, emergencies, AIS (Automatic Identification System) spoofing, etc. With global warming, the north-east and -west passages have opened up for shipping, fishing, and cruise ships in uncharted reef-infested territories littered with sea-ice and titanic icebergs.

The Special Issue entitled "Remote Sensing in Vessel Detection and Navigation", was focused on many aspects of multispectral, multi-sensor, SAR, and other sensors related to science/research, algorithm/technical development, analysis tools, synergy with sensors in multiple wavelengths of the e.m. spectrum, synergy with other measurements such as AIS, as well as reviews of the state-of-the-art in-ocean processes using multispectral and SAR imagery for oceans and sea ice, and vessel monitoring for surveillance, trafficking, and navigation. Topics for the Special Issue included the following:

- Vessel detection, classification, and identification;
- Sea-ice and iceberg detection and tracking;
- Multi-sensor data fusion;
- Autonomous ships navigation;
- Comparative (terrain reference) navigation;
- Change detection for classifying islands, reefs, and other static objects;
- Synergy with and comparison to AIS and other vessel identification data;
- Synergies between satellite sensors with airborne platforms; multiple satellite SAR; optical and thermal infrared sensors including finer resolution sensors, for example, sentinels and other satellites, and in situ measurements;
- The use of multispectral, multiple frequencies, and polarizations to interpret and quantitatively assess various ocean surfaces, currents, and sea ice phenomena for navigation;
- Interferometric and Doppler-derived SAR oceanic and sea ice applications focused on surface motion;

- Validation studies for vessel, ocean, and sea-ice parameters based on in situ and airborne data collections;
- Use of machine learning and the build-up of annotated training databases;
- Artificial Intelligence for image data processing.

In this article we provide a brief overview of the published papers, in particular the use of advanced modern technologies and data fusion techniques. These two areas seem to be the right direction for the future development of vessel detection and navigation.

2. Overview of Contributions

2.1. Convolutional Neural Networks for Detection and Classification of Vessels

Convolutional Neural Networks (CNN) are state of the art machine learning algorithms for detection and classification of objects in images. This Special Issue includes several works on CNN applied SAR images of vessels.

Dai et al. [1] propose a novel SAR ship detector, based on three subnetworks, which is designed especially for small ships in a noisy background. Using the public SAR ship dataset (SSDD) and China Gaofen-3 satellite SAR images, their method shows excellent performance for detecting the multiscale and small-size ships with respect to some competitive models and exhibits high potential in practical application.

Fu et al. [2] investigate very deep neural networks and the problem of saturation that limits the recognition accuracy. They incorporate micro convolutional modules, residual structures and other improvements. Testing their model on a simulated dataset of HRRP signals obtained from thirteen 3D CAD object models, their model is capable of achieving higher recognition accuracy and robustness than other common network structures.

Pan et al. [3] propose a new method for dealing with the arbitrary rotations of ships in noisy SAR images by removing redundancy and improving detection. A multi-stage rotational region-based network (MSR2N) is proposed which contains three modules and several other characteristics. MSR2N is more suitable and robust for ship detection in SAR images. The experimental results on their SAR ship detection dataset (SSDD) show that the MSR2N has achieved state-of-the-art performance.

2.2. Vessel Detections by Radar

Radars are historically the most common method for navigation and vessel detection. There are, however, certain limitations which are addressed by two papers in this Special Issue.

Jiang et al. [4] address the problem of R-mode AIS signals where a ship may only receive signals from one shore station. Lacking range information it cannot determine its position. They describe a novel method using several ship antennas which then can determine its position by triangulation and discuss the precision of this method. Their proposed method expands the application scope of the AIS R-mode positioning system widely.

Lisowski [5] address the problem of inaccurate and limited radar data for autonomous ships during maneuvering. They analyze the use of game theory for decision processes in order to predict the safest trajectories, and do simulation studies of a positional game algorithm. A real situation was studied as an example, where a vessel had to sail among nine other ships. The sensitivity characteristics of safe trajectories under conditions of both good and limited visibility at sea were presented.

2.3. Vessel Detections by Video Sensors

Video surveillance is becoming increasingly popular in the process of detecting ships at short distances. It is particularly useful in lintel systems both on sea and inland waterways. There are also articles devoted to this issue in the SI.

Wawrzyniak et al. [6] present a method that allows for the detection and tracking of ships using the video streams of existing monitoring systems for ports and rivers. The article presents the way and results of experiments conducted on selected groups of data using selected cameras with different working parameters. Experiments were carried out in different locations, variable lighting and weather conditions. The detection targets were variable vessels. The obtained results confirmed the correctness of the adopted solution, although minor problems were encountered in difficult weather conditions.

Polap and Wlodarczyk-Sielicka [7] address the problem of vessel classification, not by classifying whole input data, but smaller parts of the data. A duplicate way is to divide the image of the ship into smaller parts and classify them into vectors that can be identified as features by means of a convolutional neural network (CNN). The concept presented in the article reflects the title mechanism of the word sack, where the created feature vectors can be called words, and with their help the solution can assign images a specific class. The results of two tests are presented, in which, first, two classes were analyzed, and second, the authors used much larger sets of images belonging to five types of ships. The method described in the paper allowed to obtain better results by about 5% every year, which is good for future research. The given solution may be an alternative approach to the classification of unconventional ships.

2.4. Object Detections by Different Sensors

A number of articles are devoted to the processing of data from various other sensors in the process of detection and eventual identification of floating objects.

Ru et al. [8] present a detection and tracking algorithm based on the Gibbs Generalized Labeled Multi-Bernoulli (GLMB) filter for estimating recognizable target groups. Once the state of a group target has been estimated, we extract relevant information from the estimation data to assess whether the structure or state of concentration of the target group has changed. Finally, they conduct several experiments to verify the algorithm.

Shan et al. [9] address the problem of accurate detection of the coastline and nearby ships with the available sensors due to the complex environment and six degrees of freedom of movement. The article combines data from cameras and inertial sensors and proposes an innovative algorithm for detecting sea targets based on camera position in relation to movement. The developed algorithm consists of semiconductor lighting estimation, semiconductor lighting detection and target weakness detection. Experimental results have been compared to the results of the experiments indicating that the proposed algorithm achieves a higher accuracy of detection near ships and may be of significant importance in the development of unmanned ships.

Willburger et al. [10] demonstrate of the project AMARO (Autonomous Real-Time Detection of Moving Maritime Objects) running at DLR. AMARO is a feasibility study of an on-board ship detection system involving on-board processing and real-time communication. The article describes the scope, purpose and design of the AMARO system, as well as detailed results of the flight experiment. The operation of the prototype system was successfully demonstrated on an on-board platform in spring 2018. Ground users could be informed about the detected ships within minutes after their detection without a direct communication link.

Dong et al. [11] address the problem of passive sonar detection and identification in low signal to noise rate conditions. They proposed adaptive intrawell matched stochastic resonance Adaptive Linear Amplifiers (AIMSR) method, aiming to break through the limitation of the conventional adaptive line enhancers (ALE) method by nonlinear filtering effects. The verification of the application was carried out experimentally in a tank with an autonomous submarine vehicle (AUV) in order to confirm the feasibility and effectiveness of the proposed method. The results indicate that the proposed method is superior to the conventional adaptive line enhancers (ALE) method.

2.5. Multispectral Techniques for Remote Sensing

The optical spectrum was also investigated in two papers, where special spectrum sensor features could be exploited.

Kim et al. [12] propose a new method for ground-based ship detection exploiting the strong CO_2 absorption in the mid-infrared band. Since ship emissions are hot, their CO_2 emission spectrum is broader than the subsequent absorption band in the colder atmosphere. As a result, a double peak remains in the IR spectrum that allow for robust remote detection of ships in the mid-IR band.

Heiselberg [13] demonstrated for the first time that the time delay between recording the 13 images for the Sentinel-2 multispectral imager can be exploited to determine ship and aircraft velocities from space. The movement in the image during recording is substantial for aircrafts but only just measurable for the much slower ships. The velocities (and altitudes) can be accurately determined for aircrafts and reasonably for ships without wakes. When strong wakes are visible, they distort the spectrum spatially, which can partially be corrected for.

2.6. Autonomous Ships' Navigation

The next two articles concern unmanned surface vehicles for hydrographic tasks, which have been developing rapidly in recent years pushing out more and more manned ships.

Specht et al. [14] address the problem of steering precision on low-cost Unmanned Surface Vessel (USV) during automatic survey. One of the challenges of modern hydrography are measurements in shallow waters less than 1 m deep. A potential solution seems to be the use of USV of low draught as an alternative method for such bathymetric measurements. The article presents the modernization of USV allowing for the realization of automatic mode in low-budget USV. It presents an evaluation of the usefulness of the popular autopilot and cooperating with its cheap satellite navigation system (GNSS) receiver in the automatic process of bathymetric measurements. The results obtained in terms of distance from the planned trajectory were evaluated statistically.

Stateczny et al. [15] address the issue of precise navigation of the hydrographic unit on the measuring profile with the use of a nonlinear adaptive autopilot. Precise navigation in relation to the measurement profile avoids registration of unnecessary data and saves time and costs of hydrographic measurements both during the registration process and during data processing. This article discusses the results of experiments concerning hydrographic measurements performed in real conditions with a new generation HydroDron vehicle.

3. Conclusions

The Special Issue entitled "Remote Sensing in Vessel Detection and Navigation" comprised 15 articles on many topics related to remote sensing with navigational sensors. In this paper, we have presented short introductions of the published articles.

It can be said that navigation and vessel detection still remain important and hot topics, and a lot of work will continue to be done worldwide. New techniques and methods for analyzing and extracting information from navigational sensors and data have been proposed and verified. Some of these will provoke further research, and some are already mature and can be considered for industrial implementation and development.

Author Contributions: A.S. wrote the first draft, H.H. revised and rewrite article, H.H. and A.S. read and corrected the final version. All authors have read and agreed to the published version of the manuscript.

Funding: This research was not funded by external funding.

Acknowledgments: We would like to thank all the authors who contributed to the Special Issue and the staff in the editorial office.

References

1. Dai, W.; Mao, Y.; Yuan, R.; Liu, Y.; Pu, X.; Li, C. A Novel Detector Based on Convolution Neural Networks for Multiscale SAR Ship Detection in Complex Background. *Sensors* **2020**, *20*, 2547. [CrossRef] [PubMed]

2. Fu, Z.; Li, S.; Li, X.; Dan, B.; Wang, X. A Neural Network with Convolutional Module and Residual Structure for Radar Target Recognition Based on High-Resolution Range Profile. *Sensors* **2020**, *20*, 586. [CrossRef] [PubMed]

3. Pan, Z.; Yang, R.; Zhang, Z. MSR2N: Multi-Stage Rotational Region Based Network for Arbitrary-Oriented Ship Detection in SAR Images. *Sensors* **2020**, *20*, 2340. [CrossRef] [PubMed]

4. Jiang, Y.; Zheng, K. The Single-Shore-Station-Based Position Estimation Method of an Automatic Identification System. *Sensors* **2020**, *20*, 1590. [CrossRef] [PubMed]

5. Lisowski, J. Sensitivity of Safe Trajectory in a Game Environment on Inaccuracy of Radar Data in Autonomous Navigation. *Sensors* **2019**, *19*, 1816. [CrossRef] [PubMed]

6. Wawrzyniak, N.; Hyla, T.; Popik, A. Vessel Detection and Tracking Method Based on Video Surveillance. *Sensors* **2019**, *19*, 5230. [CrossRef] [PubMed]

7. Polap, D.; Wlodarczyk-Sielicka, M. Classification of Non-Conventional Ships Using a Neural Bag-Of-Words Mechanism. *Sensors* **2020**, *20*, 1608. [CrossRef] [PubMed]

8. Ru, X.; Chi, Y.; Liu, W. A Detection and Tracking Algorithm for Resolvable Group with Structural and Formation Changes Using the Gibbs-GLMB Filter. *Sensors* **2020**, *20*, 3384. [CrossRef] [PubMed]

9. Shan, X.; Zhao, D.; Pan, M.; Wang, D.; Zhao, L. Sea–Sky Line and Its Nearby Ships Detection Based on the Motion Attitude of Visible Light Sensors. *Sensors* **2019**, *19*, 4004. [CrossRef] [PubMed]

10. Willburger, K.; Schwenk, K.; Brauchle, J. AMARO—An On-Board Ship Detection and Real-Time Information System. *Sensors* **2020**, *20*, 1324. [CrossRef] [PubMed]

11. Dong, H.; He, K.; Shen, X.; Ma, S.; Wang, H.; Qiao, C. Adaptive Intrawell Matched Stochastic Resonance with a Potential Constraint Aided Line Enhancer for Passive Sonars. *Sensors* **2020**, *20*, 3269. [CrossRef] [PubMed]

12. Kim, S.; Shin, J.; Ahn, J.; Kim, S. Extremely Robust Remote-Target Detection Based on Carbon Dioxide-Double Spikes in Midwave Spectral Imaging. *Sensors* **2020**, *20*, 2896. [CrossRef] [PubMed]

13. Heiselberg, H. Aircraft and Ship Velocity Determination in Sentinel-2 Multispectral Images. *Sensors* **2019**, *19*, 2873. [CrossRef] [PubMed]

14. Specht, M.; Specht, C.; Lasota, H.; Cywiński, P. Assessment of the Steering Precision of a Hydrographic Unmanned Surface Vessel (USV) along Sounding Profiles Using a Low-Cost Multi-Global Navigation Satellite System (GNSS) Receiver Supported Autopilot. *Sensors* **2019**, *19*, 3939. [CrossRef] [PubMed]

15. Stateczny, A.; Burdziakowski, P.; Najdecka, K.; Domagalska-Stateczna, B. Accuracy of Trajectory Tracking Based on Nonlinear Guidance Logic for Hydrographic Unmanned Surface Vessels. *Sensors* **2020**, *20*, 832. [CrossRef] [PubMed]

Publisher's Note: MDPI stays neutral with regard to jurisdictional claims in published maps and institutional affiliations.

Article

A Novel Detector Based on Convolution Neural Networks for Multiscale SAR Ship Detection in Complex Background

Wenxin Dai [1], Yuqing Mao [2], Rongao Yuan [1], Yijing Liu [1], Xuemei Pu [2,3,*] and Chuan Li [1,*]

[1]   College of Computer Science, Sichuan University, Chengdu 610065, China;
      2017223045183@stu.scu.edu.cn (W.D.); rgyuan@stu.scu.edu.cn (R.Y.); liuyijing@scu.edu.cn (Y.L.)
[2]   College of Cybersecurity, Sichuan University, Chengdu 610065, China; maoyuqing@stu.scu.edu.cn
[3]   College of Chemistry, Sichuan University, Chengdu 610065, China
*   Correspondence: xmpuscu@scu.edu.cn (X.P.); lcharles@scu.edu.cn (C.L.); Tel.: +86-28-8541-0765 (X.P.);
      +86-28-8546-6105 (C.L.)

Received: 13 March 2020; Accepted: 23 April 2020; Published: 30 April 2020

Abstract: Convolution neural network (CNN)-based detectors have shown great performance on ship detections of synthetic aperture radar (SAR) images. However, the performance of current models has not been satisfactory enough for detecting multiscale ships and small-size ones in front of complex backgrounds. To address the problem, we propose a novel SAR ship detector based on CNN, which consist of three subnetworks: the Fusion Feature Extractor Network (FFEN), Region Proposal Network (RPN), and Refine Detection Network (RDN). Instead of using a single feature map, we fuse feature maps in bottom–up and top–down ways and generate proposals from each fused feature map in FFEN. Furthermore, we further merge features generated by the region-of-interest (RoI) pooling layer in RDN. Based on the feature representation strategy, the CNN framework constructed can significantly enhance the location and semantics information for the multiscale ships, in particular for the small ships. On the other hand, the residual block is introduced to increase the network depth, through which the detection precision could be further improved. The public SAR ship dataset (SSDD) and China Gaofen-3 satellite SAR image are used to validate the proposed method. Our method shows excellent performance for detecting the multiscale and small-size ships with respect to some competitive models and exhibits high potential in practical application.

Keywords: convolutional neural network (CNN); ship detection; synthetic aperture radar (SAR); multiscale and small ship detection; complex background

---

1. Introduction

Synthetic aperture radar (SAR) can provide high-resolution images under all-weather and all-day conditions [1–4], thus playing an important role in marine monitoring and maritime traffic supervision [5–8]. Ship detections of the SAR images have attracted considerable interests [9–13], which usually consist of four steps: land masking [14], preprocessing, prescreening, and discrimination [15]. The purpose of the land masking is to eliminate adverse effects of the lands, while the preprocessing aims at improving the detection precision in subsequent stages. The prescreening step is used to locate candidate areas as ship region proposals. Among the prescreening methods, constant false alarm rate (CFAR) prescreening is the most widely used [10,16–18]. The discrimination is designed to eliminate false alarms and obtain real targets [19–21]. Traditional methods rely on hand-crafted features. Consequently, they are not promising for ship discrimination in front of complex backgrounds, which generally contain inshore or offshore locations (ship-like interferences, such as roofs, container piles, and so on), or distractions caused by sea clutter [14,15]. Therefore, it is urgent to develop new detection methods to improve the detection performance for the SAR ships.

Convolution neural network can learn deep features from the data itself [22]. Its feature extraction performs much better than the hand-crafted one for target detections [23–26]. Thus, convolution neural network (CNN)-based detectors have been applied to detect ships in the SAR images. Among the CNN methods, Faster RCNN (F-RCNN) [27] based on the region proposal is a typical detection algorithm. F-RCNN consists of the shared convolution network used for extracting features, the region proposal network (RPN) for predicting candidate regions, and the detection network for classifying ship proposals and refining their spatial locations. In F-RCNN, RPN uses an anchor mechanism to generate the region proposal directly from the topmost feature map. However, the detection performance of the F-RCNN algorithms has not been satisfactory for the small-size ships with pixels less than 30 px [28]. Thus, Li et al. [29] proposed several strategies such as transfer learning and hard negative mining to improve the standard F-RCNN algorithm. They used CNN with five layers to detect the public SAR ship dataset (SSDD), which contains different-size ships covering offshore and inshore areas. Experimental results showed that the average precision of the improved F-RCNN is 78.8%, which is 8.7% higher than that of the previous one. Although the precision of the ship detection was improved to some extent, it is still not satisfactory. This may be attributed to the small number of CNN layers, the complex background of the SAR images, and the variable sizes of ships.

As known, the feature map from each layer has differences in semantic distinction and spatial resolution for CNN. Thus, CNN has a tradeoff between them [11]. In general, shallow layers of CNN have higher spatial resolutions than the other layers. Feature maps of intermediate layers are complementary with a passable resolution, while feature maps of high layers are abstract and semantic, which could distinguish target categories. Consequently, the shallow layers are more suitable for the location while the high layers are conducive to classification [30,31]. To deal with the detection of the variable-size ships, Zhao et al. [15] proposed a coupled CNN detector, which was based on an idea of fusion feature map from the Single Shot Detector (SSD) [32] algorithm. They used a VGG16 network with 16 convolution layers and merged the last three-layer feature maps to improve semantic information. Compared with the F-RCNN method, the average precision was improved from 71.3% to 79.5% for collected Gaofen-3 datasets, which contain many small and densely clustered ships. In addition, Ji et al. [8,11] proposed a multilayer fusion convolutional neural network for the SAR ship detection, in which three shallow layers were combined. Compared with the F-RCNN method, the detection precisions were improved from 73.9%/67.2% to 83.6%/87.3% on a collected Sentinal-1 dataset. Gui et al. [22] merged shallow layers and high layers (discarding intermediate layers) to detect multiscale objects, based on a light-head detector. They achieved 84.4% of detection precision for SSDD, which was 7.8% higher than that of F-RCNN under the same experimental configurations. These observations verify that the feature merging is beneficial for improving the detection performance of the multiscale ships. However, the detection results are not very satisfactory, which are desired to be further improved either for the feature fusion or for the model construction.

Based on all the considerations above, we propose a novel SAR ship detection framework to identify the multiscale ships against complex backgrounds. In order to improve the detection performance of the multiscale ships, we also fuse feature maps in bottom–up and top–down forms, and we generate proposals from each fused feature map in order to make full use of the semantic information and the spatial one. Different from other related works, we change the convolution network to a residual learning framework [33] in order to further improve the detection performance and avoid overfitting of high network depth. In general, the RoI pooling layer extracts a fixed-length feature vector from the coarse region proposal generated by RPN in order to predict the target. As pointed out, the small size object lacks information for the location optimization and classification [11]. Thus, in order to improve their detection performance, we further merge each feature map generated by the RoI pooling layer to enhance the feature information, which is also different from the previous models with inclusion of the feature merging [8,11,22]. As expected, our experiments on the public SAR Ship Detection Dataset (SSDD) and the Chinese Gaofen-3 dataset show that the proposed framework could significantly improve the detection performance on the ship targets with different sizes in front of complex backgrounds.

The rest of this paper is organized as follows. Section 2 describes the framework of our method in detail. Section 3 introduces the datasets used in the work and the experimental results. The final section gives the conclusion.

2. Methodology

Figure 1 illustrates the detailed architecture of our proposed method, including three subnetworks: Fusion Feature Extractor Network (FFEN), Region Proposal Network (RPN), and Refine Detection Network (RDN). Firstly, FFEN extracts features from the SAR images and fuses features through the bottom–up and top–down ways, which are shared by the following two subnetworks. Next, RPN is used to predict the region proposals at each feature fusion layer. Finally, RDN implements the target detection, based on the region proposals and the feature maps from FFEN. Detailed introductions for the three subnetworks are shown in the following sections. In addition, we also test the computational costs of the three subnetworks after the whole framework is constructed.

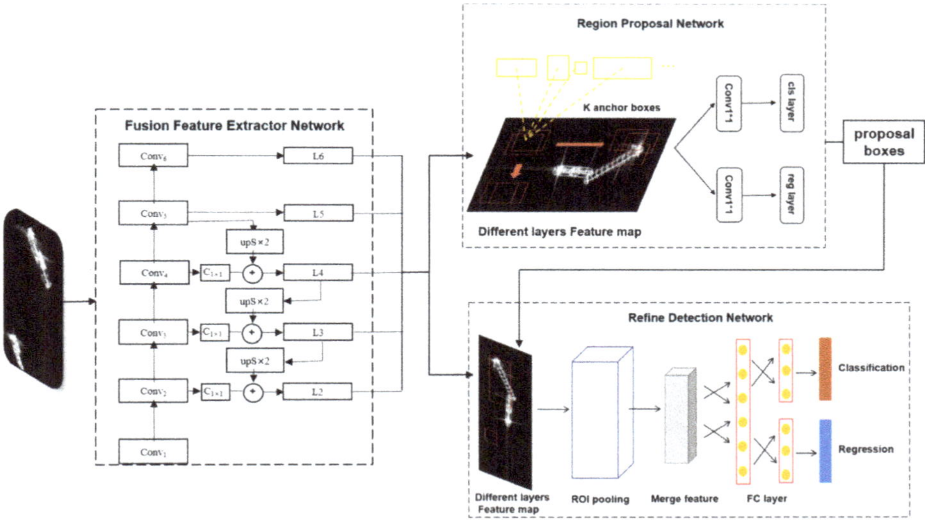

Figure 1. The architecture of our proposed method, which consists of the Fusion Feature Extractor Network (FFEN), Region Proposal Network (RPN), and Refine Detection Network (RDN).

2.1. Fusion Feature Extractor Network

As known, convolution neural networks are generally composed of multiple convolution layers and pooling layers, through which CNN can extract features from the input image. In order to reduce the number of parameters in the neural network, CNN always shrinks its feature maps after the convolutions by means of the max pooling operation. Herein, we take VGG16 for example to visualize the feature maps of different convolution layers. In Figure 2, *convi* (i = 1, 2, 3, 4, 5) denotes different convolution layers from shallow to high in VGG16. It can be seen that the shallow layers (*conv1* and *conv2*) present higher spatial resolutions but are scarce in the semantic information. One pixel on *conv1* almost corresponds to one pixel in the input image; thus, it is similar in size to the input image. After the pooling layer reduces the number of the training parameters and the dimension of the feature vectors from the convolution layers, the feature map will become small, thus showing lower resolution. As depicted by Figure 2, the feature semantic information of higher layers such as *conv4* and *conv5* is rich but abstract, in which one pixel corresponds to several pixels of the input image. Thus, the object location in the high layers is rough. Overall, the shallow layers can achieve more accurate location, and the high layers are conducive to classify in a wide range. Thus, we construct FFEN by fusing feature information of all the convolution layers in order to make full use of the semantic and spatial information.

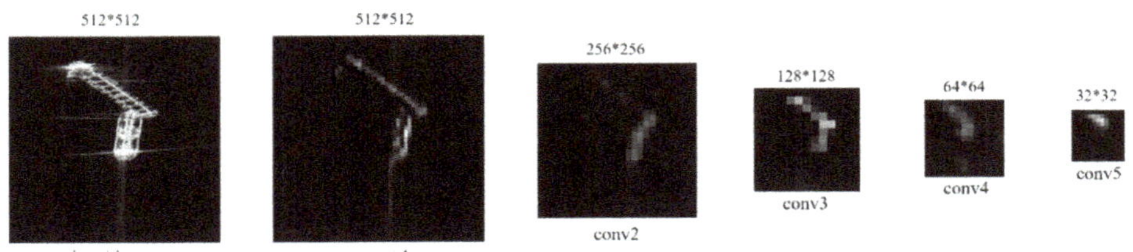

Figure 2. Visualization of feature maps from different convolution layers in VGG16, *convi* (i = 1, 2, 3, 4, 5) denotes different convolution layers from shallow to high.

Herein, we use the idea of Feature Pyramid Networks (FPN). Specifically, the structure includes bottom–up and top–down processes, as shown in the left side of Figure 1. In the bottom–up feedforward network, there are often many layers producing output maps with the same sizes, which are taken as one feature mapping layer. In total, we select such five feature mapping layers $Conv_i$ (i = 2, 3, 4, 5, 6), and $Conv_6$ is a stride two max-pooling of $Conv_5$. The feature extracted from each feature mapping layer is the output of its last layer with strong semantic information. A top–down approach is adopted, which first undergoes a 1 × 1 convolutional layer (vide $C_{1\times1}$ in Figure 1) to reduce the dimension of corresponding $Conv_i$ (i = 2, 3, 4), and uses the nearest neighbor up-sampling to up-sample the fused feature maps higher than it to its size. Then, the up-sampled map is merged with the corresponding bottom–up one, as shown in Figure 1. For example, the up-sampled map $Conv_5$ is merged with $Conv_4$, which generates L_4. Then, the up-sampled map L_3 is merged with $Conv_3$, outputting L_3. Finally, the fusion of the up-sampled map L_3 and $Conv_2$ generates L_2. This process is iterated until the finest resolution map is obtained. In addition, a 3 × 3 convolution filter is appended to each fused feature map to generate the fusion feature mapping layer L_i (i = 2, 3, 4, 5) so that the aliasing effect of the upper sampling could be reduced. Consequently, the merged feature mapping layer could enhance integrity of the location and semantics information, which is beneficial for the multiscale ship detection.

Ren et al. [34] pointed out that the CNN depth is very important to improve the performance of the feature representation. However, as the depth increases, the training of the network becomes difficult due to an explosion of parameters and disappearance of gradients, which leads to a drop in the precision of the network. To solve the problem, a residual learning depth network based on ResNet was proposed to ease the training process and improve the detection accuracy [34]. Instead of stacking convolution layers directly, ResNet connects these layers to fit a residual mapping. Formally, x denotes the input SAR image, and $H(x)$ represents the underlying output mapping. We let the stacked nonlinear layers fit another mapping of $F(x) := H(x) - x$. Then, the original mapping is recast into $F(x) + x$. The process could be realized by feedforward networks with shortcut connections, as shown in Figure 3. The shortcut connections do not add additional parameters and computational complexity. Based on the strategy, the entire network could propagate signals with more layers. Herein, ResNet50 is used as the residual network [34].

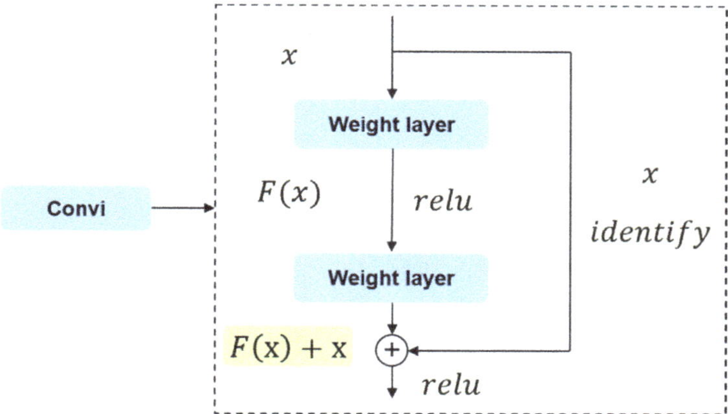

Figure 3. The shortcut connection of ResNet.

2.2. Region Proposal Network

The region proposal network is utilized to classify the ships and the background in the SAR images and to generate coarse region proposals through using the fusion feature mapping layer Li ($i = 2, 3, 4, 5, 6$) provided by FFEN as inputs. The feature maps of different layers represent different feature semantic information and spatial resolutions. For the F-RCNN detector, only the top-level features of the network are used for prediction (see Figure 4a). This may be attributed to the fact that it cannot detect the multiscale ships well. Single Shot Detector (SSD) uses multiscale feature fusion to extract features from the middle and top layers for prediction, as shown in Figure 4b. Although these methods utilized the feature fusion, they ignored the low-level feature information, which is useful for the accurate location. Thus, in order to make full use of the feature semantic information, we design a hierarchical prediction structure of feature fusion, in which RPN is attached to each fusion feature map Li so that it could achieve high performance for the detection of the multiscale ships in the complex background, as shown in Figure 4c.

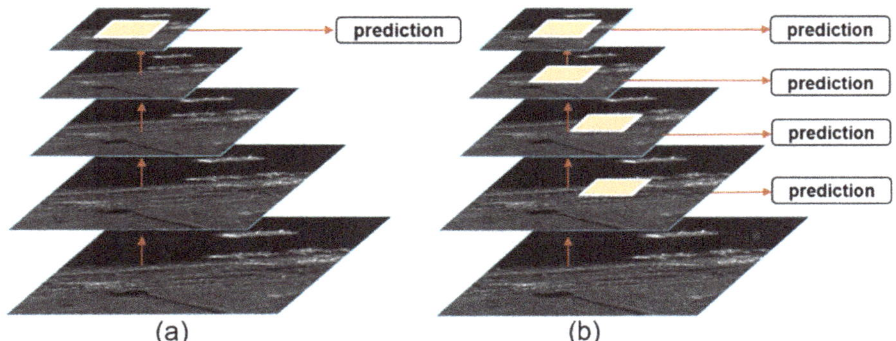

Figure 4. *Cont.*

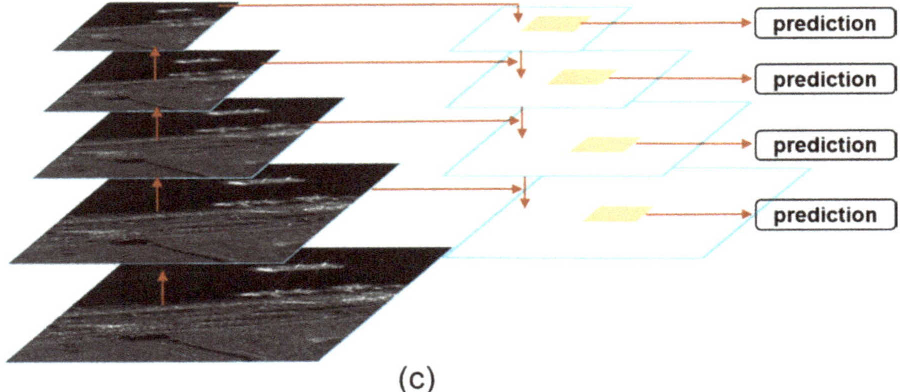

(c)

Figure 4. Different strategies for the multiscale detection. (**a**) Prediction from the top feature map such as Faster RCNN (F-RCNN); (**b**) Prediction from multiple feature maps such as Single Shot Detector (SSD); (**c**) RPN of our framework is a hierarchical prediction structure of feature fusion.

For RPN, we use anchors (a set of reference boxes, also called as region proposals) to measure the ship position and predict whether it is a ship target. The anchors are involved in multiple predefined scales and aspect ratios in order to cover ship targets of different scales. All the anchors have the same center points. We assign five different-scale ($Scale_i$ (i = 2, 3, 4, 5, 6) = {32 × 32, 64 × 64, 128 × 128, 256 × 256, 512 × 512}) anchors to each fusion feature mapping layer L_i (i = 2, 3, 4, 5, 6). The aspect ratios of the anchors of the fusion feature mapping layers L_i (i = 2, 3, 4, 5, 6) are {1:1, 1:2, 2:1}. Consequently, 15 (5 scales and 3 aspect ratios) anchors are generated for each Li (i = 2, 3, 4, 5, 6). As shown in Figure 1, these anchors are transmitted to the cls_layer and reg_layer in RPN (cls_layer for the ship target classification and reg_layer for the anchor regression). The cls_layer outputs 2K(K=15) scores, which are used to estimate probability of the object for each proposal. The reg_layer has 4K outputs encoding coordinates of boxes. Since this stage produces a large number of coarse anchors and many of them overlap each other, we use non-maximum suppression (NMS) [35] to reduce the number of coarse anchors. The retention of the anchors is measured by Intersection-Over-Union (IoU) between each anchor and the corresponding ground-truth. IoU is generally defined as:

$$\text{IoU} = (Area_{bbox} \cap Area_{gt}) \,/\, (Area_{bbox} \cup Area_{gt}) \tag{1}$$

where $Area_{bbox}$ and $Area_{gt}$ represent the prediction box and the ground-truth box, respectively. If the IoU of an anchor is higher than 0.7, it is considered as a positive anchor. An anchor with IoU less than 0.3 is taken as a negative anchor. The anchors with IoU in the range of 0.3–0.7 are ignored and do not participate in the training. For each image, we sample 512 anchors to train, in which a ratio of 1:1 is used for the sampled positive and negative anchors.

2.3. Refine Detection Network

As reflected by Figure 1, the Refine Detection Network (RDN) is the third stage of our algorithm framework, which uses the characteristics provided by FFEN and the coarse anchors of RPN as inputs. Its main function is to refine the coarse anchors and get the final prediction result. In RDN, the RoI pooling layer extracts a fixed-length feature vector with a 7 × 7 × 512 size from the coarse region proposal generated by RPN. In order to enhance the semantic information about the small-size objects, we further merge the features generated by the RoI pooling layer. Then, the merged features are fed back to the fully connected layers to obtain the final detection result, as shown by Figure 1. The impact of the feature merging in RDN will be evaluated in the following experiment section.

2.4. Computational Costs

Herein, we test the computational cost of the whole network. Table 1 shows the structure of ResNet-50, the number of parameters, and the multiply–add computational cost (MAC), which was derived from the 224 × 224 size of the input image block. The parameters and MAC of each layer are computed in terms of the configuration of each layer. Table 2 summarizes the MAC and the number of parameters for the three subnetworks (FFEN, RPN, and RDN). As shown in Table 2, our method requires 53 billion MAC and 260 million parameters for an iteration. Judged from MAC, the FFEN part is the least in the computing cost. The required times for training and testing mainly depend on the RPN and RDN parts. The result also indicates that the computing cost of FFEN with inclusion of ResNet-50 is not increased despite increasing the number of convolution layers.

Table 1. The detailed structure, the number of parameters, and the multiply–add computational cost (MAC) for the FFEN network.

Name	Type	Stride	Output	Params	MAC
$Conv_1$	$[7 \times 7, 64] \times 1$	2	$112 \times 112 \times 64$	9.47 K	118.01 M
$Conv_2$	$\begin{bmatrix} 1 \times 1, 64 \\ 3 \times 3, 64 \\ 1 \times 1, 256 \end{bmatrix} \times 3$	2	$112 \times 112 \times 64$	9.47 K	118.01 M
$Conv_3$	$\begin{bmatrix} 1 \times 1, 128 \\ 3 \times 3, 128 \\ 1 \times 1, 512 \end{bmatrix} \times 4$	2	$56 \times 56 \times 256$	262.19 K	877.88 M
$Conv_4$	$\begin{bmatrix} 1 \times 1, 256 \\ 3 \times 3, 256 \\ 1 \times 1, 1024 \end{bmatrix} \times 6$	2	$28 \times 28 \times 512$	1154.1 K	1056.11 M
$Conv_5$	$\begin{bmatrix} 1 \times 1, 512 \\ 3 \times 3, 512 \\ 1 \times 1, 2048 \end{bmatrix} \times 3$	2	$14 \times 14 \times 1024$	7360.7 K	1389.24 M

Table 2. The number of parameters and MAC for each part of our network with the image size of 224 × 224 as the input.

	Params	MAC
FFEN	24.286 M	4201.04 M
RPN	7.499 M	6144.39 M
RDN	228.47 M	42736.8 M
Total	260.3 M	53 B

3. Experiments and Results

In this section, experiments are carried out to evaluate the performance of the proposed method. First, we briefly describe the datasets used and experimental settings. Then, we used a standard dataset (the public synthetic aperture radar (SAR) ship detection dataset, SSDD) to evaluate the performance of the proposed framework. Finally, our model is further applied to the Gaofen-3 dataset (the first high-resolution civil SAR satellite in China) in order to test its robustness in practice. For the two types of datasets, our model is compared with some competitive methods reported and exhibits better performances.

3.1. Experimental Datasets and Settings

3.1.1. Dataset Descriptions

The public SAR Ship Detection Dataset (SSDD) [29] is used in the work, which follows a similar format to Pascal VOC [36]. SSDD includes SAR images collected from Radarsat-2, Terrasar-x, and Sentinel-1 [37] with resolutions ranging from 1 to 15 m and polarimetric modes of HH, HV, VV, and

VH. Table 3 lists specific information of the ships in SSDD. In SSDD, there are a total of 1160 images and 2456 ships, and the average number of ships per image is 2.12. Statistics for the number of the ships and the images are shown in Table 4. We divide the dataset into three parts (training set, test set, and validation set) with the ratio of 7:2:1. Figure 5 representatively shows some images of SSDD. In addition, in order to further verify the robustness of our model in practice, we also use the SAR image taken from Ganfen-3 as one independent test set, which contains 102 ships with different sizes in a complex environment. Gaofen-3 is the first C-band multi-polarization SAR satellite developed by China, and its resolution could reach 1 m. The specific information of the Gaofen-3 dataset is listed in Table 5.

Table 3. The detailed information of the synthetic aperture radar ship dataset (SSDD) dataset.

Sensors	Resolution	Polarization	Ship	Position
Sentinel-1 RadarSat-2 TerraSAR-X	1–15 m	HH,VV VH,HV	Different size and material	In the sea and offshore

Table 4. Statistics for the number of ships and images.

NoS [1]	1	2	3	4	5	6	7	8	9	10	11	12	13	14
NoI [2]	725	183	89	47	45	16	15	8	4	11	5	3	3	0

[1] NoS denotes the number of ships. [2] NoI denotes the number of images.

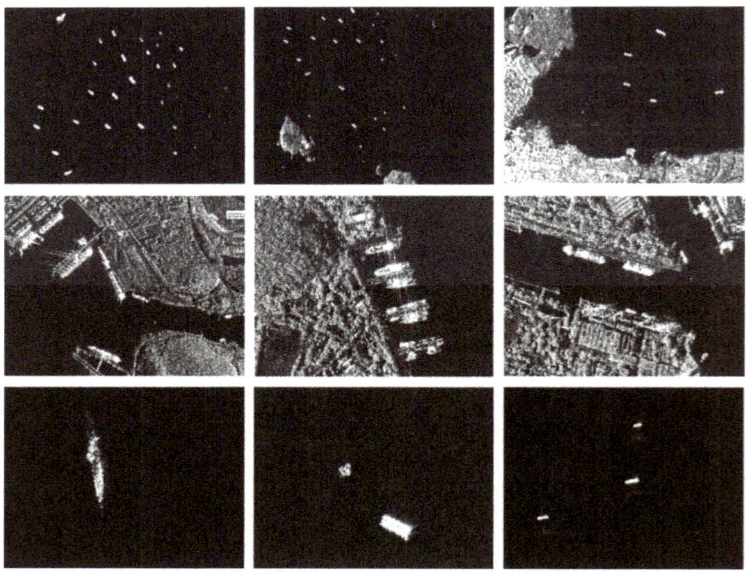

Figure 5. Some representative images of the SAR ship detection dataset (SSDD) involved in multiscale, near-shore, and small ships.

Table 5. Detailed information of the GF3 image.

Sensors	Resolution	Polarization	Ship	Position	Pixel	Imaging Time
GF-3	3M	HH	Different Size	Sea and offshore	Width: 29,986 Height: 15,648	17 July 2018

3.1.2. Experimental Settings

All experiments are implemented using the deep learning framework Caffe [38] and executed on a PC with an Intel(R) Xeon(R) CPU E3-1230 v5 @ 3.40GHz, NVIDIA GTX-1080T GPU (12 GB memory), and the PC operating system is Ubuntu 16.04. We firstly use the pretraining model ResNet-50 to initialize our network. Then, we utilize the end-to-end training strategy to train our model, in which the gradient descent algorithm is used to update the network weight. A total of 40 k iterations are performed. The learning rate of the first 20,000 iterations is 0.001, and the learning rate of the last 20,000 iterations is 0.0001. The weight decay and momentum are set to be 0.0001 and 0.9, respectively.

3.1.3. Evaluation Metrics

In this work, we utilize three criteria widely used to quantitatively evaluate the detection performance. They are precision, recall, and F1-score. The precision measures the detection fraction of true positive samples in terms of Equation (2).

$$precision = \frac{TP}{TP + FP} \tag{2}$$

The recall measures fraction of positives over the number of ground-truths, which is defined by Equation (3)

$$recall = \frac{TP}{TP + FN} \tag{3}$$

Herein, TP, FN, and FP denote true positive, false negative, and false positive, respectively. In general, a detection result is considered to be a true positive if the overlap ratio of the IoU between a detected bounding box and a ground truth bounding box is greater than 0.5. Otherwise, the detection is considered as a false positive. IoU is generally defined by Equation (1) above.

As shown in Equation (4), the F1-score combines the precision and recall metrics as a single measure; thus, it could comprehensively evaluate the quality of the ship detection model:

$$F1 = 2 \times \frac{precision \times recall}{precision + recall} \tag{4}$$

3.2. Experiments on SSDD

3.2.1. The Effect of the Number of Network Layers

As known, the depth of the convolution layers is associated with the detection precision. In order to observe the effect of the depth of the convolution layers, we test and compare three network depths (layer-5 (ZF [39]), layer-16 (VGG16 [40]), and layer-50 (ResNet-50 [34])). To eliminate the influence of other factors, we only change the network depth, not considering the other operations such as the feature fusion. Table 6 lists the detection precision, recall, and F1-score for the three types of network depths. It can be seen that the 50-layer (ResNet-50) model exhibits the best performance for recall, precision, and F1-score, indicating that the precision of SAR ship detection could be improved by increasing the depth of the network within the framework of the residual block. Thus, the 50-layer network is adopted in the subsequent experiments.

Table 6. Detection performances for three network depths, in which any feature merging is not considered.

Depth	Precision	Recall	F1-Score
5-layer	73%	84.4%	78%
16-layer	80.5%	86.4%	83.4%
50-layer	82.6%	87.1%	84.8%

3.2.2. The Effect of Feature Merging in RDN

As mentioned above, the small-size object lacks information regarding the location optimization and the classification. Thus, in order to improve their detection performances, we fully merge the features generated by the RoI pooling layer and compare the results between the model with inclusion of the feature merging (labeled as the merge model) and one without the feature merging (labeled as the no-merge model). Table 7 lists their detection precisions, detection recalls, and F1-scores. It can be seen that the recall values of the two models are similar, but the precision and the F1-score of the merge model are higher than those of the no-merge model. Therefore, the feature merging in RDN could further improve the detection performance. In order to observe the impact of the feature merging on the detection performance of the small-size ships, we further check the number of small ships detected by the two models. There are 269 small ships in total for the test set. Herein, the target with less than 30 px is considered as the small-size ship [28]. The model without the feature merging could correctly identify 242 ships, while it is increased to 256 after merging the features. Figure 6 representatively displays the detection results of the two models. It is also observed that the merge model could identify more small-size SAR ships than the no-merge one. These observations confirm the efficacy of our fusing strategy in improving the detection of the small-size ships.

Table 7. The effect of feature merging of RDN on detection performances, which include the feature merging of FFEN.

Methods	Precision	Recall	F1-Score
no-merge	86.9%	93.8%	90.2%
merge	89.9%	93.2%	91.5%

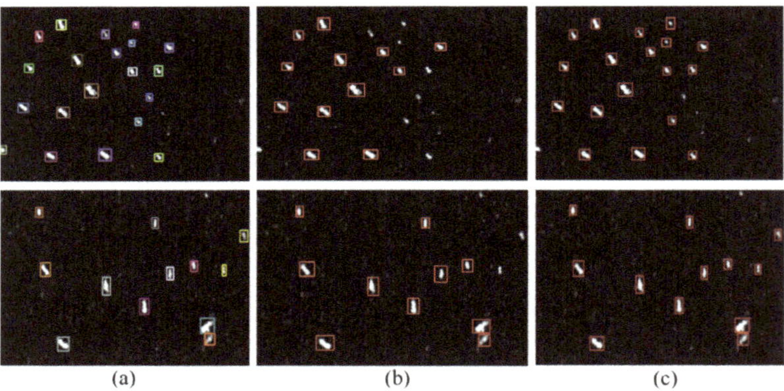

(a)   (b)   (c)

Figure 6. The detection results of the small size ships. (**a**) is the ground truth, (**b**) is the detection result of the model without merging features generated by the region-of-interest (RoI) pooling layer (called as no-merge model), (**c**) is the detection result of the model with the feature merging (called as merge model).

3.2.3. Comparisons with Other Methods

To further evaluate the detection performance of our model, some competitive methods applied to SSDD are compared, including traditional a CFAR detector [41], Faster RCNN (F-RCNN) [27], Coupled-CNN_E_A [15], SSD [32], and a multilayer fusion light-head detector (MFLHD) [22]. These comparison results are shown in Table 8. Herein, we construct an improved CFAR based on the traditional two-parameter CFAR detector through combining a morphological filter and a density filter. The Faster RCNN method was reported to be a particularly influential detector, in which 16 convolution layers (VGG16) were used. Coupled-CNN_E_A and MFLHD are detectors specially designed to detect the multiscale ships in the SAR images, which exhibited good performances for the ship detection in the complex environment. SSD is a single-stage detector and it is faster than F-RCNN, which used anchor boxes to predict bounding boxes from multiple feature maps with different resolutions. In comparison, we used the choices laid out in the original papers as soon as possible.

Table 8. Performance comparisons of several methods with our method for the SSDD dataset. The bold numbers denote the optimal values in each column. CFAR: constant false alarm rate.

Methods	Precision	Recall	F1-Score
CFAR	52.7%	64.2%	57.9%
F-RCNN	83.2%	81.9%	82.5%
Coupled-CNN_E_A	83.1%	85.7%	84.4%
SSD	80.4%	74.8%	77.5%
MFLHD [22] [1]	87.5%	81.6%	84.4%
Our method	**89.9%**	**93.2%**	**91.5%**

[1] The results come from Ref. [22].

It can be seen from Table 8 that the traditional CFAR exhibits the poorest performance for the multiscale ship detection in the complex environment, while our method significantly improves the detection performance compared with the other methods for the SSDD dataset, as evidenced by the precision, recall, and F1-score. In addition, Li et al. [29] used an improved F-RCNN to perform the ship detection for the SSDD dataset. In the work, they utilized AP to evaluate the detection performance, rather than the three evaluation metrics used in the work. In order to compare, we also calculate the AP value (89.4%), which is significantly higher than 78.8% reported by the work [29]. These comparisons above further confirm that our proposed method has excellent performance in the ship detection.

On the other hand, we also compute recall and precision at different IoU ratios with the ground truth boxes for the four representative methods (SSD, F-RCNN, Coupled-CNN_E_A, and our method) in order to diagnose models, as shown in Figure 7. It can be seen from Figure 7a that the recall rate of each method decreases with increasing IoU. The recall rate of SSD detector is the lowest, and our method is superior to the other methods for recall-IoU. As reflected by Figure 7a, the recall values begin to drop when the IoU is higher than 0.5. Thus, it should be reasonable to set IoU to be 0.5 for calculating prediction results. In addition, Figure 7b further displays the precision-recall curve. A good model should possess high precision and high recall. However, the precision rate would present a drop when the recall rate is increased up to a point. As reflected by Figure 7b, the other three methods present sudden precision drops when the recall rate gets higher than 0.6, while our method begins to decrease when the recall rate is greater than 0.8. These observations further show the superiority of our model over the other three methods.

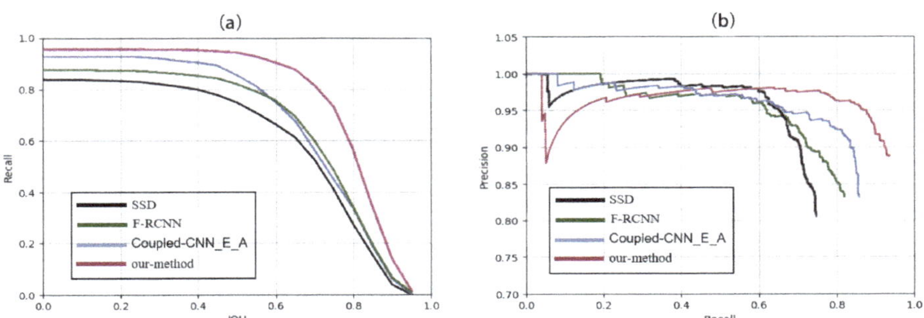

Figure 7. Performance curves for the four methods. (**a**) Recall vs. Intersection-Over-Union (IoU) curve, (**b**) Precision vs. recall curve.

3.3. Robustness Testing on the GF-3 Dataset

3.3.1. Detection Results and Comparisons

As mentioned above, our method exhibits better performance on the SSDD dataset containing multiscale SAR ships. In order to further evaluate the application of our model in practice, it is applied to detect a large Ganfen-3 SAR image, which includes 102 ships with different sizes in the complicated environment (see Figure 8). Due to the large size of the whole Ganfen-3 SAR image, a 512 × 512 pixel sliding window is used without any overlapping. Similarly, the performance of our model is compared with the four representative detectors (CFAR, F-RCNN, Coupled-CNN_E_A, and SSD), as shown in Table 9. It can be seen that our method still exhibits better performance than the other methods for the independent GF-3 dataset. Figure 8 representatively shows the detection results from our method and F-RCNN, since F-RCNN has been recognized as a very influential detector. It is clear that our method almost detects all the ships on the ocean, including ships in offshore or inshore areas, while the F-RCNN method misses many ships. The result confirms that our method is effective for detecting the multiscale ships in practice.

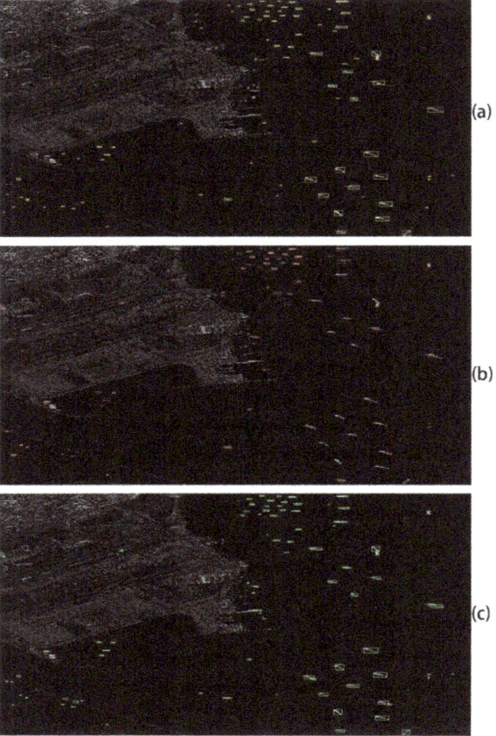

Figure 8. Detection results on the large GF3 SAR ship imagery (**a**) the ground truth; (**b**) detection results of F-RCNN; (c) detection results of our method. Yellow, red, and green rectangles represent the ground truth, the detection result of F-RCNN, and the detection result of our method, respectively.

Table 9. Performance comparisons of several methods with our method for the GF-3 dataset. The bold numbers denote the optimal values in each column.

Methods	Precision	Recall	F1-Score
CFAR	50.4%	62.9%	56.0%
F-RCNN	78.2%	83.5%	80.8%
Coupled-CNN_E_A	84.7%	82.2%	83.5%
SSD	79.2%	71.3%	75.1%
Our Method	**91.3%**	**93.1%**	**92.1%**

3.3.2. Analysis on Missing Ships and False Alarms

Although our method achieves excellent performance for the SSDD dataset and the GF-3 image, a few missing ships and false alarms still exist. For the GF-3 image with 102 ship targets, there are seven missing ships and nine false alarms. As can be seen from Figure 9a,b, some missing ships present very weak or low intensity, so that they would induce few responses on the shallow layers, in turn leading to them being missed. Recently, a new Perceptual Generative Adversarial Net-work (Perceptual GAN) model was proposed to improve the detection of small objects through narrowing the representation differences of the small objects from the large ones, rather than learning representations of all the objects at multiple scales [42]. The introduction of the perceptual GAN should be beneficial for detecting small size ships in the future. In addition, some ships side by side are detected to be one ship due to their close distances. It may be improved by modifying the method of non-maximum suppression (NMS) such as soft-NMS [43]. The method decays the detection scores of all other objects as a continuous function of their overlaps with the detection box so that no object is eliminated in the process. Besides these missing targets, some false alarms are also observed in our prediction results. They mainly come from some building facilities on land, some harbor facilities in the open ocean area, or near the coast, which are similar to ships in shape and intensity, as reflected by Figure 9c,d. For these false alarms, they may be ruled out with sea–land segmentation in image preprocessing or the addition of environmental information into the network.

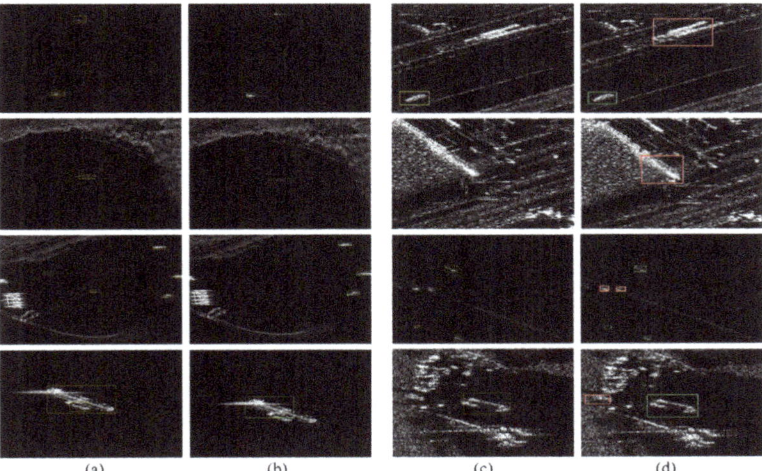

Figure 9. Some missing ships and false alarms for the detection result of the GF-3 synthetic aperture radar (SAR) image with our proposed method. (**a**) Ground truth; (**b**) detection result from our method with respect to (**a**); (**c**) ground truth; (**d**) detection result from our method with respect to (**c**). Yellow, green, and red rectangles denote the ground truth, the detection result of our method, the false alarm, respectively.

4. Conclusions

In order to improve the detection performance for the multiscale ships and small-size ones in complex environments, we construct a novel CNN-based detector composed of a Fusion Feature Extractor Network (FFEN), Region Proposal Network (RPN), and Refine Detection Network (RDN). Instead of using a single feature map, we fuse feature maps in bottom–up and top–down ways and generate proposals from each fused feature map in FFEN. In addition, we further merge features generated by the region-of-interest (RoI) pooling layer in RDN. Based on the feature fusing strategy, rich location and semantics information could be obtained for the multiscale ships, in particular for the small-size ones. On the other hand, the residual block is introduced to FFEN in order to further improve the detection accuracy. Finally, the experimental results on the public SAR ship dataset (SSDD) and the Gaofen-3 satellite SAR image verify that our method could improve the detection performance of the multiscale and small-size ships in front of complex backgrounds. Compared to some competitive methods reported, our model exhibits better performance and high potential for practical applications.

Author Contributions: W.D. finished the experiment and the manuscript. Y.M. and R.Y. helped to calculate and analyze data. Y.L. helped to discuss the proposed method and search related references. C.L. and X.P. designed the experiment and supervised the manuscript writing. All authors have read and agreed to the published version of the manuscript.

Funding: This work was supported by NSAF in China (Grant No: U1730127).

Conflicts of Interest: The authors declare no conflict of interest.

References

1. Moreira, A.; Krieger, G.; Hajnsek, I.; Papathanassiou, K.; Younis, M.; Lopez-Dekker, P.; Huber, S.; Villano, M.; Pardini, M.; Eineder, M.; et al. Tandem-L: A Highly Innovative Bistatic SAR Mission for Global Observation of Dynamic Processes on the Earth's Surface. *IEEE Geosci. Remote Sens. Mag.* **2015**, *3*, 8–23. [CrossRef]
2. Liu, L.; Gao, Y.; Wang, F.; Liu, X. Real-Time Optronic Beamformer on Receive in Phased Array Radar. *IEEE Geosci. Remote Sens. Lett.* **2018**, *16*, 387–391. [CrossRef]
3. Li, N.; Wang, R.; Deng, Y.; Chen, J.; Liu, Y.; Du, K.; Lu, P.; Zhang, Z.; Zhao, F. Waterline Mapping and Change Detection of Tangjiashan Dammed Lake After Wenchuan Earthquake From Multitemporal High-Resolution Airborne SAR Imagery. *IEEE J. Sel. Top. Appl. Earth Obs. Remote Sens.* **2014**, *7*, 3200–3209. [CrossRef]
4. Liao, M.; Tang, J.; Wang, T.; Balz, T.; Zhang, L. Landslide monitoring with high-resolution SAR data in the Three Gorges region. *Sci. China Earth Sci.* **2011**, *55*, 590–601. [CrossRef]
5. Gao, G.; Shi, G. CFAR Ship Detection in Nonhomogeneous Sea Clutter Using Polarimetric SAR Data Based on the Notch Filter. *IEEE Trans. Geosci. Remote Sens.* **2017**, *55*, 4811–4824. [CrossRef]
6. Marino, A.; Sugimoto, M.; Ouchi, K.; Hajnsek, I. Validating a Notch Filter for Detection of Targets at Sea With ALOS-PALSAR Data: Tokyo Bay. *IEEE J. Sel. Top. Appl. Earth Obs. Remote Sens.* **2013**, *7*, 4907–4918. [CrossRef]
7. Wang, S.; Jiao, L.; Wang, M.; Yang, S. New Hierarchical Saliency Filtering for Fast Ship Detection in High-Resolution SAR Images. *IEEE Trans. Geosci. Remote Sens.* **2016**, *55*, 351–362. [CrossRef]
8. Lin, Z.; Ji, K.; Leng, X.; Kuang, G. Squeeze and Excitation Rank Faster R-CNN for Ship Detection in SAR Images. *IEEE Geosci. Remote Sens. Lett.* **2018**, *16*, 751–755. [CrossRef]
9. Heiselberg, P.; Heiselberg, H. Ship-Iceberg Discrimination in Sentinel-2 Multispectral Imagery by Supervised Classification. *Remote Sens.* **2017**, *9*, 1156. [CrossRef]
10. Kang, M.; Leng, X.; Lin, Z.; Ji, K. A modified faster R-CNN based on CFAR algorithm for SAR ship detection. In Proceedings of the 2017 International Workshop on Remote Sensing with Intelligent Processing (RSIP), Shanghai, China, 19–21 May 2017; pp. 1–4. [CrossRef]
11. Kang, M.; Ji, K.; Leng, X.; Lin, Z. Contextual Region-Based Convolutional Neural Network with Multilayer Fusion for SAR Ship Detection. *Remote Sens.* **2017**, *9*, 860. [CrossRef]
12. Wang, Y.; Wang, C.; Zhang, H. Combining a single shot multibox detector with transfer learning for ship detection using sentinel-1 SAR images. *Remote Sens. Lett.* **2018**, *9*, 780–788. [CrossRef]

13. Greidanus, H.; Alvarez, M.; Santamaria, C.; Thoorens, F.-X.; Kourti, N.; Argentieri, P. The SUMO Ship Detector Algorithm for Satellite Radar Images. *Remote Sens.* **2017**, *9*, 246. [CrossRef]

14. Yang, X.; Sun, H.; Fu, K.; Yang, J.; Sun, X.; Yan, M.; Guo, Z. Automatic Ship Detection in Remote Sensing Images from Google Earth of Complex Scenes Based on Multiscale Rotation Dense Feature Pyramid Networks. *Remote Sens.* **2018**, *10*, 132. [CrossRef]

15. Zhao, J.; Guo, W.; Zhang, Z.; Yu, W. A coupled convolutional neural network for small and densely clustered ship detection in SAR images. *Sci. China Inf. Sci.* **2018**, *62*, 42301. [CrossRef]

16. Dai, H.; Du, L.; Wang, Y.; Wang, Z. A Modified CFAR Algorithm Based on Object Proposals for Ship Target Detection in SAR Images. *IEEE Geosci. Remote Sens. Lett.* **2016**, *13*, 1925–1929. [CrossRef]

17. Ao, W.; Xu, F.; Li, Y.; Wang, H. Detection and Discrimination of Ship Targets in Complex Background from Spaceborne ALOS-2 SAR Images. *IEEE J. Sel. Top. Appl. Earth Obs. Remote Sens.* **2018**, *11*, 536–550. [CrossRef]

18. Mazzarella, F.; Vespe, M.; Santamaria, C. SAR Ship Detection and Self-Reporting Data Fusion Based on Traffic Knowledge. *IEEE Geosci. Remote Sens. Lett.* **2015**, *12*, 1685–1689. [CrossRef]

19. Li, T.; Liu, Z.; Xie, R.; Ran, L. An Improved Superpixel-Level CFAR Detection Method for Ship Targets in High-Resolution SAR Images. *IEEE J. Sel. Top. Appl. Earth Obs. Remote Sens.* **2018**, *11*, 184–194. [CrossRef]

20. Jiang, S.; Wang, C.; Zhang, B.; Zhang, H. Ship detection based on feature confidence for high resolution SAR images. In Proceedings of the 2012 IEEE International Geoscience and Remote Sensing Symposium, Munich, Germany, 22–27 July 2012; pp. 6844–6847. [CrossRef]

21. Gambardella, A.; Nunziata, F.; Migliaccio, M. A Physical Full-Resolution SAR Ship Detection Filter. *IEEE Geosci. Remote Sens. Lett.* **2008**, *5*, 760–763. [CrossRef]

22. Gui, Y.; Li, X.; Xue, L. A Multilayer Fusion Light-Head Detector for SAR Ship Detection. *Sensors* **2019**, *19*, 1124. [CrossRef]

23. Wang, Y.; Wang, C.; Zhang, H.; Dong, Y.; Wei, S. Automatic Ship Detection Based on RetinaNet Using Multi-Resolution Gaofen-3 Imagery. *Remote Sens.* **2019**, *11*, 531. [CrossRef]

24. Guo, Y.; Liu, Y.; Oerlemans, A.; Lao, S.; Wu, S.; Lew, M.S. Deep learning for visual understanding: A review. *Neurocomputing* **2016**, *187*, 27–48. [CrossRef]

25. Wang, H.; Li, S.; Zhou, Y.; Chen, S. SAR Automatic Target Recognition Using a Roto-Translational Invariant Wavelet-Scattering Convolution Network. *Remote Sens.* **2018**, *10*, 501. [CrossRef]

26. Xu, Z.; Wang, R.; Zhang, H.; Li, N.; Zhang, L. Building extraction from high-resolution SAR imagery based on deep neural networks. *Remote Sens. Lett.* **2017**, *8*, 888–896. [CrossRef]

27. Ren, S.; He, K.; Girshick, R.; Sun, J. Faster R-CNN: Towards Real-Time Object Detection with Region Proposal Networks. *IEEE Trans. Pattern Anal. Mach. Intell.* **2017**, *39*, 1137–1149. [CrossRef]

28. Li, Q.; Mou, L.; Liu, Q.; Wang, Y.; Zhu, X. HSF-Net: Multiscale Deep Feature Embedding for Ship Detection in Optical Remote Sensing Imagery. *IEEE Trans. Geosci. Remote Sens.* **2018**, *56*, 7147–7161. [CrossRef]

29. Li, J.; Qu, C.; Shao, J. Ship detection in sar images based on an improved faster r-cnn. In Proceedings of the 2017 SAR in Big Data Era: Models, Methods and Applications (BIGSARDATA), Beijing, China, 13–14 November 2017; pp. 1–6.

30. Jiao, J.; Zhang, Y.; Sun, H.; Yang, X.; Gao, X.; Hong, W.; Fu, K.; Sun, X. A Densely Connected End-to-End Neural Network for Multiscale and Multiscene SAR Ship Detection. *IEEE Access* **2018**, *6*, 20881–20892. [CrossRef]

31. Lin, T.; Dollár, P.; Girshick, R.; He, K.; Hariharan, B.; Belongie, S. Feature pyramid networks for object detection. In Proceedings of the 2017 IEEE Conference on Computer Vision and Pattern Recognition (CVPR), Honolulu, HI, USA, 21–26 July 2017; pp. 936–944.

32. Liu, W.; Anguelov, D.; Erhan, D.; Szegedy, C.; Reed, S.; Fu, C.Y.; Berg, A.C. SSD: Single Shot MultiBox Detector. *Eur. Conf. Comput. Vis.* **2016**, 21–37.

33. He, K.; Zhang, X.; Ren, S.; Sun, J. Identity Mappings in Deep Residual Networks. In Proceedings of the Lecture Notes in Computer Science, Hong Kong, China, 16–18 March 2016; Volume 9908, pp. 630–645.

34. He, K.; Zhang, X.; Ren, S.; Sun, J. Deep Residual Learning for Image Recognition. In Proceedings of the 2016 IEEE Conference on Computer Vision and Pattern Recognition (CVPR), Las Vegas, NV, USA, 27–30 June 2016; pp. 770–778. [CrossRef]

35. Neubeck, A.; Van Gool, L. Efficient Non-Maximum Suppression. In Proceedings of the 18th International Conference on Pattern Recognition (ICPR'06), Hong Kong, China, 20–24 August 2006; Volume 3, pp. 850–855. [CrossRef]

36. Everingham, M.; Zisserman. The 2005 pascal visual object classes challenge. In Proceedings of the Machine Learning Challenges Workshop, Southampton, UK, 11–13 April 2005; pp. 117–176.
37. Huang, L.; Liu, B.; Li, B.; Guo, W.; Yu, W.; Zhang, Z.; Yu, W. OpenSARShip: A Dataset Dedicated to Sentinel-1 Ship Interpretation. *IEEE J. Sel. Top. Appl. Earth Obs. Remote Sens.* **2018**, *11*, 195–208. [CrossRef]
38. Jia, Y.; Shelhamer, E.; Donahue, J.; Karayev, S.; Long, J.; Girshick, R.; Guadarrama, S.; Darrell, T. Caffe: Convolutional Architecture for Fast Feature Embedding. In Proceedings of the Proceedings of the 22nd ACM international conference on Multimedia, Mountain View, CA, USA, 18–19 June 2014; pp. 675–678.
39. Zeiler, M.D.; Fergus, R. Visualizing and Understanding Convolutional Networks. In Proceedings of the European Conference on Computer Vision, Zurich, Switzerland, 6–12 September 2014; pp. 818–833.
40. Simonyan, K.; Zisserman, A. Very deep convolutional networks for large-scale image recognition. *arXiv* **2014**, arXiv:1409.1556.
41. Novak, L.M.; Owirka, G.J.; Netishen, C.M. Performance of a high-resolution polarimetric SAR automatic target recognition system. *Linc. Lab. J.* **1993**, *6*, 11–24.
42. Li, J.; Liang, X.; Wei, Y.; Xu, T.; Feng, J.; Yan, S. Perceptual Generative Adversarial Networks for Small Object Detection. In Proceedings of the 2017 IEEE Conference on Computer Vision and Pattern Recognition (CVPR), Honolulu, HI, USA, 22–25 July 2017; pp. 1951–1959.
43. Bodla, N.; Singh, B.; Chellappa, R.; Davis, L.S. Soft-NMS—Improving Object Detection with One Line of Code. In Proceedings of the 2017 IEEE International Conference on Computer Vision (ICCV), Venice, Italy, 22–29 October 2017; pp. 5562–5570.

Article

A Neural Network with Convolutional Module and Residual Structure for Radar Target Recognition Based on High-Resolution Range Profile

Zhequan Fu *, Shangsheng Li, Xiangping Li, Bo Dan and Xukun Wang

Coast Defense College, Naval Aviation University, Yantai 264001, China; liss1965@163.com (S.L.);
lixiangping401@126.com (X.L.); lovelin19841204@163.com (B.D.); wang17616244926@163.com (X.W.)
* Correspondence: fuzq2413@163.com; Tel.: +86-186-6008-1572

Received: 14 December 2019; Accepted: 19 January 2020; Published: 21 January 2020

Abstract: In the conventional neural network, deep depth is required to achieve high accuracy of recognition. Additionally, the problem of saturation may be caused, wherein the recognition accuracy is down-regulated with the increase in the number of network layers. To tackle the mentioned problem, a neural network model is proposed incorporating a micro convolutional module and residual structure. Such a model exhibits few hyper-parameters, and can extended flexibly. In the meantime, to further enhance the separability of features, a novel loss function is proposed, integrating boundary constraints and center clustering. According to the experimental results with a simulated dataset of HRRP signals obtained from thirteen 3D CAD object models, the presented model is capable of achieving higher recognition accuracy and robustness than other common network structures.

Keywords: neural network; target recognition; HRRP; residual structure; loss function

---

1. Introduction

The high-resolution range profile (HRRP) of a target refers to the projection of the target scattering center following the radar line of sight, covering numerous target characteristics (e.g., size and structure). HRRP can be acquired, processed and stored easily; it has a simplified computation and robust real-time performance. For this reason, it has constantly been a critical data source for target recognition. Researchers are able to harvest separable features from HRRP to classify and identify a range of targets. Previous HRRP-based radar target classification and recognition placed primary emphasis on feature extraction on the basis of researchers' prior knowledge and experience, as well as optimization and fusion of classification algorithms. Common features consist of time domain characteristics [1,2] (as manifested by original image, central moment, structure contour, strong scattering points, etc.), while power spectrum, polarization ratio, polarization matrix and other frequency domain [3,4] and polarization domain [5,6] characteristics are also covered.

Fueled by advances in computer technology, and in accordance with deep learning theory, deep learning has become a hotspot in research in various fields [7–10]. It has been extensively employed in radar target detection, recognition, and classification. At the same time, HRRP and CNN also have significant applications in the field of unmanned aerial vehicles and unmanned surface vehicles [11,12]. Deep learning-based object recognition refers to feature extraction using a neural network. HRRP-based radar recognition can also be achieved using a deep learning algorithm. This field has aroused a great deal of attention from researchers, and considerable new achievements have been made, which will be presented below. There are many methods for enhancing the recognition accuracy of neural networks, such as improving the structure of the neural network, optimizing the loss function, and increasing the training data. To be specific, the neural network structure for HRRP target recognition consists of an autoencoder (AE) and a convolutional neural network (CNN).

CNN refers to a critical deep learning structure, capable of automatically extracting the effective separable characteristics of HRRP and addressing the susceptibility to amplitude, translation and orientation that results from the structural similarity between different ships. Compared with the conventional classification algorithm, it exhibits a better recognition effect. In [13], CNN was employed to identify aircraft based on HRRP, and an analysis was conducted on the effects of activation function, convolution kernel size, learning rate and weight decay coefficient on recognition accuracy. It exhibited higher recognition accuracy than those of back propagation (BP) neural network, support vector machine (SVM), and K-nearest neighbor (KNN). Their dataset originated from actual measurements of four aircraft scale models. In [14], CNN was also adopted to achieve target recognition, with the dataset being derived from the simulation calculation of 10 ship targets. In [15], when CNN was employed for target recognition, white noise was added to expand the dataset. Moreover, the recognition results of different radars were fused, and the threshold was set to determine whether the target was known or unknown. Furthermore, the effect of radar numbers and SNR on recognition accuracy was analyzed. In [16], an algorithm that integrates HRRP with polarization information was proposed. With the use of the polarization matrix, Pauli decomposition and Freeman decomposition, 12 eigenvectors were achieved to form the dataset. According to the simulation result, the recognition accuracy based on fully polarized datasets was 5 percentage higher than that of single polarized datasets. Similar work was also conducted in [17], wherein the dataset originated from full-polarization measurements of the model using 77 GHz electromagnetic waves in a microwave anechoic chamber. In [18], two CNN models with various structures were built, and the difference in recognition effect was studied.

AE refers to a type of data compression algorithm, capable of reproducing the input signal to the greatest extent by harvesting the crucial features of the input data. The vital features extracted can be exploited to identify the target. In [19], a deep network, termed a sparse convolution autoencoder (S_1C_1AE), was presented to deal with HRRP target recognition. The model was employed to identify 3 vehicle models. While data were being preprocessed, amplitude normalization and centroid alignment were performed. As compared with other models, the recognition accuracy was enhanced noticeably. Comparison models covered linear discriminate analysis (LDA), principal component analysis (PCA), linear support vector machine (LSVM), denoise sparse autoencoder (D_1S_1AE), and deep belief network (DBN). In [20], a stacked corrective autoencoder (S_2C_2AE) was built, and the data preprocessing was the same as the process in [19]. The correction was achieved by averaging HRRP for respective frame. The model exhibits better generalization performance. To form a loss function based on the Mahalanobis distance, the covariance matrix of each HRRP was employed. Experimental results suggested that the deeper layers of the model, the better the recognition effect could be. In [21], a HRRP recognition model was proposed, combining S_1C_1AE and multiple classifiers. First, the S_1C_1AE was employed to extract features, and subsequently the random forest (RF), naive Bayes (NB) and minimum classifier were fused for features classification. According to experimental results, the model exhibited good noise robustness. In [22–24], the recognition method of fusion neural networks and classifiers was also studied.

There are no publicly available datasets for deep learning-based HRRP target recognition. Most of the datasets adopted by the various researchers of HRRP target recognition are derived from measurements in a microwave anechoic chamber, as well as from their simulation calculation. Nevertheless, different studies have suggested that deep learning-based HRRP target recognition exhibits higher accuracy than conventional classification methods. At present, the enhancement of HRRP target recognition accuracy based on neural networks is primarily focused on the enhancement and fusion of the mentioned structures. In contrast, only rare studies have sought to enhance the accuracy of HRRP target recognition by optimizing the loss function. Most studies that optimize the loss function to enhance the recognition effect are aiming towards face recognition, and many of them can be applied to HRRP target recognition.

Designing a good neural network structure is one of the most efficient and challenging approaches to enhancing classification performance. Under the premise of sufficient datasets, the learning ability of the model can be enhanced by up-regulating the depth and width of the neural network. AlexNet [25] and VGG [26] have both demonstrated that model recognition accuracy displays a positive correlation with the network depth in a certain range. Nevertheless, with the increase in network depth, gradient explosion, disappearance, and saturation of network recognition accuracy may take place in the back propagation of CNN in the training process. By introducing a residual learning framework, Kaiming He and Xiangyu Zhang [27] addressed the degradation problem. Accordingly, the problem whereby the accuracy reaches saturation and subsequently degrades rapidly with the rise in the network depth was avoided. However, to enhance the recognition effect, the residual learning framework requires further increases in the network depth.

In this study, an efficient and extensible convolutional module is presented by optimizing the residual learning framework. The convolutional module contains left and right branches. Among them, based on the left branch structure of convolutional module, the effect of network deepening and widening can be simulated. The skip structure of the right branch is capable of transferring features and gradients more effectively. The convolutional module is capable of achieving the recognition effect of a deep network with fewer network parameters. Additionally, a novel loss function is proposed to enhance the recognition accuracy by combining central clustering and additive margin strategy. The features extracted by the novel loss function are characterized by larger inter-class variations, smaller intra-class variations, and stronger separability. In the meantime, by combining convolutional module that exhibits the same topology, the presented model can be extended to adapt to various difficulty classification tasks. According to the experimental results with a simulated dataset of HRRP, the presented model is capable of achieving higher recognition accuracy than the conventional algorithm. The rest of this paper is organized as follows. Section 2 presents the composition and structure of one-dimensional convolutional network. The design of convolutional module and loss function are elucidated in Section 3. The experimental effect of the model is demonstrated in Section 4 from different aspects. Lastly, the concluded remarks are drawn in Section 5.

2. One-Dimensional Convolutional Neural Network

CNN refers to a type of feedforward neural network that covers convolution calculation. For its translation invariance in the calculation process, it is capable of avoiding complex preprocessing (e.g., HRRP data alignment) and exhibits higher robustness. The model employed in this study complies with the convolutional neural network. First, the basic structure of CNN is introduced, which covers five parts, namely, input layer, convolutional layer, pooling layer, fully connected layer, and output layer. The CNN structure for HRRP is illustrated in Figure 1.

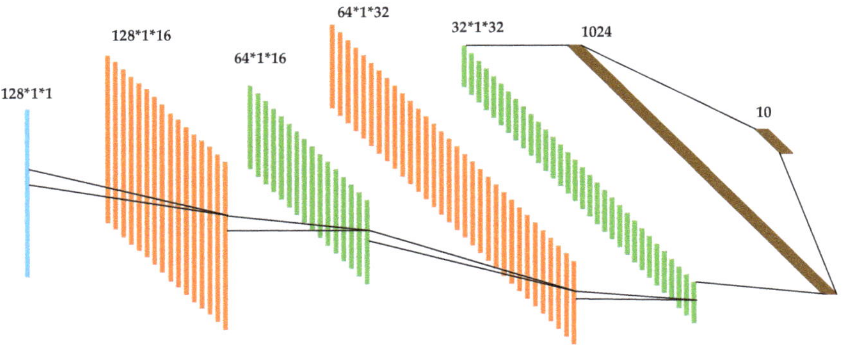

Figure 1. Schematic diagram of the CNN structure for HRRP. It shows CNN's classification process of 10 types of targets for 128-length HRRP data.

The input layer acts as the start of the neural network, generally requiring simple preprocessing of data to make the data have the identical dimension and satisfy the same distribution characteristics. Preprocessing is capable of down-regulating the effect of amplitude perturbation on the extraction characteristics of different HRRP data and enhancing the robustness of the model. It is also convenient to find the minimum value more directly in the iterative process of the gradient descent method, so the model can converge faster. It can be performed in the two steps below:

1. Normalize the amplitude of HRRP. The data after amplitude normalization of the nth HRRP is expressed as $x'_n = x_n / \max(|x_n|)$, where $\max(|x_n|)$ denotes the maximum absolute value of all elements in HRRP.
2. Subtract the mean value of the normalized HRRP data from the respective element.

The major function of the convolutional layer is to extract the features of the input data. In Figure 1, the first convolutional layer covers 16 convolution kernels, and the second convolutional layer consists of 32 convolution kernels. Each convolution kernel element is composed of weight coefficient and bias. In deep learning, the weight coefficient initialization method of the neural network plays an important role in the convergence speed and performance of the model. Common weight coefficient initialization methods include random initialization, Xavier initialization [28], and He initialization [29]. Random initialization may cause gradient disappearance when the neural network layers are deep. To solve this problem, the Xavier initialization method was proposed. When used in conjunction with the Tanh activation function, the Xavier initialization method makes the output value of the activation function of the network layer obey the Gaussian distribution. The generation of gradient disappearance is avoided. However, when used with the Relu activation function, the problem of gradient disappearance still exists. The He initialization method proposed in [29] solves the problem of gradient disappearance when the Relu activation function is used in combination with it. The convolution kernel calculates the input data by convolution, adds the bias, and then activates it by means of the activation function. The output of the convolutional layer is the extracted feature. The calculation process can be written as:

$$x_j^l = f\left(\sum_{i \in M_j} x_i^{l-1} * k_{ij}^l + b_j^l \right),$$

(1)

where x_j^l denotes the output of the jth channel, belonging to the lth convolutional layer. $f(\cdot)$ refers to the activation function, employing the Relu function. k_{ij}^l is the convolution kernel vector of the jth channel of the convolutional layer l that corresponds to the ith input vector. b_j^l is the bias of the jth channel of the convolutional layer l, * represents the convolution operation. The parameters of the convolutional layer consist of convolution kernel size, step size, filling category, as well as activation function. The common activation functions cover the Sigmoid function, the Relu function, etc. For various parameters, the convolutional layer exhibits different characteristics.

The function of the pooling layer aims to select the features extracted by the convolutional layer and down-regulate the dimension by down-sampling. Max-pooling, mean-pooling and mix-pooling are the common pooling layers.

On the whole, the fully connected layer is placed on the back side of the neural network. The major function is to arrange the features extracted from the previous layer to yield the one-dimensional vector. The whole CNN outputs target-related outcomes through the output layer classifier. The common classifiers are softmax and SVM. In the task of target recognition, the output of CNN can cover the category, size and central coordinates of the target. The learning process of CNN usually updates parameters iteratively by back propagation, and stable identification results are obtained by minimizing the error calculated by the loss function.

3. Model Analysis and Design

3.1. Design of Convolutional Module

The depth of neural networks is critical. The deep convolutional neural network is capable of extracting and fusing features of different levels for end-to-end target recognition. Nevertheless, the deepening of network layers will cause saturated recognition accuracy. To address this problem, residual structure is introduced, as illustrated in Figure 2.

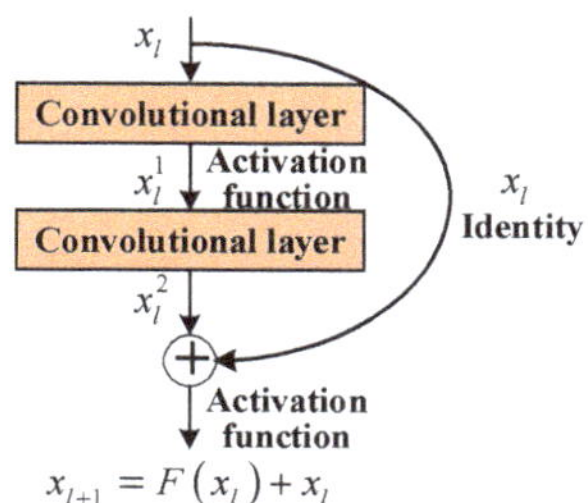

Figure 2. Schematic diagram of residual block.

The residual block in the residual structure consists of convolutional layers, and the number of convolutional layers in Figure 2 is 2. The residual structure outputs the sum of the input feature, and the output of the last convolutional layer is expressed by

$$x_{l+1} = F(x_l) + x_l, \qquad (2)$$

where x_l and x_{l+1} represent the input and output feature vector of the residual block, respectively. $F(x_l)$ denotes the mapping of residual blocks.

Research results reveal that the saturated recognition accuracy of deep network can be effectively addressed by replacing the required fitting mapping $F(x_l) + x_l$ with the fitting mapping $F(x_l)$ [27]. In particular, if the network has extracted the optimal features required for classification, the residual structure should only carry out identity mapping of skip connections to ensure the maximal recognition accuracy. For neural networks, zero residual block is more efficient than the use of multilayer neural networks to fit identity mapping. Figure 3 presents the structure of the convolutional module promoted in this study based on the residual structure, where conv denotes the convolutional layer.

The convolutional module proposed in this article is set up as a highly modular network structure that exhibits high expansibility. The features extracted by the upper layer network act as the input of this layer, and the input will pass through two branches, as shown in Figure 3. In the left branch, the convolution kernel of 1×1 is adopted to fuse the features between layers first. Subsequently, the fused features are split into x branches according to the number of layers. Each branch contains 3 layers of features, and all branches adopt a convolution kernel of 3×1 to extract features; the step size is 2. Since the step size of the convolution kernel is 2, the number of layers of the output feature remains unchanged, and the dimension is halved. Next, the features of all branches are concatenated. Moreover, the size of x can be ascertained according to the complexity of the classification tasks. The larger x is, the easier it is to extract stronger separable features, and the better the recognition effect is in a more complex classification task. Such structure is similar to Inception [30]. Nevertheless, the size and number of the convolution kernel for each branch in Inception are customized step by step. In the convolutional module proposed in this article, a small-scale convolution kernel of 3×1 is uniformly chosen to simplify the structure design and ensure the recognition effect in the meantime. After concatenation, the features are fused again with the convolution kernel of 1×1, and the number of feature layers is then up-regulated. In the left branch in Figure 3, the number of feature layers increases from N to 4N/3. Then, according to the number of layers, the features are split into two parts to prepare for the subsequent fusion of the features of the two branches, where the number of layers of features for add is N, and the number of layers of features for concatenate is N/3, as shown in Figure 3. The right branch directly uses the convolution kernel of 1×1 to fuse the input features and rises the number of feature layers. In the meantime, the features are also separated into two parts according to the number of layers. The number of layers of features for add is N, and the number of layers of features for concatenate is 2N/3. Lastly, the corresponding features in the left and right branches are added or concatenated, as illustrated in Figure 3.

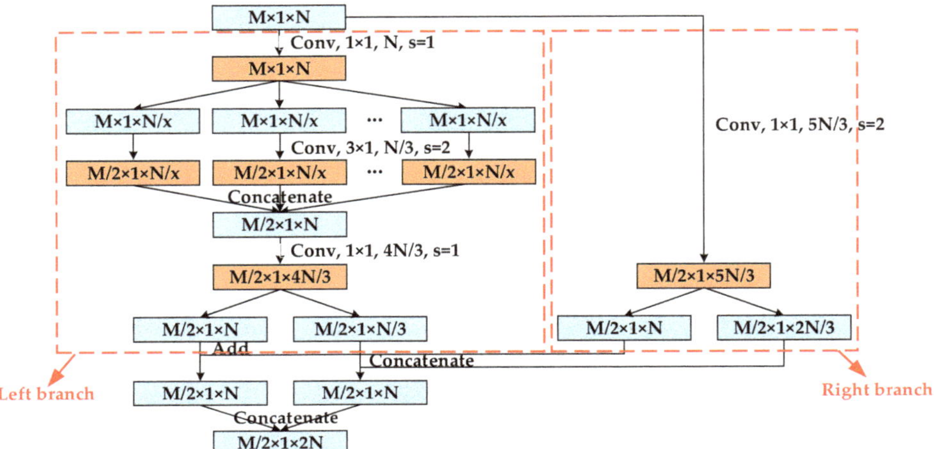

Figure 3. Structure of the convolutional module. $M \times 1 \times N$ represents one-dimensional data with input characteristics of $M \times 1$, N feature layers, s is the moving step size of the convolution kernel, and the unmarked step size is 1 by default.

Compared with the input of the convolutional module, the dimension of the output features is halved, and the number of layers is doubled. The right branch exerts similar effects as the residual network, making the transfer of features and gradients more efficient. Because of the right branch, each layer of convolutional module is capable of acquiring information from the loss function and the original input, and the exploitation of shallow features is facilitated. Then, the problem that the recognition accuracy decreases with the rise in the number of network layers is avoided.

3.2. Design of Loss Function

The loss function is adopted to identify the difference between the predicted value and the real value. Softmax loss commonly acts as the loss function for multi-classification convolutional neural network. However, from the clustering perspective, the feature extracted from softmax loss will display larger intra-class variations than inter-class variations. In the meantime, the features extracted by softmax loss are not discriminative enough, since they still display significant intra-class variations. Under too many target types, the features will overlap, which is not conducive to object classification. To solve this problem, numerous solutions have been proposed in face recognition [31–35]. They primarily focus on promoting inter-class variations and lowering intra-class variations.

For softmax loss, features can be brought closer by enhancing the boundary constraints between various targets. It also promotes the inter-class variations of targets. The formulation of the original softmax loss is defined as:

$$
\begin{aligned}
L_S &= -\frac{1}{m}\sum_{i=1}^{m}\log\frac{e^{W_{y_i}^T x_i}}{\sum_{j=1}^{n}e^{W_j^T x_j}}\\
&= -\frac{1}{m}\sum_{i=1}^{m}\log\frac{e^{\|W_{y_i}\|\|x_i\|\cos(\theta_{y_i})}}{\sum_{j=1}^{n}e^{\|W_j\|\|x_i\|\cos(\theta_j)}}
\end{aligned}
\tag{3}
$$

where x represents the input of the last fully connected layer. $x_i \in \mathbb{R}^d$ denotes the ith deep feature, belonging to the y_ith class. d indicates the feature dimension. $W_j \in \mathbb{R}^d$ refers to the jth column of the weights $W \in \mathbb{R}^{d \times n}$ in the last fully connected layer. $W_{y_i}^T x_i$ denotes the target logit of the ith sample. m and n represent the size of mini-batch and the number of class, respectively.

The design of the loss function proposed refers to the additive margin softmax loss (AM-softmax), which is used in face recognition [35]. In the meantime, considering the constraint of intra-class variations of features, a loss function named margin center (Referred to MC), integrating additive margin and center constraint, is proposed. The loss function uses the additive margin to increase the inter-class variations of features; the center constraint is also employed to reduce the intra-class variations of features. As a result, the inter-class variations of features are larger, the intra-class variations are smaller, and the separability of features is enhanced. The formulation of the loss function proposed in this study is given by

$$
\begin{aligned}
L_{AMSC} &= L_{AMS} + \lambda L_C\\
&= -\frac{1}{m}\sum_{i=1}^{m}\log\frac{e^{s\cdot(W_{y_i}^T x_i - \mu)}}{e^{s\cdot(W_{y_i}^T x_i - \mu)}+\sum_{j=1,j\neq y_i}^{n}e^{sW_j^T x_i}} + \frac{\lambda}{2}\sum_{i=1}^{m}\|x_i - c_{y_i}\|_2^2\\
&= -\frac{1}{m}\sum_{i=1}^{m}\log\frac{e^{s\cdot(\cos\theta_{y_i} - \mu)}}{e^{s\cdot(\cos\theta_{y_i} - \mu)}+\sum_{j=1,j\neq y_i}^{n}e^{s\cdot\cos\theta_j}} + \frac{\lambda}{2}\sum_{i=1}^{m}\|x_i - c_{y_i}\|_2^2
\end{aligned}
\tag{4}
$$

where the hyper-parameter s is adopted to scale the cosine values, and cosine values represent the similarity between the features. μ is applied for the control of the distance between the edges of the feature. $c_{y_i} \in \mathbb{R}^d$ denotes the y_ith class center of features, and c_{y_i} can constantly update with the variation of the features of each batch. L_{AMS} proposes a specific $\psi(\theta) = \cos\theta - \mu$ to introduce the additive margin property and enhances the recognition effect by promoting the inter-class variations of features. L_C constructs a class center for the features of each class of target and punishes the features far away from the class center. Accordingly, the intra-class variations of features becomes more compact, the intra-class variations are lowered, and the inter-class variations are promoted. The gradients of L_C with respect to x_i and update equation of c_{y_i} are computed as:

$$
\frac{\partial L_C}{\partial x_i} = x_i - c_{y_i},
\tag{5}
$$

$$
\Delta c_j = \frac{\sum_{i=1}^{m}\delta(y_i = j)\cdot(c_j - x_i)}{1 + \sum_{i=1}^{m}\delta(y_i = j)},
\tag{6}
$$

where $\delta(\cdot) = 1$ if the identification is correct; otherwise, $\delta(\cdot) = 0$. Under the constraint of joint loss function L_{AMSC}, the learning details in network can be summarized in the following (Algorithm 1):

Algorithm 1. The neural network algorithm with convolution module and residual structure

---

Input: Training samples $\{x_i\}$. Initialized parameters θ_c in convolution kernel. Weight matrix W. The jth class center c_j of features. Hyper-parameter s, μ in L_{AMS}. Learning rate α for feature center in L_C. Weight λ and learning rate lr in network. The number of iteration $t \leftarrow 0$.
Output: The parameters θ_c.

Step 1: **while** not converge **do**
Step 2: $t \leftarrow t + 1$.
Step 3: compute the joint loss by $L^t_{AMSC} = L^t_{AMS} + \lambda L^t_C$.
Step 4: compute the backpropagation error $\dfrac{\partial L^t_{AMSC}}{\partial x^t_i}$ for each i by $\dfrac{\partial L^t_{AMSC}}{\partial x^t_i} = \dfrac{\partial L^t_{AMS}}{\partial x^t_i} + \lambda \cdot \dfrac{\partial L^t_C}{\partial x^t_i}$.
Step 5: update the parameters W by $W^{t+1} = W^t - lr \cdot \dfrac{\partial L^t_{AMSC}}{\partial W^t} = W^t - lr \cdot \dfrac{\partial L^t_{AMS}}{\partial W^t}$.
Step 6: update the parameters c_j by $c^{t+1}_j = c^t_j - \alpha \cdot \Delta c^t_j$.
Step 7: update the parameters θ_c by $\theta^{t+1}_c = \theta^t_c - lr \sum\limits_i^m \dfrac{\partial L^t_{AMSC}}{\partial x^t_i} \cdot \dfrac{\partial x^t_i}{\partial \theta^t_c}$.

Step 8: **end while**

3.3. Design of Model Structure

The block diagram of the presented model in this study is shown in Figure 4, which primarily includes an initial convolutional layer, several convolutional modules with the same topology connected sequentially, and the last two fully connected layers. The dimension of the latter fully connected layer is 2, which is conducive to visualizing the features extracted by the model and analyze the clustering effect of the features.

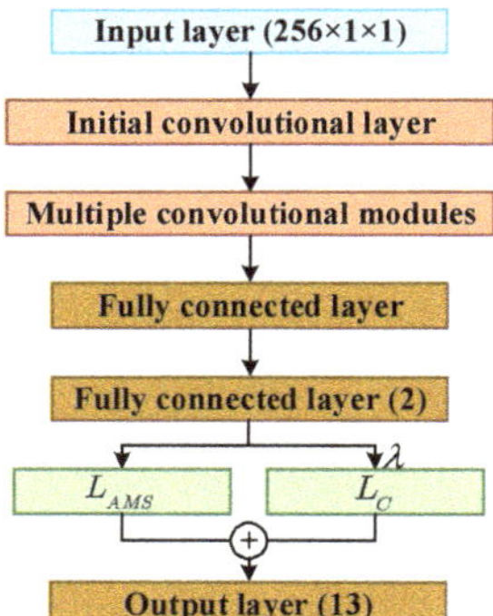

Figure 4. The structure of the presented model.

In Figure 4, the numbers in brackets represent the data dimension after the data passes through this layer, consistent with Figure 3. The output data dimensions of each convolutional module and the first fully connected layer are determined according to the number of convolutional modules. Lastly, the result of the output layer is one-dimensional data that represents the target types. The number of target types in this study is 13. In the presented model, a one-dimensional convolution kernel with a scale of 7×1 is taken for the initial convolutional layer. The selection of convolution kernel with relatively large scale in the first layer of the network is conducive to the extraction of the features (e.g., contour and texture in the HRRP). After each convolution operation in this model, batch normalization and Relu activation are performed on the extracted features. Since the Relu activation function is used, the He initialization method is chosen for all weight initialization of the model proposed.

4. Experimental Simulation and Analysis

4.1. Data Set Construction

On the whole, there are two ways to obtain the target echo signal, namely the measured method and the theoretical calculation method. Since most ship targets are non-cooperative targets, it is very difficult to obtain the HRRP from field measurement. In this study, 13 ship models were built by 3D Max, and HRRP was calculated by FEKO. FEKO is 3D electromagnetic field simulation software, and is an abbreviation of "FEldberechnung für Körper mit beliebiger Oberfläche", in German. When calculating the HRRP of a ship, the ship is stationary, and the HRRP of the ship in different directions is obtained by changing the incident direction of the electromagnetic wave. Since the ship is stationary when calculating HRRP, we do not apply three-dimensional rotation around the different Cartesian axes. The set simulation parameters include the center frequency of the radar as 10 GHz, the bandwidth as 80 MHz, the number of frequency sampling points as 256, the calculated azimuth range as 0–360°, and the interval as 1°. The grazing angle is 10°. The obtained HRRP has 256 range cells, with the corresponding length of each range cell as 1.875 m. The model and amplitude normalized HRRP of one of the ships are illustrated in Figure 5. Models of all of the ship targets are presented in Figure 6.

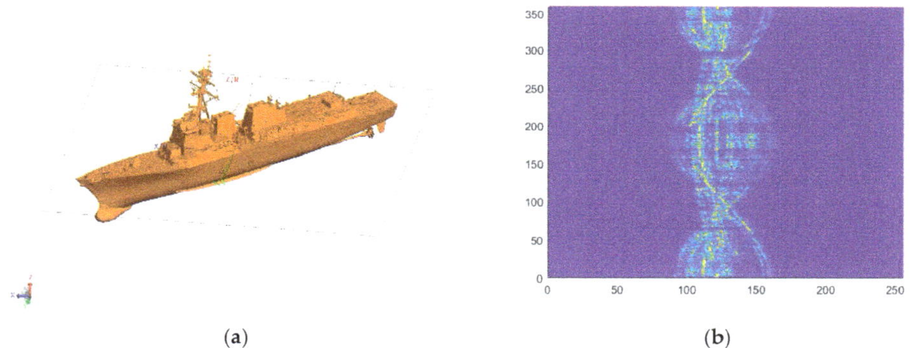

(a)   (b)

Figure 5. Ship model and the graph of HRRP after amplitude normalization. (a) Ship model; (b) The graph of HRRP after amplitude normalization.

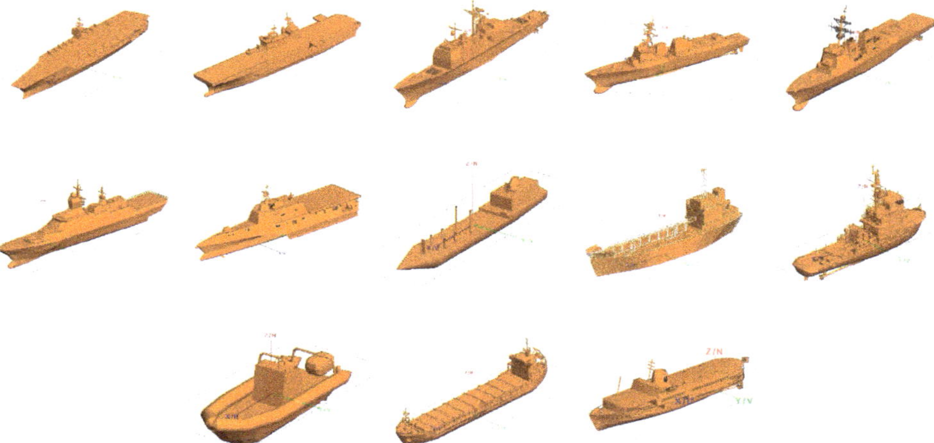

Figure 6. Models of all the ship targets.

In Figure 5b, the horizontal axis and the vertical axis represent HRRP length and azimuth angle, respectively. Each ship acquires 360 HRRP data. To meet the requirement of the data amount of the sample during neural network training and prevent over-fitting, the dataset should be expanded. The process is as follows:

1. Translation interception of HRRP. As revealed by Figure 5, when HRRP is calculated, the coordinate axis coincides with the center of the ship, so the effective HRRP information is generally in the middle region. However, when the radar detects the target, the echo signal may be incomplete or partially missing. Accordingly, the first step of data expansion is the translation interception of HRRP. Since each HRRP is one-dimensional data, only a one-dimensional translation interception is applied. The HRRP is shifted to the left and right by 32 and 64 range cells in turn. The data removed is discarded, and the blank part is supplemented with 0, as presented in Figure 7. The number of samples is increased to 5 times by taking those HRRP samples that overlap but are not identical. It should be noted that the translation interception of HRRP is to simulate the partially missing echo signal, and there is no spatial transformation performed on the object during the HRRP acquisition and expansion process.

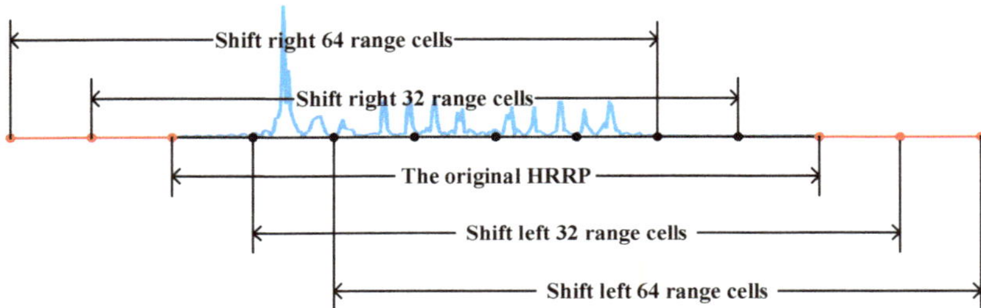

Figure 7. Interception of the HRRP.

2. Random noise is added to the translated HRRP data. Gaussian white noise was added to the data 10 times, and the data after adding noise meets a certain SNR.

2/3 of the target data of each class of ship are randomly taken as the training dataset and 1/3 as the testing dataset. In the database, the training dataset samples and the testing dataset samples were 156,000 and 78,000, respectively.

4.2. Model Identification Performance Analysis

In this section, the performance of the presented model is analyzed in three aspects. The first part primarily shows the effect of different loss functions on the recognition effect. The second part primarily analyzes the advantages of the presented model compared with the comparison model. The third part primarily analyzes the enhancement of model complexity to recognition effect.

The experiment was conducted under the following circumstances. Operating system: Windows 10. Memory: 64 GB. Video memory: 11 GB. GPU: NVIDIA GeForce RTX 2080 Ti. CPU: Intel(R) Xeon(R) w-2125 CPU @4.00GHz.

All the networks were trained from scratch. The iterations were set to 200. The learning rate began with 0.01, and it was halved every 20 training iterations. The Adam optimizer was employed to update the network weight. The batch gradient descent method was applied, and the number of training samples per batch was 512.

4.2.1. Effect of Loss Function on Recognition Effect

The hyper-parameters of the presented model are limited to three types: the number of convolutional modules, the number of left branches in the modules and the parameters of the joint loss function.

To verify the effectiveness of the structure and loss function proposed, model A, with low complexity, is built first. The number of convolutional modules in model A is 4, and the number of left branches inside the module is 3. The parameters of the joint loss function are fine-tuned in accordance with the identification effect. Table 1 elucidates the structure and parameters of each stage in model A. After each convolutional layer, there are batch normalization and Relu activation operations. The number of parameters in the respective stages covers convolution kernel parameters and batch normalization parameters. For instance, the number of parameters of the initial convolutional layer is 63 + 36, suggesting 63 convolutional kernel parameters and 36 batch normalized parameters, respectively. The total number of parameters of model A is 37,538.

Table 1. Details of structure and parameters of each stage in model A.

Stage	Output Size	Structure		Number of Parameters
		Left branch	Right branch	
Initial convolutional layer	$128 \times 1 \times 9$	$7 \times 1, 9, s = 2$		$63 + 36$
Convolutional module 1	$64 \times 1 \times 18$	$1 \times 1, 9$ $3 \times 1, 3, s = 2, x = 3$ $1 \times 1, 12$	$1 \times 1, 15, s = 2$	$405 + 180$
Convolutional module 2	$32 \times 1 \times 36$	$1 \times 1, 18$ $3 \times 1, 6, s = 2, x = 3$ $1 \times 1, 24$	$1 \times 1, 30, s = 2$	$1620 + 360$
Convolutional module 3	$16 \times 1 \times 72$	$1 \times 1, 36$ $3 \times 1, 12, s = 2, x = 3$ $1 \times 1, 48$	$1 \times 1, 60, s = 2$	$6480 + 720$
Convolutional module 4	$8 \times 1 \times 144$	$1 \times 1, 72$ $3 \times 1, 24, s = 2, x = 3$ $1 \times 1, 96$	$1 \times 1, 120, s = 2$	$25,920 + 1440$
Fully connected layer 1	144	Global max pooling and global average pooling		0
Fully connected layer 2	2			288
Output layer	13	Joint loss function		26
Total number of parameters			37,538	

First, the effect of different loss functions on the recognition effect is compared under the structure of model A. The loss functions participating in the comparison refer to L_S, L_{AMS} and L_{SC}.

Classification Effect Comparison of Loss Function L_{AMS} and Loss Function L_S

The hyper-parameter μ in L_{AMS} constrains the boundaries between features and s scales the cosine values. In [35], it was reported that the s will not increase, and the network converges in a relatively slow manner if the s is set to be learned. Thus, s is fixed at 30, which is a sufficiently large value. Thus, experiments are performed to delve into the sensitivity of parameter μ.

In the dataset with SNR of 0, 5, 10 and 15 dB, respectively. s is fixed to 30 and μ varies from 0 to 1 to compare the recognition accuracy of model A using loss function L_{AMS} and L_S. The recognition accuracy is obtained by calculating the percentage of correctly classified samples in the testing dataset in the total number of samples, and the simulation results are presented in Figure 8. As suggested by Figure 8, compared with the conventional loss function L_S, the use of loss function L_{AMS} improves the model recognition accuracy under different SNR conditions to a certain extent. In addition, the lower the SNR of the dataset is, the greater the enhancement in recognition accuracy. At different SNR, with the rise in the boundary constraint strength μ, the enhancement of recognition accuracy generally presents a downward trend. It is also noted that the effective range of boundary constraint strength is small when the SNR is low. In Figure 8a, the effective range of μ is only from 0 to 0.25 at SNR of 0 dB. Furthermore, the recognition accuracy after exceeding the range is lower than that with the use of loss function L_S only. When the loss function L_{AMS} is adopted for ship target recognition in our dataset, the value of boundary constraint strength should not be too large; 0.05 is generally appropriate, applying to a larger SNR range.

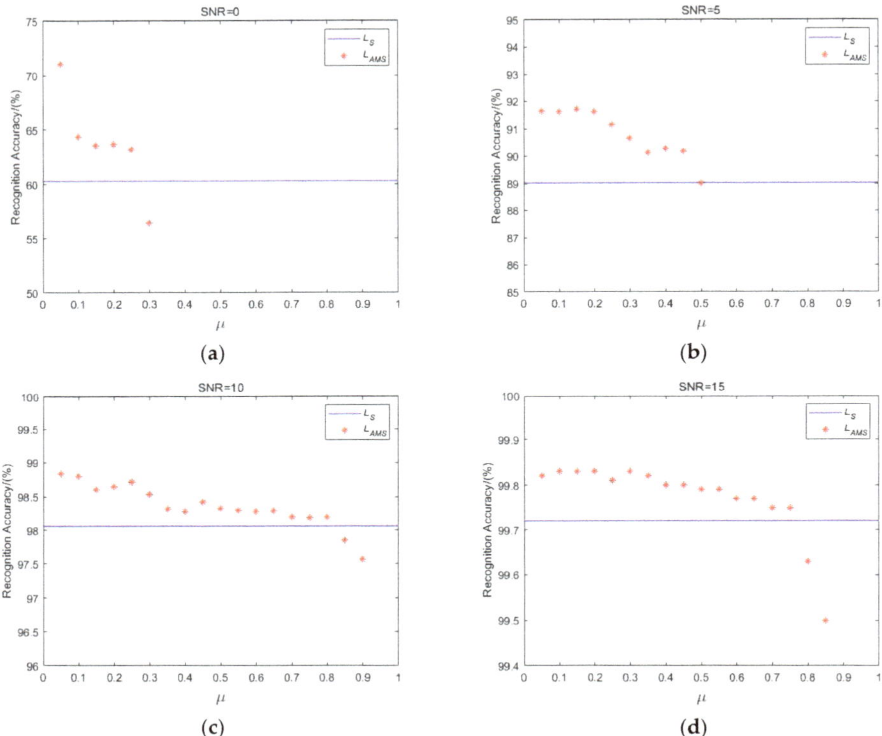

Figure 8. Recognition accuracy of model A with the dataset under different SNR conditions. (**a**) SNR = 0 dB; (**b**) SNR = 5 dB; (**c**) SNR = 10 dB; (**d**) SNR = 15 dB. The blue line suggests the recognition accuracy of model A using the loss function L_S, and the discrete red points indicate the recognition accuracy of model A using the loss function L_{AMS} under different μ.

To show the effect of loss function L_{AMS} on the separability of features extracted from model A more intuitively, when the SNR is 15 dB, the testing dataset is visualized with the 2d features of the second full-connection layer in model A, as shown in Figure 9. It can be seen that after the loss function L_{AMS} is used, the corner space occupied by the extracted features in sample of each class becomes smaller, the inter-class variations of features become larger, and the features are more separable. It is also noted that the scale of the feature increases with the use of the loss function L_{AMS}. That is, the features of the same class become more slender in terms of spatial distribution.

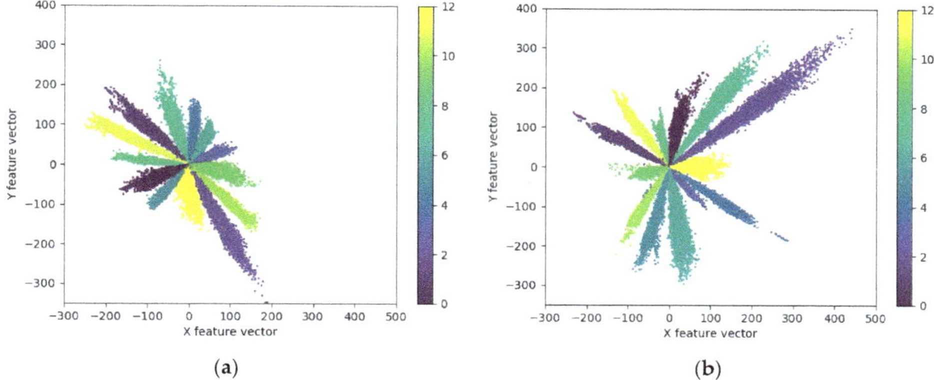

Figure 9. Visualization of the feature extracted by model A with a variety of loss functions at the SNR of 15 dB. (**a**) With the loss function L_S; (**b**) With the loss function L_{AMS} and $\mu = 0.5$. Each data point represents the two-dimensional feature extracted from HRRP data by the model A, and different colors represent different target types. The total number of types is 13.

Classification Effect Comparison of Loss Function L_{SC} and Loss Function L_S

When the weight λ is introduced to fuse the loss function L_S with the loss function L_C, it yields L_{SC}. The hyper-parameter α in L_{SC} is adopted to control the learning rate of center for the features, and λ is applied for the balance of the two functions. Experimental results reveal that when the learning rate α varies, the recognition accuracy fluctuates slightly. Here, to simplify model design and optimization, the learning rate α is directly fixed at 0.6. Therefore, we conduct experiments to investigate the sensitivity of parameter λ while the dataset under different SNR conditions. The simulation results are listed in Table 2.

Table 2. Recognition accuracy of model A when the λ in loss function L_{SC} is different while the dataset is under different SNR conditions.

Loss Function and Parameter	Recognition Accuracy(%)			
	SNR = 0 dB	SNR = 5 dB	SNR = 10 dB	SNR = 15 dB
L_S	60.32	89.03	98.06	99.72
$L_{SC}, \lambda = 0.001$	60.45	89.10	98.07	99.73
$L_{SC}, \lambda = 0.005$	60.38	89.08	98.06	99.75
$L_{SC}, \lambda = 0.01$	60.42	89.08	98.07	99.75
$L_{SC}, \lambda = 0.05$	60.46	89.06	98.08	99.74
$L_{SC}, \lambda = 0.1$	60.40	89.11	98.09	99.73
$L_{SC}, \lambda = 0.2$	60.35	89.10	98.09	99.73
$L_{SC}, \lambda = 0.4$	60.37	89.08	98.07	99.74
$L_{SC}, \lambda = 0.6$	60.40	89.08	98.08	99.72
$L_{SC}, \lambda = 0.8$	60.38	89.09	98.07	99.74
$L_{SC}, \lambda = 1$	60.35	89.09	98.07	99.75

Comparing the recognition accuracy in Table 2 with the results in Figure 8 using the loss function L_{AMS}, the loss function L_{SC} is suggested to be more robust to noise, whereas it has a limited effect on the enhancement of recognition accuracy, indicating that reducing the intra-class variations of features alone cannot significantly enhance the recognition effect of the model. To show the process of establishing the center of features extracted by the loss function L_{SC}. When SNR of the dataset is 15 dB and the weight λ is 0.6, the 2d features of the second fully connected layer of the dataset in model A are visualized for every 50 iterations, as shown in Figure 10.

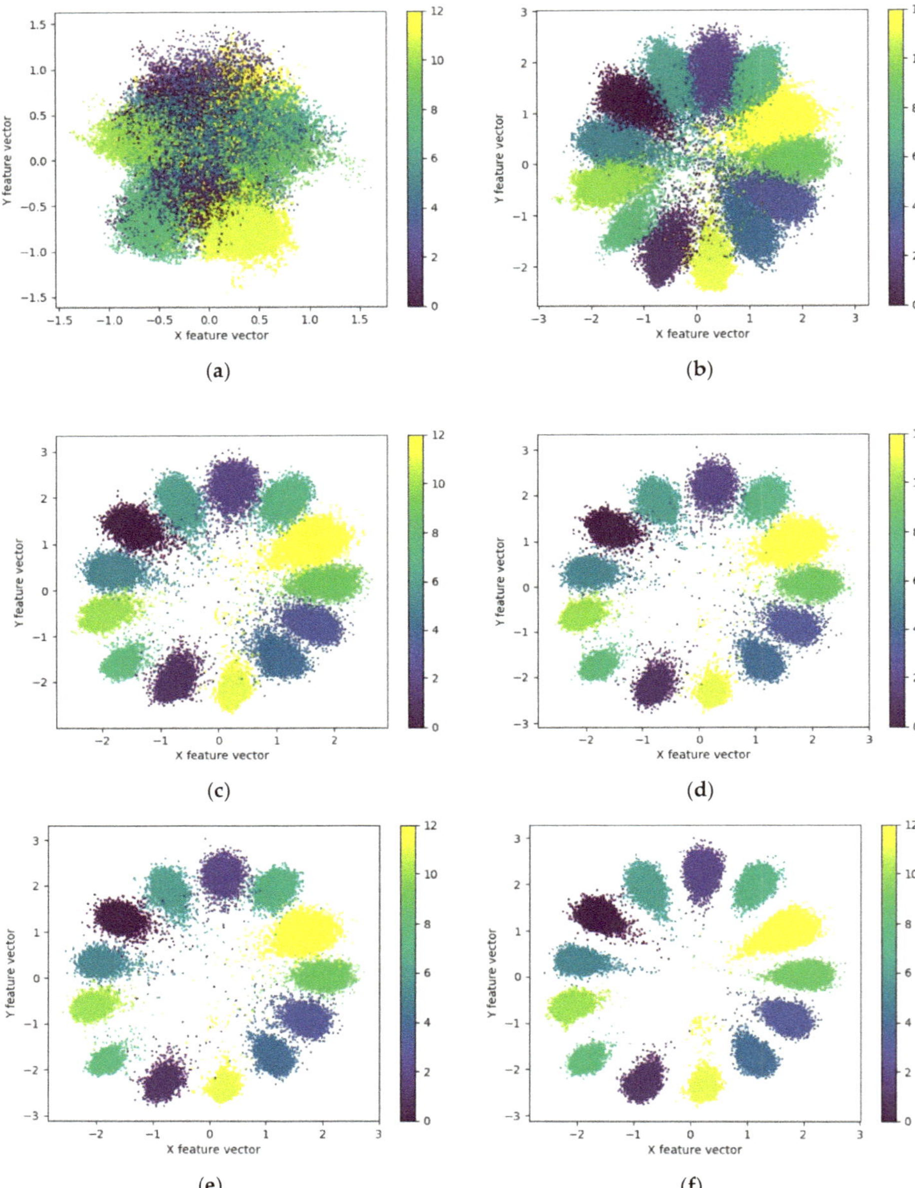

Figure 10. Visualization of the feature extracted by model A with the loss function L_{SC} during training. (**a**–**e**) Feature visualization of testing dataset, and (**f**) feature visualization of training dataset. (**a**) Iterations = 1 and recognition accuracy = 0.3274; (**b**) Iterations = 50 and recognition accuracy = 0.9489; (**c**) Iterations = 100 and recognition accuracy = 0.9968; (**d**) Iterations = 150 and recognition accuracy = 0.9968; (**e**) Iterations = 200 and recognition accuracy = 0.9972; (**f**) Iterations = 200 and recognition accuracy = 0.9988.

As suggested by Figure 10a, the initial features of each class are inseparable, and the initial recognition accuracy is only about 0.3274. With the increase in iteration times and the constant updating of parameters, the features of various samples are gradually separated and concentrated in their category centers. With the enhancement of feature separability, the model recognition accuracy rises. The comparison between (c) and (d) in Figure 10 suggests that though the recognition accuracy of the model is not improved between 100 and 150 iterations, the features of various samples are more clustered, and the intra-class variations of features are gradually decreased. As suggested by the comparison of Figure 10e,f, although the training dataset exhibits stronger feature separability and higher recognition accuracy, the testing dataset have similar feature distribution and recognition accuracy. It is, therefore, revealed that the model has no obvious overfitting and the extracted features have good generalization performance. Compared with the visualization of features extracted by model A when loss function L_{AMS} and L_S are used in Figure 9, the feature scale extracted by loss function L_{SC} is smaller, and the distribution range is narrowed from $[-400,400]$ to $[-3,3]$. The distribution of features in space varies from divergence to aggregation by class, and the intra-class difference is smaller.

Classification Effect Comparison between Loss Function L_{SC} and Others

By analyzing the described results, it can be concluded that the boundary constraint strength μ of the loss function L_{AMS} can significantly improve the recognition accuracy. However, when the SNR is low, the value of μ should not be overly large. The weight λ of loss function L_{SC} has better adaptability and can improve the intra-class aggregation effect of features within a larger value range, but the enhancement of recognition accuracy is limited.

In this section, we verify the enhancement of recognition accuracy with the joint loss function L_{AMSC}, where s, α and μ are fixed at 30, 0.6 and 0.05, respectively. When λ is taken to have different values, the recognition accuracy of model A under different SNR conditions is listed in Table 3.

Table 3. Recognition accuracy of model A when the λ in loss function L_{AMSC} is different while the dataset is under different SNR conditions.

Loss Function and Parameter	Recognition Accuracy(%)			
	SNR = 0 dB	SNR = 5 dB	SNR = 10 dB	SNR = 15 dB
L_{AMSC}, $\lambda = 0.001$, $\mu = 0.05$	71.26	**93.39**	98.91	99.89
L_{AMSC}, $\lambda = 0.01$, $\mu = 0.05$	70.58	93.16	99.03	99.89
L_{AMSC}, $\lambda = 0.1$, $\mu = 0.05$	72.28	92.90	99.08	99.91
L_{AMSC}, $\lambda = 0.2$, $\mu = 0.05$	69.70	92.46	98.89	99.91
L_{AMSC}, $\lambda = 0.3$, $\mu = 0.05$	70.61	93.09	99.00	99.90
L_{AMSC}, $\lambda = 0.4$, $\mu = 0.05$	71.14	92.36	98.99	99.90
L_{AMSC}, $\lambda = 0.6$, $\mu = 0.05$	69.78	92.49	99.00	99.88
L_{AMSC}, $\lambda = 0.8$, $\mu = 0.05$	70.92	92.85	99.08	99.91
L_{AMSC}, $\lambda = 1$, $\mu = 0.05$	71.41	93.26	99.06	99.91
L_{AMS}	65.03	91.72	98.84	99.84
L_{SC}	60.46	89.11	98.09	99.75
L_S	60.32	89.03	98.06	99.72

In Table 3, the recognition accuracy of model A is given when the λ in the joint loss function L_{AMSC} is taken to have different values. In the meantime, it also shows the recognition accuracy when using loss function L_{AMS}, L_{SC} and L_S. Among them, the loss function L_{AMS} and L_{SC} show the best recognition accuracy when assuming different values of parameters. As suggested by Table 3, when the joint loss function L_{AMSC} is used, the recognition accuracy is improved stably under different SNR conditions. In addition, when the SNR of the dataset is relatively low, the enhancement is greater. When the value of λ is 0.001, 0.01, 0.1 and 1, respectively. we visualize the features extracted by model A, as shown in Figure 11.

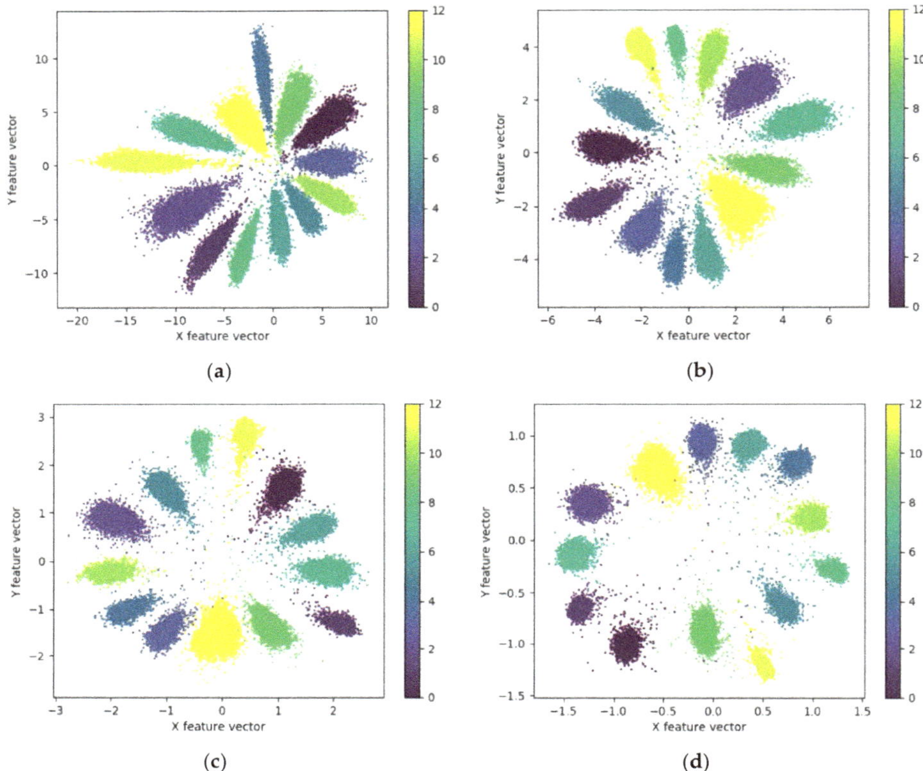

Figure 11. Visualization of the feature extracted by model A with different value of λ. (**a**) $\lambda = 0.001$; (**b**) $\lambda = 0.01$; (**c**) $\lambda = 0.1$; (**d**) $\lambda = 1$.

As suggested by Figure 11, with increasing value of λ, the intra-class differences of features gradually become smaller, and the features of different types gradually converge to the center of the class. The spatial distribution range of features is narrowed from [−20, 15] to [−1.5, 1.5]. As suggested by the recognition accuracy and the intra-class aggregation of features, the recognition effect is identified the optimal at the value of λ as 0.1.

4.2.2. Analysis of the Recognition Effect of the Presented Model and the Comparison Model

In this section, the common target recognition algorithm based on HRRP is selected as the comparison model to verify the effectiveness of the presented model and loss function. Conventional comparison algorithms based on machine learning include: KNN [36], LSVM [37], RBF-SVM [38], RF [39] and NB [40]. The comparison algorithms based on neural network includes: CNN [18], Stack Sparse Auto Encoder and K-Nearest Neighbor (sDSAE&KNN) [24], Stack Convolutional Auto Encoder (SCAE) [41]. For the highest recognition accuracy, the hyper-parameters in the comparison algorithm are fine-tuned. Tables 4–6 elucidate the structure and parameters of each comparison model based on the neural network. The pooling layer in each model is max-pooling, and batch normalization is performed after the convolutional layer in CNN.

Table 4. Details of structure and parameters of CNN.

Stage	Output Size	Structure	Number of Parameters
Convolutional layer 1	$256 \times 1 \times 8$	$3 \times 1, 8, s = 1$	$32 + 32$
Pooling layer 1	$128 \times 1 \times 8$	$2 \times 1, s = 2$	0
Convolutional layer 2	$128 \times 1 \times 16$	$3 \times 1, 16, s = 1$	$400 + 64$
Pooling layer 2	$64 \times 1 \times 16$	$2 \times 1, s = 2$	0
Convolutional layer 3	$64 \times 1 \times 32$	$3 \times 1, 32, s = 1$	$1568 + 128$
Pooling layer 3	$32 \times 1 \times 32$	$2 \times 1, s = 2$	0
Convolutional layer 4	$32 \times 1 \times 64$	$3 \times 1, 64, s = 1$	$6208 + 256$
Pooling layer 4	$16 \times 1 \times 64$	$2 \times 1, s = 2$	0
Convolutional layer 5	$16 \times 1 \times 64$	$1 \times 1, 64, s = 1$	$4160 + 256$
Pooling layer 5	$8 \times 1 \times 64$	$2 \times 1, s = 2$	0
Fully connected layer 1	64		32,832
Fully connected layer 2	2		130
Output layer	13	L_s	39
Total number of parameters			46,105

Table 5. Details of structure and parameters of sDSAE&KNN.

Stage	Output Size	Number of Parameters
Hidden layer 1	150×1	38,550
Hidden layer 2	100×1	15,100
Hidden layer 3	50×1	5050
Hidden layer 4	10×1	510
Total number of parameters		59,210

Table 6. Details of structure and parameters of SCAE.

Stage	Output Size	Structure	Number of Parameters
Convolutional layer 1	$256 \times 1 \times 128$	$5 \times 1, 128, s = 1$	768
Pooling layer 1	$128 \times 1 \times 128$	$2 \times 1, s = 2$	0
Convolutional layer 2	$128 \times 1 \times 64$	$5 \times 1, 64, s = 1$	41,024
Pooling layer 2	$64 \times 1 \times 64$	$2 \times 1, s = 2$	0
Convolutional layer 3	$64 \times 1 \times 32$	$3 \times 1, 32, s = 1$	6176
Pooling layer 3	$32 \times 1 \times 32$	$2 \times 1, s = 2$	0
Convolutional layer 4	$32 \times 1 \times 16$	$3 \times 1, 16, s = 1$	1552
Pooling layer 4	$16 \times 1 \times 16$	$2 \times 1, s = 2$	0
Convolutional layer 5	$16 \times 1 \times 8$	$1 \times 1, 8, s = 1$	136
Pooling layer 5	$8 \times 1 \times 8$	$2 \times 1, s = 2$	0
Output layer	13	L_s	845
Total number of parameters			50,501

Since the complexity of the model is associated with the recognition accuracy, the number of parameters of each model is similar to model A when the comparison model based on neural network is designed. The total parameters of the model are employed here to represent the complexity of the model. As suggested by the table above, the complexity of each model based on neural network is shown in descending sequence: sDSAE&KNN, SCAE, CNN, model A.

First, the recognition effect of all models was compared using the dataset under the condition SNR = 5 dB. The recognition accuracy of each model is shown in Figure 12.

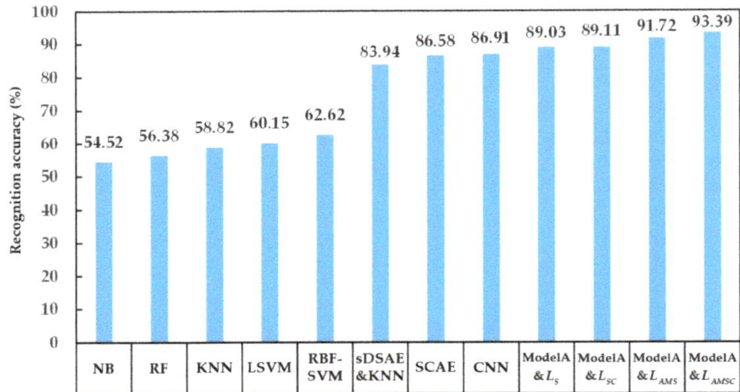

Figure 12. Recognition accuracy of different models when the dataset SNR is 5 dB.

Figure 12 shows the best recognition accuracy of the comparison model and the model A with a variety of loss functions. As suggested in Figure 12, each model based on the proposed structure (model A) achieves better recognition effect. Additionally, the recognition of model A combined with the joint loss function L_{AMSC} exhibits the highest accuracy among all models. The effectiveness of the proposed structure and the joint loss function is verified, respectively.

In the meantime, the recognition effect based on neural network model appears to be generally better than that based on conventional machine learning model. In the neural network models, the model including the convolution kernel (model A, CNN, SCAE) can achieve a prominent recognition effect. In the meantime, the model based on the convolutional neural network (model A, CNN) outperforms the model based on the auto-encoder (SCAE, sDSAE&KNN). During the expansion of the dataset, translation interception is performed to simulate target occlusion and information loss in the echo signal to some extent. The convolutional neural network-based recognition exhibits higher accuracy, revealing that the convolution kernel helps the model extract the effective separable features of different target echo signals, achieve better recognition effect, and avoid being adversely affected by incomplete echo signal information.

Under different SNR conditions, the optimal recognition results of each model based on neural network are listed in Table 7.

Table 7. Recognition accuracy of model A and the comparison model under different SNR conditions.

Model Name	Number of Parameters	Computational Time for Each HRRP (us)	Recognition Accuracy (%)			
			SNR = 0 dB	SNR = 5 dB	SNR = 10 dB	SNR = 15 dB
Model A&L_{AMSC}	37538	258	72.28	93.39	99.08	99.91
CNN	46105	69	58.22	86.91	95.51	98.79
SCAE	50501	47	54.78	86.58	94.44	98.78
sDSAE&KNN	59210	68	46.50	83.94	93.44	98.65

As suggested by the recognition results in Table 7, the recognition accuracy of each model noticeably impacts SNR. Additionally, the recognition accuracy of each model is enhanced with the rise in SNR of the dataset. Compared with the comparison model, model A exhibits the least number of parameters and the least complexity, whereas the highest recognition accuracy is achieved under different SNR datasets. By enhancing the network structure and loss function, the presented model achieves better recognition effect with less model complexity and exhibits higher generalization performance and noise robustness. It should be noted that the calculation process of the model proposed is more complicated, so it takes more time to identify the target.

4.2.3. Effect of Model Complexity on Recognition Effect

The mentioned experimental results verify the effectiveness of the structure and loss function proposed. Since the recognition accuracy of the model displays a positive correlation with the depth and width of the model within a certain range. In the present section, different parameters will be selected, three models with different complexity will be designed, and their recognition effects will be compared. Model A refers to the model adopted in Section 4.2.1. Model B is developed by up-regulating the number of convolutional modules in model A to 5, and Model C is obtained by up-regulating the number of branches in the left branch of model A to 6. The details of the structure and parameters of each stage in model B and C are listed in Tables 8 and 9.

Table 8. Details of structure and parameters of each stage in model B.

Stage	Output Size	Structure		Number of Parameters
Initial convolutional layer	$128 \times 1 \times 9$	$7 \times 1, 9, s = 2$		$63 + 36$
		Left branch	Right branch	
Convolutional module 1	$64 \times 1 \times 18$	$1 \times 1, 9$ $3 \times 1, 3, s = 2, x = 3$ $1 \times 1, 12$	$1 \times 1, 15, s = 2$	$405 + 180$
Convolutional module 2	$32 \times 1 \times 36$	$1 \times 1, 18$ $3 \times 1, 6, s = 2, x = 3$ $1 \times 1, 24$	$1 \times 1, 30, s = 2$	$1620 + 360$
Convolutional module 3	$16 \times 1 \times 72$	$1 \times 1, 36$ $3 \times 1, 12, s = 2, x = 3$ $1 \times 1, 48$	$1 \times 1, 60, s = 2$	$6480 + 720$
Convolutional module 4	$8 \times 1 \times 144$	$1 \times 1, 72$ $3 \times 1, 24, s = 2, x = 3$ $1 \times 1, 96$	$1 \times 1, 120, s = 2$	$25,920 + 1440$
Convolutional module 5	$4 \times 1 \times 288$	$1 \times 1, 144$ $3 \times 1, 48, s = 2, x = 3$ $1 \times 1, 192$	$1 \times 1, 240, s = 2$	$103,680 + 2880$
Fully connected layer 1	144	Global max pooling and global average pooling		0
Fully connected layer 2	2			578
Output layer	13	Joint loss function		26
Total number of parameters		144,353		

Table 9. Details of structure and parameters of each stage in model C.

Stage	Output Size	Structure		Number of Parameters
Initial convolutional layer	$128 \times 1 \times 18$	$7 \times 1, 18, s = 2$		$126 + 72$
		Left branch	Right branch	
Convolutional module 1	$64 \times 1 \times 36$	$1 \times 1, 18$ $3 \times 1, 3, s = 2, x = 6$ $1 \times 1, 24$	$1 \times 1, 30, s = 2$	$1458 + 360$
Convolutional module 2	$32 \times 1 \times 72$	$1 \times 1, 36$ $3 \times 1, 6, s = 2, x = 6$ $1 \times 1, 48$	$1 \times 1, 60, s = 2$	$5832 + 720$
Convolutional module 3	$16 \times 1 \times 144$	$1 \times 1, 72$ $3 \times 1, 12, s = 2, x = 6$ $1 \times 1, 96$	$1 \times 1, 120, s = 2$	$23{,}328 + 1440$
Convolutional module 4	$8 \times 1 \times 288$	$1 \times 1, 144$ $3 \times 1, 24, s = 2, x = 6$ $1 \times 1, 192$	$1 \times 1, 240, s = 2$	$93{,}312 + 2880$
Fully connected layer 1	288	Global max pooling and global average pooling		0
Fully connected layer 2	2			578
Output layer	13	Joint loss function		26
Total number of parameters			13,0132	

Under different SNR conditions, the optimal recognition results of each model are listed in Table 10. The proposed joint loss function L_{AMSC} is used in all models, and the values of each parameter of the loss function are $s = 30$, $\alpha = 0.6$, $\mu = 0.05$, $\lambda = 0.1$.

Table 10. Recognition accuracy of different complexity models under different SNR conditions.

Model Name	Number of Parameters	Computational Time for Each HRRP (us)	Recognition Accuracy (%)			
			SNR = 0 dB	SNR = 5 dB	SNR = 10 dB	SNR = 15 dB
Model A	37538	258	72.28	92.90	99.08	99.91
Model B	144353	326	77.12	95.28	99.49	99.93
Model C	130132	323	76.31	95.50	99.43	99.93

As revealed by Table 10, the depth and width of the model directly impact the recognition effect. Compared with model A, models B and C both enhance the recognition accuracy noticeably. In particular, when the SNR is low, the enhancement becomes more obvious. Meanwhile, the time required for model A, B and C to calculate each HRRP is also listed in Table 10. It can be seen that compared with model A, the computational times for model B and model C increase due to the increased complexity of the model. However, compared with the increased number of parameters, the increase in calculation time is not large. To compare and delve into the convergence speed of various complexity models, at the dataset SNR of 15 dB, the recognition accuracy and loss curves in the training process are plotted in Figure 13.

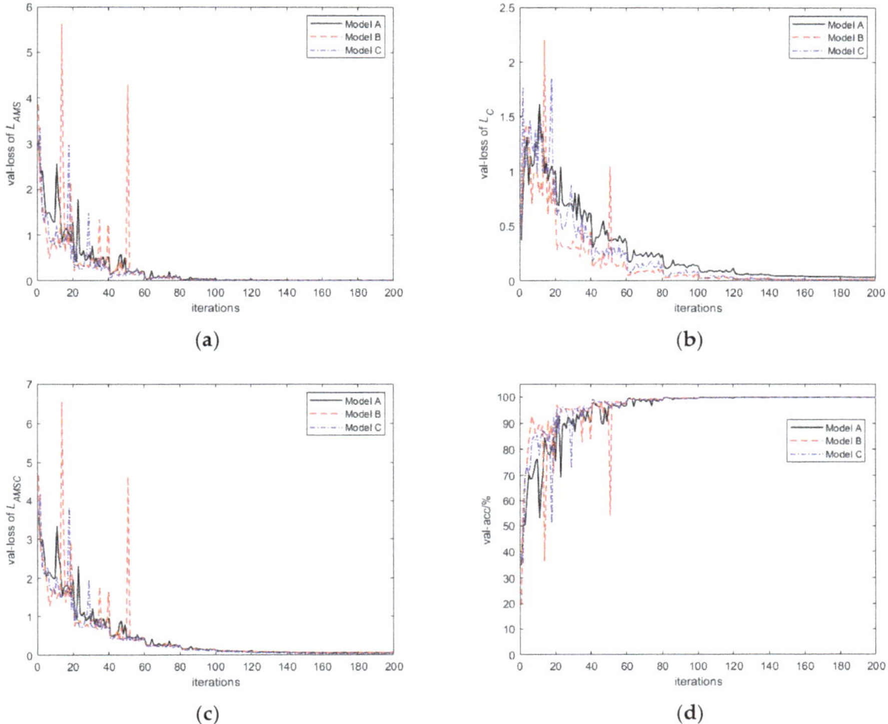

Figure 13. The recognition accuracy and loss curves in the training process at the dataset SNR of 15 dB, where val-loss and val-acc refer to the loss and recognition accuracy of the testing dataset, respectively. L_{AMS} and L_C refer to the combined parts of the joint loss function L_{AMSC}, as shown in Equation (4). (**a**) The loss curve of L_{AMS}; (**b**) The loss curve of L_C; (**c**) The loss curve of L_{AMSC}; (**d**) The accuracy curve.

Figure 13 reveals that the recognition accuracy curve and the loss curve of model B and C fluctuate more dramatically during the training process, whereas they converge faster, as compared with those of model A. In the initial 60 iterations, the loss and the recognition accuracy curves of the testing dataset decline and increase rapidly. After 60 iterations, the model comes to exhibit a relatively high training effect. Subsequently, until the end of training, the loss and recognition accuracy gradually converge to stable values. In the meantime, the loss curve of model A in Figure 13b is always higher than that of models B and C, and a certain gap remains until the end of the training. The L_C in the joint loss function L_{AMSC} indicates the intra-class difference of the features extracted by the model, which suggests that the features extracted by model B and C undergo intra-class aggregation more effectively. The visualization of features also verifies this conclusion. The feature visualization of each model is illustrated in Figures 14 and 15, respectively.

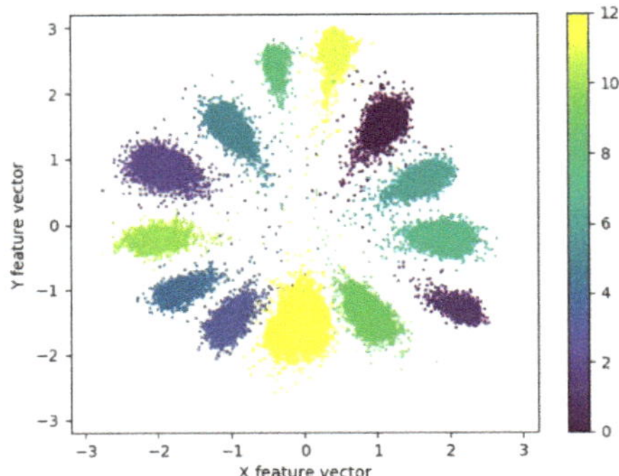

Figure 14. Visualization of the feature extracted by model A.

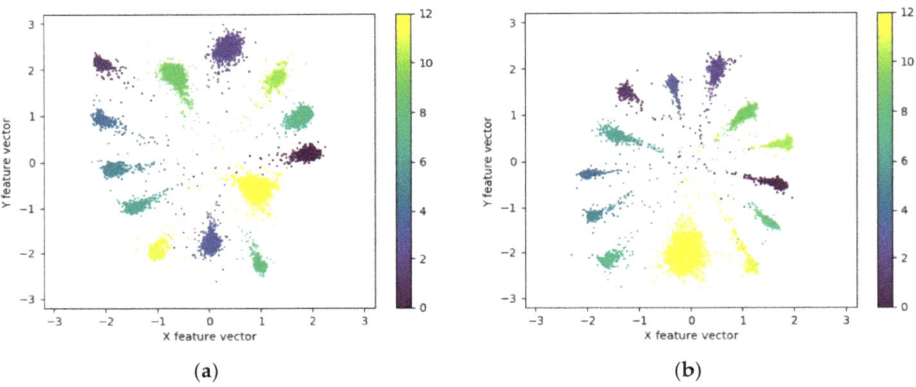

Figure 15. Visualization of the feature extracted by model B and C. (**a**) Model B; (**b**) Model C.

Though the model becomes more complex with the increases in depth and width, the presented model is capable of extracting deeper and more stable separable features in HRRP data for identification, thereby making the model more adaptable to SNR.

5. Conclusions

In this study, a neural network model integrating micro convolutional module and residual structure is proposed to classify ship targets based on HRRP. The model is characterized by few hyper-parameters, has easy to expand properties, and high recognition accuracy. The convolutional module is set as a simple and highly modular network structure that exhibits strong scalability. Based on the left branch structure of convolutional module, the effect of network deepening and widening can be simulated. The skip structure of the right branch is capable of transferring features and gradients more effectively. The presented model can up-regulate the utilization rate of shallow features while lowering the risk of gradient disappearance and recognition rate saturation. In the meantime, a novel loss function combining boundary constraint and center clustering is developed. The features extracted by the novel loss function are characterized by larger inter-class variations, smaller intra-class variations, as well as stronger separability. The effects of loss function and model complexity on recognition accuracy are analyzed by simulation experiments. Compared with other commonly used network structures, the presented model in this study exhibits higher recognition accuracy with fewer model parameters, good generalization performance and robustness.

Author Contributions: Conceptualization, Z.F., S.L. and X.L.; methodology, Z.F. and B.D.; software, Z.F. and X.W.; validation, Z.F. and X.L.; resources, Z.F.; data curation, B.D.; writing—original draft preparation, Z.F. and B.D.; writing—review and editing, Z.F.; visualization, Z.F. and X.W.; project administration, S.L. All authors have read and agreed to the published version of the manuscript.

Funding: This research received no external funding.

Acknowledgments: The authors would like to thank the editors and anonymous reviewers for the valuable comments and suggestions.

Conflicts of Interest: The authors declare no conflict of interest.

References

1. Li, Y.; Yu, L.; Yang, Y. Method of Aerial Target Length Extraction Based on High Resolution Range Profile. *Mod. Radar* **2018**, *07*, 32–35. [CrossRef]
2. Wei, C.; Duan, F.; Liu, X. Length estimation method of ship target based on wide-band radar's HRRP. *Syst. Eng. Electron.* **2018**, *40*, 1960–1965. [CrossRef]
3. He, S.; Zhao, H.; Zhang, Y. Signal Separation for Target Group in Midcourse Based on Time-frequency Filtering. *J. Radars* **2015**, *05*, 545–551. [CrossRef]
4. Chen, M.; Wang, S.; Ma, T.; Wu, X. Fast analysis of electromagnetic scattering characteristics in spatial and frequency domains based on compressive sensing. *Acta Phys. Sin.* **2014**, *17*, 50–54. [CrossRef]
5. Liu, S. Research on Feature Extraction and Recognition Performance Enhancement Algorithms Based on High Range Resolution Profile. Ph.D. Dissertation, National University of Defense Technology, Changsha, China, 2016.
6. Wu, J.; Chen, Y.; Dai, D.; Chen, S.; Wang, X. Target Recognition for Polarimetric HRRP Based on Fast Density Search Clustering Method. *J. Electron. Inf. Technol.* **2016**, *10*, 2461–2467. [CrossRef]
7. Zhou, Y.; Wang, H.; Xu, F.; Jin, Y. Polarimetric SAR Image Classification Using Deep Convolutional Neural Networks. *IEEE Geosci. Remote Sens. Lett.* **2016**, *13*, 1935–1939. [CrossRef]
8. Pei, J.; Huang, Y.; Sun, Z.; Zhang, Y.; Yang, J.; Yeo, T. Multiview synthetic aperture radar automatic target recognition optimization: Modeling and implementation. *IEEE Trans. Geosci. Remote. Sens.* **2018**, *56*, 6425–6439. [CrossRef]
9. Fu, H.; Li, Y.; Wang, Y.; Li, P. Maritime Ship Targets Recognition with Deep Learning. In Proceedings of the 37th Chinese Control Conference (CCC), Wuhan, China, 25–27 July 2018; pp. 9297–9302.
10. Xing, S.; Zhang, S. Ship model recognition based on convolutional neural networks. In Proceedings of the 2018 IEEE International Conference on Mechatronics and Automation (ICMA), Changchun, China, 5–8 August 2018; pp. 144–148.
11. Luo, C.; Yu, L.; Ren, P. A Vision-Aided Approach to Perching a Bioinspired Unmanned Aerial Vehicle. *IEEE Trans. Ind. Electron.* **2018**, *65*, 3976–3984. [CrossRef]
12. Li, C.; Hao, L. High-Resolution, Downward-Looking Radar Imaging Using a Small Consumer Drone. In Proceedings of the 2016 IEEE Antennas and Propagation Society International Symposium (APSURSI), Fajardo, Puerto Rico, 26 June–1 July 2016; pp. 2037–2038.
13. Yin, H.; Guo, Z. Radar HRRP target recognition with one-dimensional CNN. *Telecommun. Eng.* **2018**, *58*, 1121–1126. [CrossRef]
14. Karabayir, O.; Yucedag, O.M.; Kartal, M.Z.; Serim, H.A. Convolutional neural networks-based ship target recognition using high resolution range profiles. In Proceedings of the 18th International Radar Symposium (IRS), Prague, Czech Republic, 28–30 June 2017; pp. 1–9.
15. Lunden, J.; Koivunen, V. Deep learning for HRRP-based target recognition in multistatic radar systems. In Proceedings of the 2016 IEEE Radar Conference, Philadelphia, PA, USA, 2–6 May 2016; pp. 1–6.
16. Gai, Q.; Han, Y.; Nan, H.; Bai, Z.; Sheng, W. Polarimetric radar target recognition based on depth convolution neural network. *Chin. J. Radio Sci.* **2018**, *33*, 575–582. [CrossRef]
17. Visentin, T.; Sagainov, A.; Hasch, J.; Zwick, T. Classification of objects in polarimetric radar images using CNNs at 77 GHz. In Proceedings of the 2017 IEEE Asia Pacific Microwave Conference (APMC), Kuala Lumpur, Malaysia, 13–16 November 2017; pp. 356–359.
18. Yang, Y.; Sun, J.; Yu, S.; Peng, X. High Resolution Range Profile Target Recognition Based on Convolutional Neural Network. *Mod. Radar* **2017**, *39*, 24–28. [CrossRef]

19. Yu, S.; Xie, Y. Application of a convolutional autoencoder to half space radar hrrp recognition. In Proceedings of the 2018 International Conference on Wavelet Analysis and Pattern Recognition (ICWAPR), Chengdu, China, 15–18 July 2018; pp. 48–53.

20. Feng, B.; Chen, B.; Liu, H. Radar HRRP target recognition with deep networks. *Pattern Recogn.* **2017**, *61*, 379–393. [CrossRef]

21. Wang, C.; Hu, Y.; Li, X.; Wei, W.; Zhao, H. Radar HRRP target recognition based on convolutional sparse coding and multi-classifier fusion. *Syst. Eng. Electron* **2018**, *11*, 2433–2437. [CrossRef]

22. Zhang, H. RF Stealth Based Airborne Radar System Simulation and HRRP Target Recognition Research. Master's Dissertation, Nanjing University of Aeronautics and Astronautics, Nanjing, China, 2016.

23. Zhang, J.; Wang, H.; Yang, H. Dimension reduction method of high resolution range profile based on Autoencoder. *J. PLA Univ. Sci. Technol. (Nat. Sci. Ed.)* **2016**, *17*, 31–37. [CrossRef]

24. Zhao, F.; Liu, Y.; Huo, K. Radar Target Recognition Based on Stacked Denoising Sparse Autoencoder. *J. Radars* **2017**, *6*, 149–156. [CrossRef]

25. Krizhevsky, A.; Sutskever, I.; Hinton, G.E. ImageNet classification with deep convolutional neural networks. In Proceedings of the 26th Annual Conference on Neural Information Processing Systems (NIPS), Lake Tahoe, NV, USA, 3–6 December 2012; pp. 1097–1105.

26. Simonyan, K.; Zisserman, A. Very deep convolutional networks for large-scale image recognition. In Proceedings of the 3rd International Conference on Learning Representations (ICLR), San Diego, CA, USA, 7–9 May 2015; pp. 1–14.

27. He, K.; Zhang, X.; Ren, S.; Sun, J. Deep residual learning for image recognition. In Proceedings of the 29th IEEE Conference on Computer Vision and Pattern Recognition (CVPR), Las Vegas, NV, USA, 26 June–1 July 2016; pp. 770–778.

28. Glorot, X.; Bengio, Y. Understanding the difficulty of training deep feedforward neural networks. In Proceedings of the 13th International Conference on Artificial Intelligence and Statistics (AISTATS), Sardinia, Italy, 13–15 May 2010; pp. 249–256.

29. He, K.; Zhang, X.; Ren, S.; Sun, J. Delving Deep into Rectifiers: Surpassing Human-Level Performance on ImageNet Classification. In Proceedings of the 2015 IEEE International Conference on Computer Vision (ICCV), Santiago, Chile, 7–13 December 2015; pp. 1026–1034.

30. Szegedy, C.; Ioffe, S.; Vanhoucke, V.; Alemi, A.A. Inception-v4, Inception-ResNet and the Impact of Residual Connections on Learning. In Proceedings of the 31st AAAI Conference on Artificial Intelligence (AAAI), San Francisco, CA, USA, 4–10 February 2017; pp. 4278–4284.

31. Liu, W.; Wen, Y.; Yu, Z.; Li, M.; Raj, B.; Song, L. SphereFace: Deep hypersphere embedding for face recognition. In Proceedings of the 30th IEEE Conference on Computer Vision and Pattern Recognition (CVPR), Honolulu, HI, USA, 21–26 July 2017; pp. 6738–6746.

32. Liu, W.; Wen, Y.; Yu, Z.; Yang, M. Large-Margin Softmax Loss for Convolutional Neural Networks. In Proceedings of the 33th International Conference on Machine Learning (ICML), New York, NY, USA, 19–24 June 2016; pp. 1–10.

33. Wang, F.; Xiang, X.; Cheng, J.; Yuille, A.L. NormFace: L2 Hypersphere Embedding for Face Verification. In Proceedings of the 25th ACM International Conference on Multimedia, Mountain View, CA, USA, 23–27 October 2017; pp. 1041–1049.

34. Liu, Y.; Liu, Q. Convolutional neural networks with large-margin softmax loss function for cognitive load recognition. In Proceedings of the 36th Chinese Control Conference (CCC), Dalian, China, 26–28 July 2017; pp. 4045–4049.

35. Wang, F.; Cheng, J.; Liu, W.; Liu, H. Additive Margin Softmax for Face Verification. *IEEE Signal Process. Lett.* **2018**, *25*, 926–930. [CrossRef]

36. Chen, F.; Du, L.; Bao, Z. Modified KNN rule with its application in radar HRRP target recognition. *J. Xidian Univ.* **2007**, *34*, 681–686.

37. Bao, Z. Study of Radar Target Recognition Based on Continual Learning. Master Dissertation, Xidian University, Xi'an, China, 2018.

38. Huang, Y.; Zhao, K.; Jiang, X. RBF-SVM feature selection arithmetic based on kernel space mean inter-class distance. *Appl. Res. Comput.* **2012**, *29*, 4556–4559. [CrossRef]

39. Yao, L.; Wu, Y.; Cui, G. A New Radar HRRP Target Recognition Method Based on Random Forest. *J. Zhengzhou Univ. (Eng. Sci.)* **2014**, *35*, 105–108. [CrossRef]

40. Yang, K.; Li, S.; Zhang, K.; Niu, S. Research on Anti-jamming Recognition Method of Aerial Infrared Target Based on Naïve Bayes Classifier. *Flight Control Detect.* **2019**, *2*, 62–70.
41. Liu, X. The application of a multi-layers pre-training convolutional neural network in image recognition. Master Dissertation, South–Central University for Nationalities, Wuhan, China, 2018.

Article

MSR2N: Multi-Stage Rotational Region Based Network for Arbitrary-Oriented Ship Detection in SAR Images

Zhenru Pan [1,2,*], Rong Yang [1,2] and Zhimin Zhang [1]

[1] Space Microwave Remote Sensing System, Aerospace Information Research Institute, Chinese Academy of Sciences, Beijing 100190, China; yangrong16@mails.ucas.ac.cn (R.Y.); zmzhang@mail.ie.ac.cn (Z.Z.)
[2] School of Electronic, Electrical and Communication Engineering, University of Chinese Academy of Sciences, Beijing 100039, China
*   Correspondence: panzhenru16@mails.ucas.ac.cn

Received: 25 March 2020; Accepted: 17 April 2020; Published: 20 April 2020

Abstract: In synthetic aperture radar (SAR) images, ships are often arbitrary-oriented and densely arranged in complex backgrounds, posing enormous challenges for ship detection. However, most existing methods detect ships with horizontal bounding boxes, which leads to the redundancy of detected regions. Furthermore, the high Intersection-over-Union (IoU) between two horizontal bounding boxes of densely arranged ships can cause missing detection. In this paper, a multi-stage rotational region based network (MSR2N) is proposed to solve the above problems. In MSR2N, the rotated bounding boxes, which can reduce background noise and prevent missing detection caused by high IoUs, are utilized to represent ship regions. MSR2N consists of three modules: feature pyramid network (FPN), rotational region proposal network (RRPN), and multi-stage rotational detection network (MSRDN). First of all, the FPN is applied to combine high-resolution features with semantically strong features. Second, in RRPN, a rotation-angle-dependent strategy is employed to generate multi-angle anchors which can represent arbitrary-oriented ship regions more felicitously than horizontal anchors. Finally, the MSRDN with three sub-networks is proposed to regress proposals of ship regions stage by stage. Meanwhile, the incrementally increasing IoU thresholds are selected for resampling positive and negative proposals in sequential stages of MSRDN, which eliminates close false positive proposals successively. With the above characteristics, MSR2N is more suitable and robust for ship detection in SAR images. The experimental results on SAR ship detection dataset (SSDD) show that the MSR2N has achieved state-of-the-art performance.

Keywords: synthetic aperture radar (SAR) ship detection; multi-stage rotational region based network (MSR2N); rotated anchor generation; multi-stage rotational detection network (MSRDN)

---

1. Introduction

Ship detection is one of the most significant missions of marine surveillance. With the characteristics of working all-weather, all-time [1], and imaging relatively wide areas at constant resolution [2], synthetic aperture radars (SAR) such as Terra-X, COSMOS-SkyMed, RADARSAT-2, Sentinel-1, and GF-3 are widely applied in ship detection [3–7].

Traditional ship detection methods are mainly based on the following three aspects: (1) statistics characteristics [8–11]; (2) wavelet transform [12,13]; and (3) polarization information [14,15]. Among these methods, constant false alarm rate (CFAR) and variants thereof [8–10] are most widely utilized. CFAR detectors adaptively calculate the detection thresholds by estimating the statistics of background clutter and maintain a constant probability of false alarm. However, the determination of detection thresholds depends on the distribution of sea clutter, which is not robust enough for the detection of multi-scale ships in multi-scenes. On the other hand, CFAR based methods require land masking and post-processing to reduce false alarms, which is insufficiently automated.

Recently, with the rapid development of deep convolutional networks [16–20], great progress has been made in deep convolutional neural networks (CNN)-based object detection [21–33]. Generally, the current CNN-based detectors can be divided into one-stage [25,26,29,31] and two-stage detectors [22–24,28,32,33]. One-stage detectors include you only look once (YOLO) [25] and its derivative versions [34,35], single shot detector (SSD) [26] and RetinaNet [29], et al. YOLO reframes object detection as a regression problem. The input images are divided into $S \times S$ grid cells and then YOLO predicts bounding boxes and class probabilities for each grid cell straightly. SSD generates a set of default boxes over different sizes per feature map location to match the shape of objects better. RetinaNet proposed the focal loss to overcome the extreme foreground-background class imbalance. On the other hand, faster region-based CNN (Faster R-CNN) [23] and region-based fully convolutional networks (R-FCN) [28] are representative two-stage detectors. Faster R-CNN generates anchors of different scales and aspect ratios through the region proposal network (RPN). In addition, then the feature map and proposals rescaled from anchors are fed into the Fast R-CNN sub-network to predict the location and class probabilities of bounding boxes. Different from the per-region sub-network of Faster R-CNN, R-FCN is a fully convolutional network with the shared computation on the entire image. It proposed position-sensitive score maps to address the problem between translation-invariance in image classification and object detection [28]. Feature pyramid network (FPN) [24] combines low-level and high-level features for more comprehensive feature expression, which has outstanding performance on multi-scale object detection. In summary, one-stage detectors show superiority in detection speed benefits from the single network of detection pipeline. However, for accuracy, the two-stage detectors are better than that of one-stage, especially for small dense object detection.

For SAR ship detection, Deep CNNs have been widely applied in recent years. As a typical one-stage detection method, YOLOv2 was utilized to detect ships in SAR imagery [36]. Wang et al. [37] utilized RetinaNet for automatic ship detection of multi-resolution GF-3 imagery. Zhang et al. [38] proposed a lightweight feature optimizing network with lightweight feature extraction and attention mechanism for better feature representation. On the other hand, many two-stage detectors were proposed for higher detection accuracy. Ref. [39] proposed an improved Faster R-CNN for SAR ship detection. A multilayer fusion light-head detector [40] was proposed to improve the detection speed. Jiao et al. [41] proposed a densely connected neural network, which utilizes a modified FPN, for multi-scale and multi-scene ship detection. Ref. [42–45] added attention mechanisms into CNNs as attention mechanisms adaptively recalibrate feature responses to increase representation power [19,46].

Though many CNN-based methods have been proposed for SAR ship detection, they still encounter bottlenecks on the following issues: (1) Quite different from natural images, in SAR images, strip-like ships are often presented under bird's eye perspectives with various rotation angles and densely arranged in an inshore complex background, as shown in Figure 1a. A ship with an inclined angle leads to a relatively large redundancy region, which would introduce background noise. Moreover, two horizontal bounding boxes of densely arranged ships have a high Intersection-over-Union (IoU) leading to missing detection after the non-maximum suppression (NMS) operation [47,48]. Under these circumstances, the limited capacity of detection with horizontal bounding boxes would be exposed. (2) In R-CNN based object detection methods, an IoU threshold is utilized to distinguish positive and negative samples in the Fast R-CNN sub-network. A relatively low IoU threshold will result in a high recall but a low precision due to the generation of noisy bounding boxes. On the contrary, a relatively high IoU threshold leads to inadequate positive samples. In this case, the overfitting model will cause missing detection. Figure 2 shows the detection results under IoU thresholds of 0.5 and 0.7, respectively. Some noisy background regions are detected as ship regions in Figure 2a, and the missing detection appears in Figure 2b.

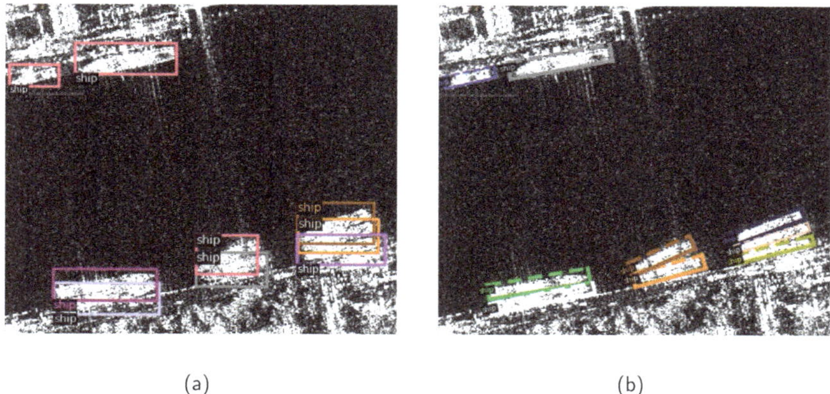

(a)   (b)

Figure 1. Densely arranged ships in an inshore complex background. (**a**) marking ship regions with horizontal bounding boxes; (**b**) marking ship regions with rotated bounding boxes.

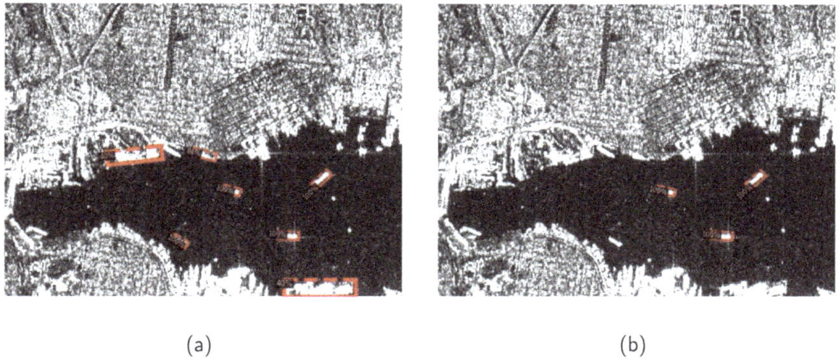

(a)   (b)

Figure 2. Detection results under different IoU thresholds; (**a**) detection results under a IoU threshold of 0.5; (**b**) detection results under a IoU threshold of 0.7.

Due to the inherent drawback of horizontal bounding boxes, rotated bounding boxes were gradually developed in optical remote sensing [47–50]. Ref. [47] proposed a rotation dense feature pyramid network (R-DFPN) in which the dense FPN and multiscale region of interest (ROI) align are used to detect ships in different scenes. In [48], a multi-category rotation detector was proposed for small, cluttered and rotated objects. Li et al. [49] proposed a rotatable region-based residual network (R3-Net) for multi-oriented vehicle detection. The ROI Transformer [50] that is lightweight was proposed to decrease the computational complexity.

To address the problems in SAR ship detection, a multi-stage rotational region based network (MSR2N) is proposed for arbitrary-oriented ship detection. As shown in Figure 1b, rotated bounding boxes can locate ships more accurately with less redundant noise background, and would not overlap with each other even in a dense arrangement. Therefore, the rotated bounding box representation is utilized in this paper. In the feature extraction module, we apply the FPN to fuse the high-resolution features and semantically strong features from the backbone network, which enhances feature representation. To generate rotated anchors and proposals, a rotational RPN (RRPN) is utilized. In addition, a multi-stage rotational detection network (MSRDN) is applied in MSR2N. The MSRDN, which contains an initial rotational detection network (IRDN) and two refined rotational detection networks (RRDN), is trained stage by stage. Three increasing IoU thresholds are selected to sample the positive and negative proposals in three stages, respectively. The increasing IoU thresholds guarantee sufficient positive samples to avoid overfitting, and reduce close false positive proposals in the meantime. Compared to other methods, the proposed MSR2N achieves state-of-the-art performance on SAR ship detection, especially for densely arranged ships in inshore complex backgrounds. The main contributions of proposed MSR2N are enumerated as follows:

1. Alluding to the characteristics of SAR images, the MSR2N framework is proposed in this paper, which is more beneficial for arbitrary-oriented ship detection than horizontal bounding box based methods.
2. In RRPN, a rotation-angle-dependent strategy is utilized to generate anchors with multiple scales, ratio aspects, and rotation angles, which can represent arbitrary-oriented ships more adequately.
3. The MSRDN is proposed, where three increasing IoU thresholds are chosen to resample and refine proposals successively. With the proposals refined more accurately, the number of refined proposals also increases.
4. The multi-stage loss function is employed to accumulate losses of RRPN and three stages of MSRDN to train the entire network.
5. Compared to other methods, the proposed MSR2N has achieved state-of-the-art performance on SAR ship detection dataset (SSDD).

This paper is organized as follows. Section 2 describes the proposed MSR2N in detail. In Section 3, the ablation and comparative experiments are carried out on SSDD, which verifies the effectiveness of proposed MSR2N. Section 4 draws the conclusions for this paper.

2. Proposed Approach

The overall framework of the proposed MSR2N is illustrated in Figure 3. The whole framework can be divided into three modules: FPN, RRPN, and MSRDN. The FPN fuses feature maps from different layers of the backbone to generate the feature pyramid. In the RRPN module, a rotated anchor generation strategy is utilized to produce anchors with various rotation angles. The RRPN head outputs the softmax probabilities of being a ship and the regression offsets that encode the coordinates of coarse proposals. Millions of proposals are generated, but most of them are redundant noisy proposals and the amount of computation is huge. Hence, we select N_{pre} proposals with the highest probabilities per feature pyramid level and perform the skew NMS operation [51] on the selected proposals to obtain N_{post} final proposals which are fed into the next MSRDN. The MSRDN containing the IRDN and two RRDNs is trained stage by stage with three increasing IoU thresholds. During training, the MSR2N is optimized under the supervision of softmax probabilities and rotated boxes regression offsets from RRPN and each stage of MSRDN. For the inference time, the final refined proposals from the last stage of MSRDN are selected by a probability threshold and sampled by the skew NMS operation. Finally, the predicted bounding boxes are obtained.

2.1. Feature Pyramid Network

In this paper, we select ResNet50 [17] as the backbone network, as shown in Figure 3. To extract sufficient ship features in SAR images, especially for small ships, the FPN [24] is utilized to combine the high-resolution, semantically weak features with low-resolution, and semantically strong features. As illustrated in Figure 4, the FPN contains a bottom-up pathway, a top-down pathway, lateral connections, and output convolutions. The outputs of ResNet50s last residual blocks, denoted as {C2, C3, C4, C5}, are chosen as the bottom-up hierarchy. The strides of {C2, C3, C4, C5} are {4, 8, 16, 32} pixels with respect to the input image. The top-down pathway upsamples the semantically stronger features by a factor of 2 and fuses them with the spatially finer features from the bottom-up pathway by element-wise addition. The lateral connections are 1×1 convolution layers with 256-channels, which reduces channel dimensions. In addition, 3×3 convolution layers with 256-channels are utilized as the output layers to reduce the aliasing effect of upsampling. The final feature maps of FPN denoted as {P2, P3, P4, P5}, are fed into the following stages.

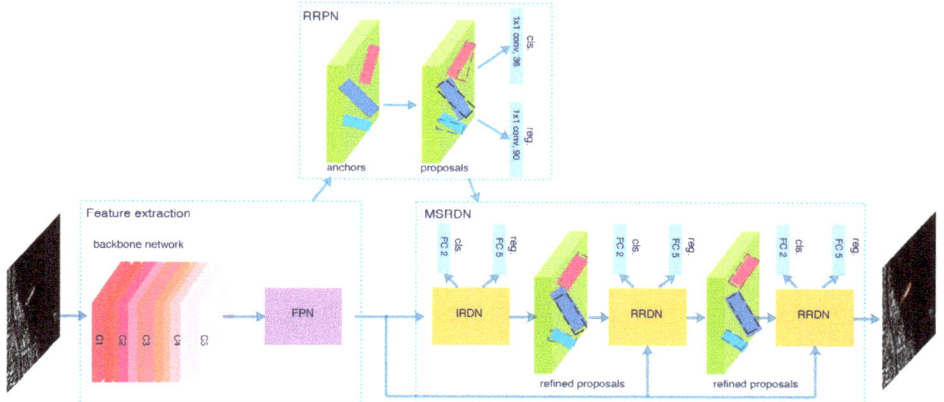

Figure 3. Overall framework of MSR2N

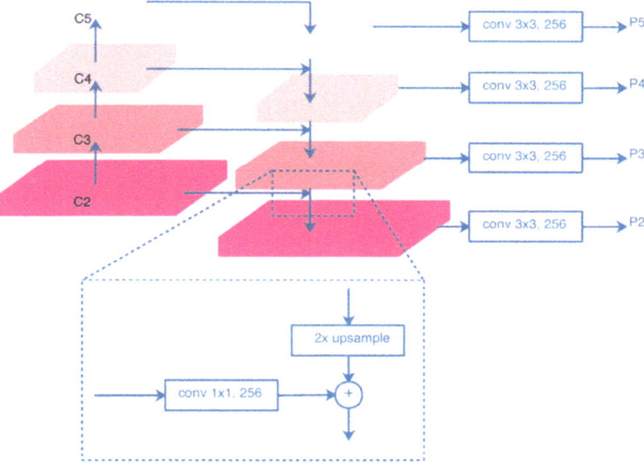

Figure 4. Architecture of FPN.

2.2. Parameterization of Rotated Bounding Box

In the conventional SAR ship detection networks, the ship region is a horizontal rectangle, which is represented by four parameters $(x_{min}, y_{min}, x_{max}, y_{max})$. (x_{min}, y_{min}) and (x_{max}, y_{max}) denote the coordinates of the top left and bottom right corners of a bounding box, respectively. For a rotated bounding box, five parameters (x, y, w, h, θ) are conducted. The coordinate (x, y) represents the center of the rotated bounding box. To avert the disorder of coordinate expression, we set w and h as the longer side and shorter side of the rotated bounding box. θ denotes the angle from the positive direction of the x-axis counterclockwise to the longer side of the rotated bounding box. As half angular space is enough for the description of the rotation angle, the range of θ is set to $[-90°, 90°)$. Figure 5 shows the geometric representation of two rotated bounding boxes with different rotation angles.

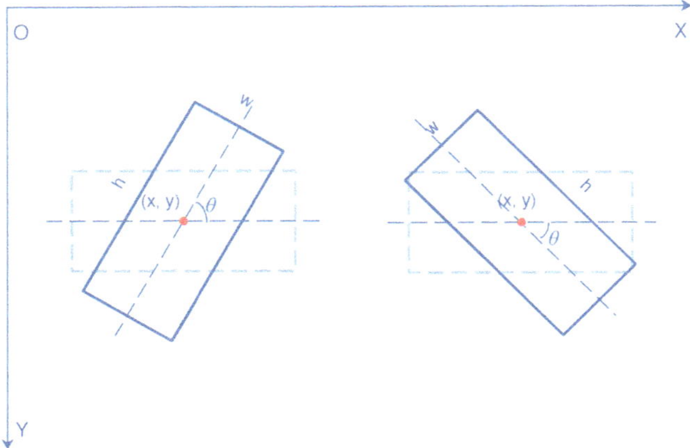

Figure 5. Geometric representation of two rotated bounding boxes. The rotation angle θ of the bounding box on the left is $60°$, and the rotation angle θ of the bounding box on the right is $-45°$.

2.3. Rotated Anchors and Proposals

In [24], multi-scale anchors with multiple aspect ratios are generated on the SAR images with respect to the feature pyramid. However, the generation of horizontal anchors is not sufficient enough for arbitrary-oriented bounding boxes. On the other hand, the characteristic of ships with large aspect ratios should be taken into consideration. Corresponding to the representation of rotated bounding boxes, we utilize a rotation-angle-dependent strategy to generate rotated anchors. As illustrated in Figure 6, a set of cell anchors are generated at the points of a SAR image with the strides of FPN, respectively. As shown in Figure 7, three aspects are taken into consideration in the cell anchor generation. First, referring to [24], the scales of anchors are set to $\{16^2, 32^2, 64^2, 128^2\}$ pixels on the feature pyramid levels of {P2, P3, P4, P5}, respectively. Second, the large aspect ratios {2, 5, 8} replace the usual aspect ratios {0.5, 1, 2} in [23,24] with a view to the characteristic of strip-like ships. Finally, we set multiple rotation angles for anchors of various scales and aspect ratios, which can represent proposals more effectively. Six rotation angles $\{-75°, -45°, -15°, 15°, 45°, 75°\}$ are set in consideration of the trade-off between half angular coverage and computational efficiency. Hence, each point per level of feature pyramid generates 18 ($1 \times 3 \times 6$) anchors. As shown in Figure 3, the classification layer outputs 36 (2×18) softmax probabilities of being a ship and background, and the regression layer outputs 90 (5×18) regression offsets that encode the coordinates (x, y, w, h, θ) for 18 proposals per position.

2.4. Multi-Stage Rotational Detection Network

2.4.1. Initial Rotational Detection Network

As shown in Figure 3, the feature pyramid {P2, P3, P4, P5} and the coarse proposals produced from RRPN are both fed into the next stage called IRDN. Different from the conventional object detection network, IRDN aims to detect rotated proposals in this paper. N_{RROI} proposals are randomly sampled as an RROI mini-batch in which an IoU threshold is chosen to distinguish the positive and negative samples. Here, we utilize the skew IoU computation [51] to compute IoU between rotated bounding boxes. If the IoU between a proposal and any ground-truth box is higher than the threshold, the proposal is assigned to a positive sample and vice versa. The ratio of positive and negative proposals is 1:3. In the RROI mini-batch, if the number of positive proposals is less than 25% of N_{RROI}, the negative proposals should be padded.

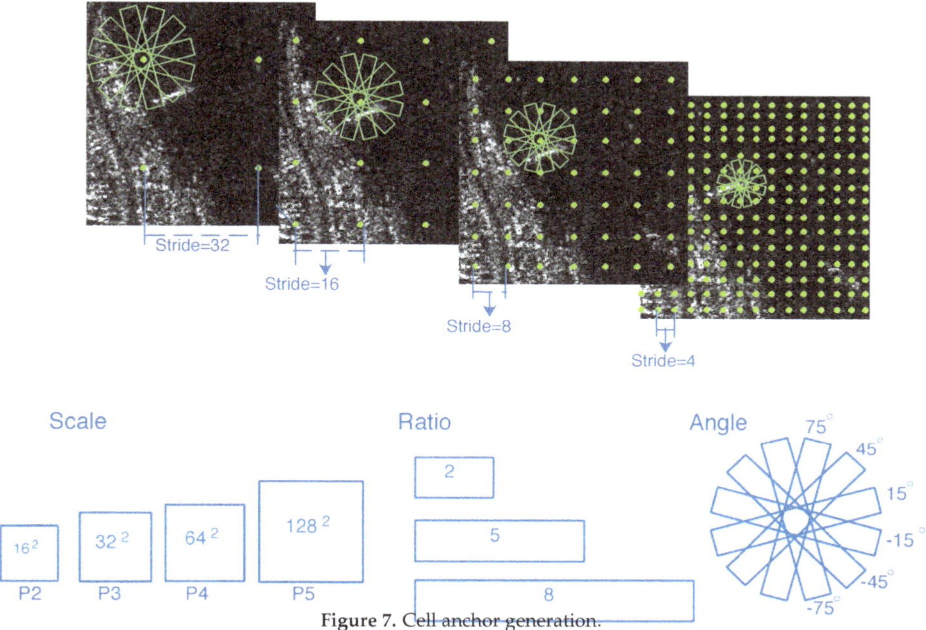

Figure 7. Cell anchor generation.

As illustrated in Figure 8, the rotational ROI (RROI) align layer is applied for arbitrary-oriented proposals. It converts the feature map of RROIs into fixed spatial feature maps. An RROI is divided into 7×7 sub-regions by the RROI align layer, and the max pooling is operated in each sub-region. Then, the feature map of an RROI with a fixed spatial extent of 7×7 can be obtained. The feature maps of RROIs are input to two successive fully connected (FC) layers with a dimension of 1024. Two 1×1 convolutional layers for classification and regression are attached to the FC layers. The classification layer outputs softmax probabilities of being a ship and background, and the regression layer outputs the regression offsets that encode the coordinates (x, y, w, h, θ) of refined proposals.

2.4.2. Multi-Stage Detector

As mentioned in Section 1, a single IoU threshold in IRDN can't separate positive and negative samples properly, which can cause false and missing detection. Inspired by [52], a multi-stage rotational detection strategy is proposed. As illustrated in Figure 3, the MSRDN contains an IRDN and two RRDNs which share the same structure with IRDN. The MSRDN is a multi-stage regression framework, which successively resamples proposals by increasing IoU thresholds. In the first stage of MSRDN, the distribution of coarse proposals is set by a low IoU threshold, which guarantees enough positive samples. The feature pyramid and sampled coarse proposals are fed into the IRDN to produce the refined proposals. In the second stage, the refined proposals from the first stage are resampled by a medium IoU threshold. These sampled refined proposals and the feature pyramid are fed into the RRDN which outputs new refined proposals. In the third stage, the repetitive procedures occur with a high IoU threshold. The process of MSRDN can be expressed as:

$$MSRDN(f, d^1) = IRDN(f, d^1) \circ RRDN_1(f, d^2) \circ RRDN_2(f, d^3) \tag{1}$$

where f denotes the feature maps from FPN, d^1 is the initial sampled distribution of proposals in IRDN, d^2 and d^3 are respectively the resampled distribution of proposals in two RRDNs, and $\circ$ represents the cascade operation. The sub-networks are optimized by the resampled distributions $\{d^1, d^2, d^3\}$, respectively. With a sequence of increasing IoU thresholds, the number of close false positive proposals are sequentially decreased, and the rotated bounding boxes regression is more accurate.

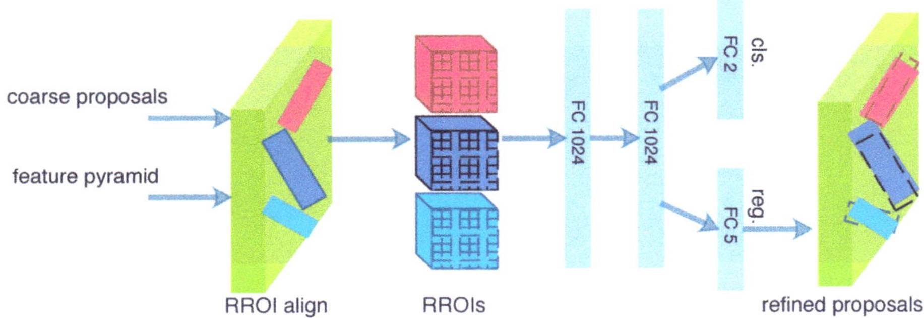

Figure 8. Structure of IRDN.

2.5. Multi-Stage Loss Computation

During the training of RRPN, a sampling strategy for rotated anchors is used. A mini-batch of N_{RRPN} anchors is selected for the loss computation, where the ratio of positive and negative samples is 1:1. In the mini-batch, if positive samples are not enough, the negative samples should be padded. The positive samples are defined by following rules: (i) the IoU between the anchor and any ground-truth is larger than 0.7; and (ii) an angular difference between the anchor and the ground-truth is smaller than $15°$. The negative samples should satisfy the following rules: (i) the IoU between the anchor and any ground-truth is lower than 0.3; or (ii) the IoU between the anchor and a ground-truth is larger than 0.7, while the angular difference is larger than $15°$. We minimize an objective function of RRPN with the multi-task loss [22] defined as:

$$L_{RRPN}(p_i, p_i^*, t_i, t_I^*) = \frac{1}{N_{cls}} \sum_i L_{cls}(p_i, p_i^*) + \lambda \frac{1}{N_{reg}} \sum_i p_i^* L_{reg}(t_i, t_i^*). \tag{2}$$

Here, i denotes the index of samples in a mini-batch, p_i^* is the label of sample i, where p_i^* equals 1 if the sample is a positive one and p_i^* equals 0 if the sample is a negative one. p_i is the predicted softmax probablility of sample i being a ship. t_i^* represents the offsets between sample i and the assigned ground-truth, and t_i represents the predicted bounding box regression offsets. In Equation (2), only if sample i is a positive one with $p_i^* = 1$ is the regression loss L_{reg} calculated. The classification loss L_{cls} is log loss which is defined as:

$$L_{cls}(p_i, p_i^*) = -log p_i p_i^*. \tag{3}$$

The regression loss is defined by the robust loss function smooth L1 as:

$$L_{reg}(t_i, t_i^*) = smooth_{L1}(t_i - t_i^*), \tag{4}$$

$$smooth_{L1}(x) = \begin{cases} 0.5x^2, & |x| < 1 \\ |x| - 0.5, & otherwise \end{cases}. \tag{5}$$

The classification loss L_{cls} and regression loss L_{reg} are respectively normalized by N_{cls} and N_{reg}, where N_{cls} and N_{reg} are both set to the mini-batch size. The regression loss L_{reg} is weighted by a balancing parameter λ, which is set to 1 in the experiments.

For rotated bounding box regression, the five parameterized coordinate regression offsets are defined as:

$$
\begin{aligned}
t_x &= (x - x_a)/w_a, t_y = (y - y_a)/h_a, \\
t_w &= log(w/w_a), t_h = log(h/h_a), \\
t_\theta &= \theta - \theta_a + k\pi, \\
t_x^* &= (x^* - x_a)/w_a, t_y = (y^* - y_a)/h_a, \\
t_w^* &= log(w^*/w_a), t_h^* = log(h^*/h_a), \\
t_\theta^* &= \theta^* - \theta_a + k\pi,
\end{aligned}
\tag{6}
$$

where x, y, w, h and θ denote the center coordinates, width, height, and rotation angle of a rotated bounding box. x, x_a, x^* are respectively for the predicted box, anchor box, and ground-truth box, likewise for y, w, h, θ. The parameter $k \in Z$ keeps θ in the range of $[-180°, 180°)$.

As described in Section 2.4, during the training of MSRDN, the proposals are resampled into a mini-batch and refined by a sequence of rotational detection sub-networks. Meanwhile, each sub-network outputs the softmax probabilities of a proposal being a ship and rotated bounding box regression offsets for loss computation. The multi-task multi-stage loss function of MSRDN is defined as:

$$
L_{MSRDN} = L_{IRDN} + L_{RRDN_1} + L_{RRDN_2}
\tag{7}
$$

where L_{IRDN}, L_{RRDN_1}, and L_{RRDN_2} represent the losses of IRDN, first RRDN and second RRDN, respectively. The losses of three stages have the same definition as the multi-task loss function of RRPN.

3. Experiments and Results

3.1. Dataset

We evaluate the proposed framework on the public SAR ship detection dataset—SSDD [39]. The SSDD contains SAR ship images of various scenes from three sensors. The details of SSDD are shown in Table 1. There are 1160 SAR images including 2456 ships in total, with an average of 2.12 ships per image. Ships in SSDD are annotated with coordinates of the four vertices of rotated bounding boxes. The SSDD is divided into a training set and a test set by a ratio of 4:1.

Table 1. Detailed description of SSDD.

Sensors	RadarSat-2, TerraSAR-X, Sentinel-1
Polarization	HH, VV, HV, VH
Sence	inshore, offshore
Resolution	1 m–15 m
Number of images	1160
Number of ships	2456

3.2. Setup and Implementation Details

The experiments are implemented in the deep learning framework Pytorch [53] on four NVIDIA TITIAN Xp GPUs with 12 GB memory. Our network is initialized by the pre-trained ResNet50 model for ImageNet classification. The whole network is trained for 12 k iterations in total, and a batch of 48 images is fed into the network in an iteration. The initial learning rate is 0.01 for the first 8k iterations;

then, learning rates of 0.001 and 0.0001 are set for the next 2.5 k iterations and the remaining 1.5 k iterations, respectively. The stochastic gradient descent (SGD) [54] optimizer with a weight decay of 0.0001 and a momentum of 0.9 is selected for the model.

Referring to faster R-CNN, we resize images such that the shorter side is 350 pixels under the premise that ensures the longer side less than 800 pixels. For data augmentation, we flip the images horizontally with a probability of 0.5.

As mentioned in Section 2, N_{pre} denotes the number of pre-selected proposals per feature pyramid level, and N_{post} represents the number of post-selected proposals. Here, we set N_{pre} to 2000 and 1000 for training and inference, respectively. In addition, N_{post} is set to 1000 for both training and inference. N_{RRPN} and N_{RROI}, which are the sizes of RRPN and RROI mini-batches, are set to 256 and 512, respectively. In inference time, an NMS threshold is used to select final predicted bounding boxes, and a predicted bounding box can be retained if its score is higher than the score threshold. The NMS threshold and score threshold are respectively set to 0.3 and 0.1 in all experiments.

3.3. Evaluation Metrics

To quantitatively evaluate the performance of proposed MSR2N, the precision, recall, mean Average Precision (mAP), and F-measure (F1) are utilized as evaluation metrics.

Precision is the rate that correctly detected ships in all detected results, and recall means the rate that correctly detected ships in all ground-truths. The definitions of precision and recall are as follows:

$$Precision = \frac{TP}{TP + FP} \tag{8}$$

$$Recall = \frac{TP}{TP + FN} \tag{9}$$

where TP, FP, and FN denote the number of correctly detected ships, false alarms, and undetected ships, respectively. In addition, a bounding box is admitted as a correctly detected ship in the case that the IoU between the bounding box and a ground-truth is higher than the threshold of 0.5. The precision–recall (PR) curve shows the precision–recall pairs at different confidence score thresholds.

The *mAP* is a comprehensive metric that calculates the average value of precision under the recall in a range of [0, 1]. The definition of mAP is as follows:

$$mAP = \int_0^1 P(R)dR \tag{10}$$

where R denotes a recall value and P represents the precision corresponding to a recall.

F1 evaluates the comprehensive performance of a detector by taking the precision and recall into consideration. *F1* is defined as:

$$F1 = \frac{2 * Precision * Recall}{Precision + Recall}. \tag{11}$$

3.4. Ablation Study

In this part, some ablation experiments are carried out to investigate the effectiveness of the main modules of the proposed MSR2N.

3.4.1. Effect of MSRDN

To illustrate the detection performance of MSRDN, the experimental results are listed in Table 2. We can find that the single-stage detectors have poor detection performance. The detector with a low IoU threshold of 0.5 achieves a high recall but a low precision due to the introduction of noisy regions. With the increase of the IoU threshold, the precision increases, but the recall decreases.

When the IoU threshold reaches 0.7, the recall decrease to 84.50% because of the overfitting caused by inadequate positive proposals. When using the two-stage detectors with an IRDN and an RRDN, they can achieve better detection performance than the single-stage ones. The detection performance of three-stage detector with the increasing IoU thresholds of {0.5, 0.6, 0.7} is significantly improved, which achieves the recall of 92.05%, precision of 86.52%, mAP of 90.11%, and F1 of 89.20%. These experiments demonstrate the effectiveness of multiple stages of MSRDN. Comparing to the three-stage detectors with uniform thresholds, the detector with the increasing IoU thresholds still has superior performance, which proves the effectiveness of increasing IoU thresholds of MSRDN. With the multi-stage regression, the number of close false positive proposals is decreased and the location of proposals is more accurate in the meantime. Therefore, the MSRDN with the increasing IoU thresholds of {0.5, 0.6, 0.7} is chosen in this paper.

Table 2. Experimental results of MSRDN.

Stage	IoU Thresholds	Recall (%)	Precision (%)	mAP (%)	F1 (%)
	{0.5}	89.53	86.03	83.66	87.75
IRDN	{0.6}	88.37	86.86	84.38	87.61
	{0.7}	84.50	**92.78**	83.09	88.44
	{0.5, 0.6}	90.70	85.56	87.57	88.05
IRDN+RRDN1	{0.5, 0.7}	89.53	86.68	84.88	88.08
	{0.6, 0.7}	89.53	88.68	86.78	89.10
	{0.5, 0.5, 0.5}	90.89	66.71	86.90	76.95
IRDN+RRDN1+RRDN2	{0.6, 0.6, 0.6}	90.70	85.25	88.55	87.89
	{0.7, 0.7, 0.7}	89.53	87.67	87.89	88.59
	{0.5, 0.6, 0.7}	**92.05**	86.52	**90.11**	**89.20**

3.4.2. Effect of Multi-Stage Loss Computation

The experimental results of multi-stage loss computation are presented in Table 3. It can be observed that the detection performance with three-stage loss computation is preferable, and the experiment only using loss from the last stage obtains the inferior detection performance. With the multi-stage loss used, the precision increases significantly, which means that the false alarms are effectively suppressed. We can conclude that the multi-stage loss computation has a strong constraint on the training of the whole network.

Table 3. Experimental results of multi-stage loss computation.

Loss Computation	Recall (%)	Precision (%)	mAP (%)	F1 (%)
RRDN2	83.91	14.26	78.97	24.37
RRDN1+RRDN2	88.57	45.84	84.43	60.41
IRDN+RRDN1+RRDN2	**92.05**	**86.52**	**90.11**	**89.20**

3.4.3. Effect of Rotation Angles

Some experiments about rotation angles are carried out, and their results are shown in Table 4. It can be seen in Table 4 that the method which only generates horizontal and vertical anchors achieves inferior detection performance. As the interval of rotation angles decreases, the detection performance improves. The method with the rotation angles of $\{-75°, -45°, -15°, 15°, 45°, 75°\}$ obtains superior detection performance, which proves the effectiveness of anchors with different rotation angles.

Table 4. Experimental results of rotation angles.

Rotation Angles	Recall (%)	Precision (%)	mAP (%)	F1 (%)
$\{-90°, 0°\}$	88.92	53.33	85.50	66.67
$\{-90°, -45°, 0°, 45°\}$	89.53	86.03	88.90	87.75
$\{-75°, -45°, -15°, 15°, 45°, 75°\}$	**92.05**	**86.52**	**90.11**	**89.20**

3.4.4. Effect of FPN

A couple of ablation experiments are performed to demonstrate the effect of FPN in the proposed MSR2N, wherein the experiment of MSR2N uses the feature pyramid {P2, P3, P4, P5} as the outputs of feature extraction module, and the experiment of MSR2N without FPN only uses C5 of ResNet50 to feed into the next stages. Table 5 shows the experimental results of the FPN. Comparing to MSR2N without FPN, our MSR2N reaches higher evaluation metrics. The significant improvement of detection performance benefits from the sufficient feature expression of FPN.

Table 5. Experimental results of FPN.

FPN	Recall (%)	Precision (%)	mAP (%)	F1 (%)
MSR2N without FPN	87.60	85.28	85.12	86.42
MSR2N	**92.05**	**86.52**	**90.11**	**89.20**

3.5. Comparison with Other Object Detection Methods

To verify the performance of our proposed MSR2N, we compare the MSR2N with the multi-stage horizontal region based network (MSHRN), rotational Faster R-CNN [23] (Faster RR-CNN), rotational FPN [24] (R-FPN), and rotational RetinaNet [29] (R-RetinaNet). MSHRN is the horizontal variant of our proposed MSR2N, where the RPN sub-network is used and the ROI pooling layer replaces the RROI align layer for predicting horizontal bounding boxes. Faster RR-CNN and R-FPN are two-stage detectors, which are respectively rotational variants of Faster R-CNN and FPN. The one-stage detector R-RetinaNet is the rotational variant of RetinaNet. In [29], the feature pyramid with levels {P3, P4, P5, P6, P7} is used as the output of the feature extraction module. For consistency, we use the feature pyramid {P2, P3, P4, P5} in the experiment of R-RetinaNet, which is the same as the experiments of R-FPN and MSR2N. Here, ResNet50 is chosen as the backbone network of all methods.

Table 6 summarizes the experimental results of different methods on SSDD, and Figure 9 shows the PR curves of different methods. From Table 6, we can find that MSR2N achieves state-of-the-art performance: 92.05% for recall, 86.52% for precision, 90.11% for mAP, and 89.20% for F1. Comparing to the other four methods, MSR2N obtains 2.27%, 7.89%, 6.45%, and 9.54% gains in mAP, respectively. In Figure 9, the PR curve of MSR2N is higher than those of Faster RR-CNN, R-FPN, and R-RetinaNet. The PR curve of MSHRN only can reach a recall of 90.36%, but MSR2N can achieve a recall of 92.05%. This phenomenon indicates that horizontal region based detection leads to more undetected ships than rotational region based detection. To further verify the capability of handling hard cases, we construct a subset by selecting SAR images, which contain densely arranged ships in inshore complex scenes, from the original test set. Table 7 shows the experimental results of different methods on hard cases. From Table 7, it can be observed that all the values of the evaluation metrics are relatively low, which indicates that these hard cases are difficult to detect correctly. For example, the mAP of R-RetinaNet only reaches 42.91%. Nevertheless, our proposed MSR2N achieves superior performance. Comparing to the other four methods, MSR2N obtains 12.43%, 8.52%, 13.08%, and 28.30% gains in mAP, respectively.

Figure 10 illustrates the detection results of four images from the test set of SSDD using different methods. From the detection results of three inshore SAR images, it can be observed that there exist undetected ships in Figure 10e–g. This is on account of the high IoU between two overlapped horizontal bounding boxes in MSHRN. On the other hand, the background noise in the horizontal bounding boxes can interfere with the detection. In addition, the detection results of Faster RR-CNN, R-FPN, and R-RetinaNet still have problems with missing detection and false detection. On the contrary, MSR2N shows superior detection performance in Figure 10u–w. All ships are successfully detected, and the position of ships is located more accurately than other methods. For the methods of MSHRN, Faster RR-CNN, and R-RetinaNet, there exist several false alarms and undetected ships in the results of fourth SAR images. Generally speaking, all methods are competent to detect ships far from shore. Based on the above discussion, we can conclude that MSR2N has state-of-the-art performance on SAR ship detection, especially for densely arranged ships in complex backgrounds.

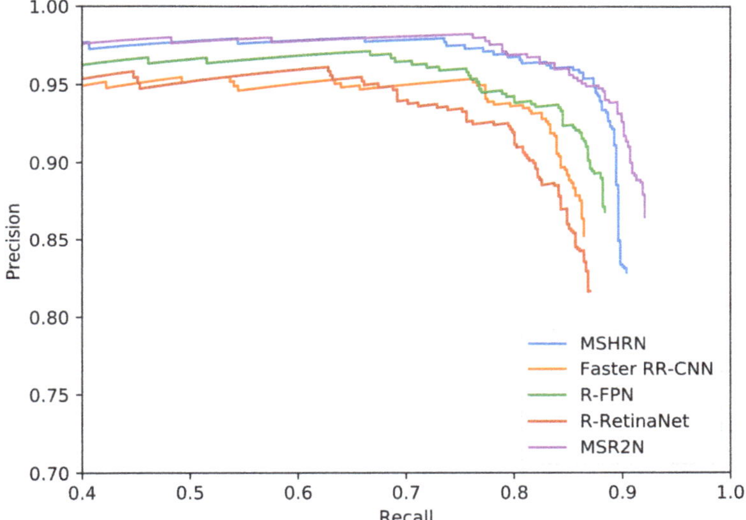

Figure 9. PR curves of different methods on SSDD.

Table 6. Experimental results of different methods on SSDD.

Method	Recall (%)	Precision (%)	mAP (%)	F1 (%)
MSHRN	90.36	82.84	87.84	86.44
Faster RR-CNN	86.43	85.28	82.22	85.85
R-FPN	88.37	**86.86**	84.38	87.61
R-RetinaNet	87.01	81.64	82.80	84.24
MSR2N	**92.05**	86.52	**90.11**	**89.20**

Table 7. Experimental results of different methods on hard cases.

Method	Recall (%)	Precision (%)	mAP (%)	F1 (%)
MSHRN	61.54	60.61	58.78	61.07
Faster RR-CNN	56.76	70.00	62.69	51.04
R-FPN	63.51	75.81	58.13	69.12
R-RetinaNet	54.05	55.56	42.91	54.79
MSR2N	**76.12**	**77.37**	**71.21**	**76.74**

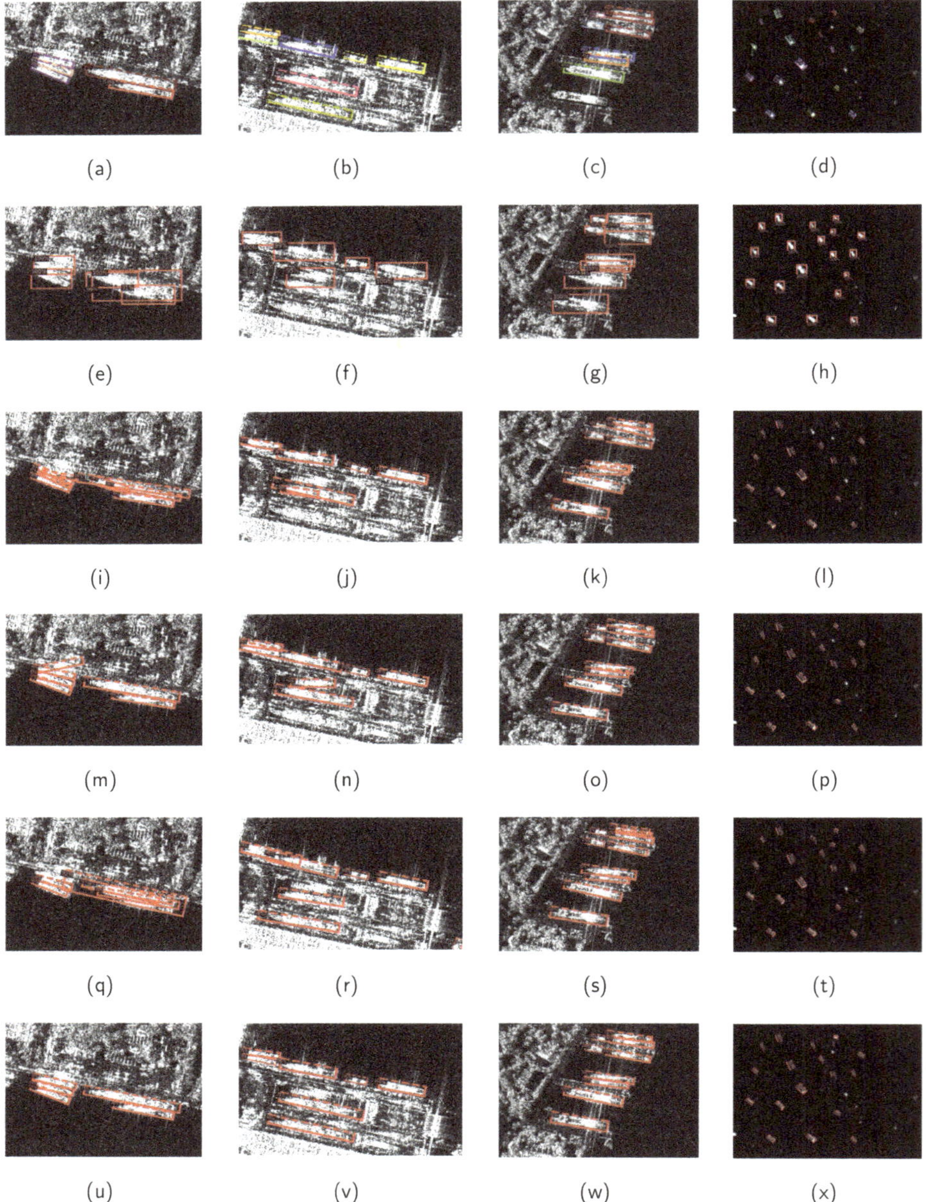

Figure 10. Detection results of different methods. (**a–d**) ground-truths; (**e–h**) detection results of MSHRN; (**i–l**) detection results of Faster RR-CNN; (**m–p**) detection results of R-FPN; (**q–t**) detection results of R-RetinaNet; (**u–x**) detection results of MSR2N.

4. Conclusions

This paper proposes an arbitrary-oriented ship detection network called MSR2N which aims at detecting ships in various complex scenes. In our MSR2N, the rotating bounding boxes are utilized to represent ship regions, which can reduce redundant noisy regions and overlaps between bounding boxes. The SSDD is employed to verify the effectiveness of the proposed MSR2N. The ablation experiments illustrate that high precision and high recall can be achieved simultaneously due to the sequential resampling of proposals and training in MSRDN sub-networks. The experiments about loss computation show the significance of multi-stage loss computation. The experiments of rotation angles and FPN are also carried out, which proves the effectiveness of rotation angles and FPN. Compared to other methods, MSR2N obtains superior detection performance, especially when ships are densely arranged in inshore complex scenes.

Author Contributions: Z.P. proposed the method and performed the experiments; R.Y. performed the experiments; and Z.Z. supervised the study and reviewed this paper. All authors have read and agreed to the published version of the manuscript.

Funding: This work was supported by the National Natural Science Foundation of China 61901442.

Conflicts of Interest: The authors declare no conflict of interest.

References

1. Moreira, A.; Prats-Iraola, P.; Younis, M.; Krieger, G.; Hajnsek, I.; Papathanassiou, K.P. A tutorial on synthetic aperture radar. *IEEE Geosci. Remote Sens. Mag.* **2013**, *1*, 6–43. [CrossRef]
2. Kanjir, U.; Greidanus, H.; Oštir, K. Vessel detection and classification from spaceborne optical images: A literature survey. *Remote Sens. Environ.* **2018**, *207*, 1–26. [CrossRef] [PubMed]
3. Brusch, S.; Lehner, S.; Fritz, T.; Soccorsi, M.; Soloviev, A.; van Schie, B. Ship surveillance with TerraSAR-X. *IEEE Trans. Geosci. Remote Sens.* **2010**, *49*, 1092–1103. [CrossRef]
4. Martorella, M.; Pastina, D.; Berizzi, F.; Lombardo, P. Spaceborne radar imaging of maritime moving targets with the Cosmo-SkyMed SAR system. *IEEE J. Sel. Top. Appl. Earth Obs. Remote Sens.* **2014**, *7*, 2797–2810. [CrossRef]
5. Touzi, R.; Hurley, J.; Vachon, P.W. Optimization of the degree of polarization for enhanced ship detection using polarimetric RADARSAT-2. *IEEE Trans. Geosci. Remote Sens.* **2015**, *53*, 5403–5424. [CrossRef]
6. Huang, L.; Liu, B.; Li, B.; Guo, W.; Yu, W.; Zhang, Z.; Yu, W. OpenSARShip: A dataset dedicated to Sentinel-1 ship interpretation. *IEEE J. Sel. Top. Appl. Earth Obs. Remote Sens.* **2017**, *11*, 195–208. [CrossRef]
7. An, Q.; Pan, Z.; You, H. Ship detection in Gaofen-3 SAR images based on sea clutter distribution analysis and deep convolutional neural network. *Sensors* **2018**, *18*, 334. [CrossRef]
8. Leng, X.; Ji, K.; Xing, X.; Zhou, S.; Zou, H. Area ratio invariant feature group for ship detection in SAR imagery. *IEEE J. Sel. Top. Appl. Earth Obs. Remote Sens.* **2018**, *11*, 2376–2388. [CrossRef]
9. Ao, W.; Xu, F.; Li, Y.; Wang, H. Detection and discrimination of ship targets in complex background from spaceborne ALOS-2 SAR images. *IEEE J. Sel. Top. Appl. Earth Obs. Remote Sens.* **2018**, *11*, 536–550. [CrossRef]
10. Pappas, O.; Achim, A.; Bull, D. Superpixel-level CFAR detectors for ship detection in SAR imagery. *IEEE Geosci. Remote Sens. Lett.* **2018**, *15*, 1397–1401. [CrossRef]
11. Lin, H.; Chen, H.; Jin, K.; Zeng, L.; Yang, J. Ship Detection With Superpixel-Level Fisher Vector in High-Resolution SAR Images. *IEEE Geosci. Remote Sens. Lett.* **2020**, *17*, 247–251. [CrossRef]
12. Tello, M.; López-Martínez, C.; Mallorqui, J.J. A novel algorithm for ship detection in SAR imagery based on the wavelet transform. *IEEE Geosci. Remote Sens. Lett.* **2005**, *2*, 201–205. [CrossRef]
13. Tello, M.; Lopez-Martinez, C.; Mallorqui, J.; Bonastre, R. Automatic detection of spots and extraction of frontiers in SAR images by means of the wavelet transform: Application to ship and coastline detection. In Proceedings of the 2006 IEEE International Symposium on Geoscience and Remote Sensing, Denver, CO, USA, 31 July–4 August 2006; pp. 383–386.
14. Chen, J.; Chen, Y.; Yang, J. Ship detection using polarization cross-entropy. *IEEE Geosci. Remote Sens. Lett.* **2009**, *6*, 723–727. [CrossRef]

15. Shirvany, R.; Chabert, M.; Tourneret, J.Y. Ship and oil-spill detection using the degree of polarization in linear and hybrid/compact dual-pol SAR. *IEEE J. Sel. Top. Appl. Earth Obs. Remote Sens.* **2012**, *5*, 885–892. [CrossRef]

16. Simonyan, K.; Zisserman, A. Very deep convolutional networks for large-scale image recognition. *arXiv* **2014**, arXiv:1409.1556.

17. He, K.; Zhang, X.; Ren, S.; Sun, J. Deep residual learning for image recognition. In Proceedings of the IEEE Conference on Computer Vision and Pattern Recognition, Las Vegas, NV, USA, 27–30 June 2016; pp. 770–778.

18. Szegedy, C.; Vanhoucke, V.; Ioffe, S.; Shlens, J.; Wojna, Z. Rethinking the inception architecture for computer vision. In Proceedings of the IEEE Conference on Computer Vision and Pattern Recognition, Las Vegas, NV, USA, 27–30 June 2016; pp. 2818–2826.

19. Hu, J.; Shen, L.; Sun, G. Squeeze-and-excitation networks. In Proceedings of the IEEE Conference on Computer Vision and Pattern Recognition, Salt Lake City, UT, USA, 18–22 June 2018; pp. 7132–7141.

20. Huang, G.; Liu, Z.; Van Der Maaten, L.; Weinberger, K.Q. Densely connected convolutional networks. In Proceedings of the IEEE Conference on Computer Vision and Pattern Recognition, Honolulu, HI, USA, 21–26 July 2017; pp. 4700–4708.

21. Girshick, R.; Donahue, J.; Darrell, T.; Malik, J. Rich feature hierarchies for accurate object detection and semantic segmentation. In Proceedings of the IEEE Conference on Computer Vision and Pattern Recognition, Columbus, OH, USA, 24–27 June 2014; pp. 580–587.

22. Girshick, R. Fast r-cnn. In Proceedings of the IEEE International Conference on Computer Vision, Santiago, Chile, 11–18 December 2015; pp. 1440–1448.

23. Ren, S.; He, K.; Girshick, R.; Sun, J. Faster r-cnn: Towards real-time object detection with region proposal networks. In Proceedings of the 28th International Conference on Neural Information Processing Systems, Montreal, QC, Canada, 8–13 December 2015; pp. 91–99.

24. Lin, T.Y.; Dollár, P.; Girshick, R.; He, K.; Hariharan, B.; Belongie, S. Feature pyramid networks for object detection. In Proceedings of the IEEE Conference on Computer Vision and Pattern Recognition, Honolulu, HI, USA, 21–26 July 2017; pp. 2117–2125.

25. Redmon, J.; Divvala, S.; Girshick, R.; Farhadi, A. You only look once: Unified, real-time object detection. In Proceedings of the IEEE Conference on Computer Vision and Pattern Recognition, Las Vegas, NV, USA, 27–30 June 2016; pp. 779–788.

26. Liu, W.; Anguelov, D.; Erhan, D.; Szegedy, C.; Reed, S.; Fu, C.Y.; Berg, A.C. Ssd: Single shot multibox detector. In Proceedings of the European Conference on Computer Vision, Amsterdam, The Netherlands, 8–16 October 2016; pp. 21–37.

27. Dai, J.; Qi, H.; Xiong, Y.; Li, Y.; Zhang, G.; Hu, H.; Wei, Y. Deformable convolutional networks. In Proceedings of the IEEE International Conference on Computer Vision, Venice, Italy, 22–29 October 2017; pp. 764–773.

28. Dai, J.; Li, Y.; He, K.; Sun, J. R-fcn: Object detection via region-based fully convolutional networks. In yProceedings of the 30th International Conference on Neural Information Processing , Barcelona, Spain, 5–10 December 2016; pp. 379–387.

29. Lin, T.Y.; Goyal, P.; Girshick, R.; He, K.; Dollár, P. Focal loss for dense object detection. In Proceedings of the IEEE International Conference on Computer Vision, Venice, Italy, 22–29 October 2017; pp. 2980–2988.

30. Li, Z.; Peng, C.; Yu, G.; Zhang, X.; Deng, Y.; Sun, J. Detnet: Design backbone for object detection. In Proceedings of the European Conference on Computer Vision (ECCV), Munich, Germany, 8–14 September 2018; pp. 334–350.

31. Zhao, Q.; Sheng, T.; Wang, Y.; Tang, Z.; Chen, Y.; Cai, L.; Ling, H. M2det: A single-shot object detector based on multi-level feature pyramid network. In Proceedings of the AAAI Conference on Artificial Intelligence, Honolulu, HI, USA, 27 January–1 February 2019; Volume 33, pp. 9259–9266.

32. Li, Z.; Peng, C.; Yu, G.; Zhang, X.; Deng, Y.; Sun, J. Light-head r-cnn: In defense of two-stage object detector. *arXiv* **2017**, arXiv:1711.07264.

33. Wang, J.; Chen, K.; Yang, S.; Loy, C.C.; Lin, D. Region proposal by guided anchoring. In Proceedings of the IEEE Conference on Computer Vision and Pattern Recognition, Long Beach, CA, USA, 16–20 June 2019; pp. 2965–2974.

34. Redmon, J.; Farhadi, A. YOLO9000: Better, faster, stronger. In Proceedings of the IEEE Conference on Computer Vision and Pattern Recognition, Honolulu, HI, USA, 21–26 July 2017; pp. 7263–7271.

35. Redmon, J.; Farhadi, A. Yolov3: An incremental improvement. *arXiv* **2018**, arXiv:1804.02767 .

36. Chang, Y.L.; Anagaw, A.; Chang, L.; Wang, Y.C.; Hsiao, C.Y.; Lee, W.H. Ship Detection Based on YOLOv2 for SAR Imagery. *Remote Sens.* **2019**, *11*, 786. [CrossRef]

37. Wang, Y.; Wang, C.; Zhang, H.; Dong, Y.; Wei, S. Automatic Ship Detection Based on RetinaNet Using Multi-Resolution Gaofen-3 Imagery. *Remote Sens.* **2019**, *11*, 531. [CrossRef]

38. Zhang, X.; Wang, H.; Xu, C.; Lv, Y.; Fu, C.; Xiao, H.; He, Y. A Lightweight Feature Optimizing Network for Ship Detection in SAR Image. *IEEE Access* **2019**, *7*, 141662–141678. [CrossRef]

39. Li, J.; Qu, C.; Shao, J. Ship detection in SAR images based on an improved faster R-CNN. In Proceedings of the 2017 SAR in Big Data Era: Models, Methods and Applications (BIGSARDATA), Beijing, China, 13–14 November 2017; pp. 1–6.

40. Gui, Y.; Li, X.; Xue, L. A Multilayer Fusion Light-Head Detector for SAR Ship Detection. *Sensors* **2019**, *19*, 1124. [CrossRef] [PubMed]

41. Jiao, J.; Zhang, Y.; Sun, H.; Yang, X.; Gao, X.; Hong, W.; Fu, K.; Sun, X. A densely connected end-to-end neural network for multiscale and multiscene SAR ship detection. *IEEE Access* **2018**, *6*, 20881–20892. [CrossRef]

42. Lin, Z.; Ji, K.; Leng, X.; Kuang, G. Squeeze and Excitation Rank Faster R-CNN for Ship Detection in SAR Images. *IEEE Geosci. Remote Sens. Lett.* **2018**, *16*, 751–755. [CrossRef]

43. Zhao, J.; Zhang, Z.; Yu, W.; Truong, T.K. A cascade coupled convolutional neural network guided visual attention method for ship detection from SAR images. *IEEE Access* **2018**, *6*, 50693–50708. [CrossRef]

44. Chen, C.; He, C.; Hu, C.; Pei, H.; Jiao, L. A deep neural network based on an attention mechanism for SAR ship detection in multiscale and complex scenarios. *IEEE Access* **2019**, *7*, 104848–104863. [CrossRef]

45. Cui, Z.; Li, Q.; Cao, Z.; Liu, N. Dense Attention Pyramid Networks for Multi-Scale Ship Detection in SAR Images. *IEEE Trans. Geosci. Remote Sens.* **2019**, *57*, 8983–8997. [CrossRef]

46. Woo, S.; Park, J.; Lee, J.Y.; So Kweon, I. Cbam: Convolutional block attention module. In Proceedings of the European Conference on Computer Vision (ECCV), Munich, Germany, 8–14 September 2018, pp. 3–19.

47. Yang, X.; Sun, H.; Fu, K.; Yang, J.; Sun, X.; Yan, M.; Guo, Z. Automatic ship detection in remote sensing images from google earth of complex scenes based on multiscale rotation dense feature pyramid networks. *Remote Sens.* **2018**, *10*, 132. [CrossRef]

48. Yang, X.; Yang, J.; Yan, J.; Zhang, Y.; Zhang, T.; Guo, Z.; Sun, X.; Fu, K. Scrdet: Towards more robust detection for small, cluttered and rotated objects. In Proceedings of the IEEE International Conference on Computer Vision, Seoul, Korea, 27 October–2 November 2019; pp. 8232–8241.

49. Li, Q.; Mou, L.; Xu, Q.; Zhang, Y.; Zhu, X.X. R3-net: A deep network for multi-oriented vehicle detection in aerial images and videos. *arXiv* **2018**, arXiv:1808.05560.

50. Ding, J.; Xue, N.; Long, Y.; Xia, G.S.; Lu, Q. Learning roi transformer for detecting oriented objects in aerial images. *arXiv* **2018**, arXiv:1812.00155.

51. Ma, J.; Shao, W.; Ye, H.; Wang, L.; Wang, H.; Zheng, Y.; Xue, X. Arbitrary-oriented scene text detection via rotation proposals. *IEEE Trans. Multimedia* **2018**, *20*, 3111–3122. [CrossRef]

52. Cai, Z.; Vasconcelos, N. Cascade r-cnn: Delving into high quality object detection. In Proceedings of the IEEE Conference on Computer Vision and Pattern Recognition, Salt Lake City, UT, USA, 18–22 June 2018; pp. 6154–6162.

53. Paszke, A.; Gross, S.; Chintala, S.; Chanan, G.; Yang, E.; DeVito, Z.; Lin, Z.; Desmaison, A.; Antiga, L.; Lerer, A. Automatic differentiation in pytorch. In Proceedings of the Neural Information Processing Systems, Long Beach, CA, USA, 4–9 December 2017.

54. Bottou, L. Large-scale machine learning with stochastic gradient descent. In Proceedings of the COMPSTAT'2010, Paris, France, 22–27 August 2010; pp. 177–186.

Article

The Single-Shore-Station-Based Position Estimation Method of an Automatic Identification System

Yi Jiang [1,*] and Kai Zheng [2]

[1]   Information Science and Technology College, Dalian Maritime University, Dalian 116026, China
[2]   Marine Electrical Engineering College, Dalian Maritime University, Dalian 116026, China; kzh@dlmu.edu.cn
*   Correspondence: j_y@dlmu.edu.cn; Tel.: +86-1810-408-0985

Received: 31 January 2020; Accepted: 9 March 2020; Published: 12 March 2020

Abstract: In order to overcome the vulnerability of the Global Navigation Satellite System (GNSS), the International Maritime Organization (IMO) initiated the ranging mode (R-Mode) of the automatic identification system (AIS) to provide resilient position data. As the existing AIS is a communication system, the number of shore stations as reference stations cannot satisfy positioning requirements. Especially in the area near a shore station, it is very common that a vessel can only receive signals from one shore station, where the traditional positioning method cannot be used. A novel position estimation method using multiple antennas on shipborne equipment is proposed here, which provides a vessel's position even though the vessel can only receive signals from a single shore station. It is beneficial for solving positioning issues in proximity to the coast. Further, as the distances between different antennas to the shore station are not sufficiently independent, the positioning matrix can easily be near singularity or ill-conditioned; thus, an effective position solving method is derived. Furthermore, the proposed method is verified and evaluated in different scenarios by numerical simulation. We assessed the influencing factors of positioning performance, such as the vessel's heading angle, the relative position, and the distances between the shore station and the vessel. The proposed method widely expands the application scope of the AIS R-Mode positioning system.

Keywords: position estimation; ranging mode; single shore station; AIS

---

1. Introduction

The vulnerability of the Global Navigation Satellite System (GNSS) to both intentional and unintentional jamming and interference is an urgent problem to be solved in the e-Navigation Strategy Implementation Plan (SIP) of the International Maritime Organization (IMO) [1,2]. The automatic identification system (AIS) is widely used for maritime communication, and the IMO has mandated its installation since 2002 in order to avoid collisions [3–6]. By reusing the existing AIS infrastructure, the ranging mode (R-Mode) of AIS has been accepted as one of the alternative GNSS backup navigation systems of the future [7]. The resilient position data, which could be supplied by both satellite and terrestrial-based navigation systems, serve as the foundation for the e-Navigation SIP [8–10]. Its aim is to contribute to safe and reliable navigation at sea, especially for autonomous ship navigation.

Generally speaking, AIS is a very high frequency (VHF)-based communication system. AIS shore stations are the most critical components in a coastal AIS network. They can not only receive signals from shipborne equipment but also transmit signals within the coverage area. AIS R-Mode adds a ranging function without influencing the existing AIS communication capacity. The existing AIS shore stations are considered reference stations. A vessel can receive signals and derive ranging information to itself from shore stations, and as a consequence, the vessel's position can be estimated [11,12]. It should be noted that the positional information in the existing AIS is now derived from GNSS [13]. If GNSS were to fail, the whole AIS would break down. Therefore, AIS R-Mode presents an efficient solution to overcome the dependence of AIS on GNSS. It will make AIS a comprehensive marine radio system integrated with communication and navigation for the e-navigation strategy. Therefore, many countries and scientists are researching AIS R-Mode. A European consortium of 12 research institutions, administrations, and industries in Germany, Poland, Sweden, and Norway has currently set up an R-Mode testbed in the Baltic Sea to study the feasibility of R-Mode [14,15]. The Maritime Safety Administration (MSA) and Dalian Maritime University have established the first AIS R-Mode testbed in China [11,16]. South Korea has performed some simulations of R-Mode with the integration of eLoran [17]. Further, the International Association of Marine Aids to Navigation and Lighthouse Authorities (IALA) is drafting the standard for R-Mode according to these countries' research results [18].

In AIS R-Mode, shipborne AIS equipment receives VHF signals from AIS shore stations. If using traditional positioning methods in AIS R-Mode, such as the time of arrival (TOA) [19–21] or the time difference of arrival (TDOA) [22–24], the vessel needs to measure the signal transmission delay or relative delay. Then, the distance or distance difference can be obtained by multiplying it by the speed of light in the free space. Finally, the vessel's position can be estimated according to the positioning equation. However, both methods require receiving signals from at least three different AIS shore stations. The existing AIS, though, is set up for communication without considering the requirement of positioning. Hence, a vessel cannot receive signals from three different AIS shore stations at the same time in some sea areas [25]. Therefore, the traditional positioning methods are not feasible. In this situation, if the vessel can receive signals from two AIS shore stations, a displacement correction position estimation method can be used to estimate the vessel's position [26,27]. The principle of this method is to calculate the displacement of the vessel for a period of time according to parameters, such as the heading and speed, provided by auxiliary sensors. The relationship of positional information between the adjacent moments can be derived by the displacement vector. Finally, the vessel's position is estimated by the continuous range measurements in adjacent moments. However, when a vessel can only receive signals from one shore station, the existing position estimation methods cannot be utilized.

We have proposed a position estimation method for AIS R-Mode, especially in the area close to a shore station. This method aims to estimate the vessel's position when it can only receive signals from a single AIS shore station. The contributions of this study are as follows.

- Traditional positioning in AIS R-Mode needs at least three reference stations. This paper proposes a position estimation method for a single shore station by using multiple antennas on a vessel.
- Due to the limited size of the vessel, the distances between different antennas to the shore station are approximately the same and not sufficiently independent. The positioning matrix is prone to being near singularity or ill-conditioned. The second significant finding is an effective position solving method for this near-singular positioning matrix.
- The proposed method was verified and evaluated in different scenarios using MATLAB simulations. The positioning performance of the proposed method was found to be influenced by the relative position between the antennas and the shore station. Further, position errors increased as the distance increased.

The remainder of this paper is organized as follows: The architecture of the AIS R-Mode positioning system is given, and different position estimation methods that can be used in AIS R-Mode are discussed in Section 2. In Section 3, we present a novel position estimation method for when a vessel can only receive signals from a single AIS shore station. Furthermore, an effective position solving method to avoid singularity is derived for the conditions of three antennas and more than three antennas. Then, four different simulation scenarios are introduced to verify and evaluate the proposed method in Section 4. Simulation results are analyzed, and the influencing factors of positioning performance are discussed in Section 5. Finally, conclusions are put forth in Section 6.

2. AIS R-Mode Positioning

The AIS R-mode positioning system is comprised of AIS shore stations and vessels. The system framework is illustrated in Figure 1. The AIS shore stations, as positioning reference stations, transmit VHF ranging signals periodically. The circles in Figure 1 indicate the coverage areas of shore stations. Only the vessels in the coverage area of the shore station can receive its signals. Then, the vessel estimates its position according to its signals received from shore stations.

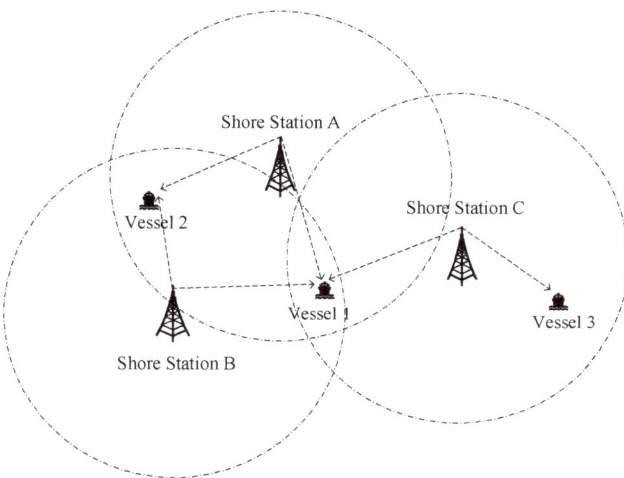

Figure 1. Coverage areas of automatic identification system (AIS) stations: different cases.

In Figure 1, if the vessel, such as Vessel 1, can receive signals from at least three AIS shore stations, either the TOA method or the TDOA method can be utilized to estimate the vessel's position in AIS R-mode positioning [25]. However, the layout of existing AIS shore stations was originally designed to satisfy communication requirements. A single shore station is sufficient for the vessel to communicate. However, as the AIS traffic load increased, more shore stations were established to increase the signal coverage rate. Still, areas exist where there are insufficient numbers of shore stations that can act as the reference stations for positioning [25]. For instance, some vessels can only receive signals from two shore stations, such as Vessel 2 in Figure 1. In this situation, the TOA method can be enhanced with the displacement correction so that the vessel's position can be estimated [26]. However, some vessels can only receive signals from a single shore station, shown in Figure 1, such as Vessel 3. Previous studies have based their criteria for at least two shore stations, so they were not able to estimate vessels' positions in this condition. This paper mainly focuses on the position estimation method based on a single shore station in AIS R-Mode.

3. Positioning Method for a Single Shore Station

The proposed positioning method based on a single shore station in AIS R-Mode uses multiple antennas on a vessel. Figure 2 presents the global coordinate system for R-Mode positioning and the local coordinate system for the proposed positioning method. O is the origin of the global system, the Y-axis is directed towards the north, and the X-axis is perpendicular to the X-axis. (X, Y) indicates the coordination in the global system. In Figure 2, B and M denote the shore station and the vessel, respectively. An inverted triangle depicts an antenna on the vessel. Multiple antennas on the vessel are shown in Figure 2. The global coordinates of the shore station B are (X_B, Y_B). (X_i, Y_i) are the global position coordinates of the ith antenna of the vessel. In the local coordinate system, o is the origin, and it is the center of the vessel; the *x*-axis is directed towards the heading direction, and the *y*-axis is perpendicular to the *x*-axis. (x_i, y_i) are the position coordinates of the *i*th antenna on the vessel in the local coordinate system.

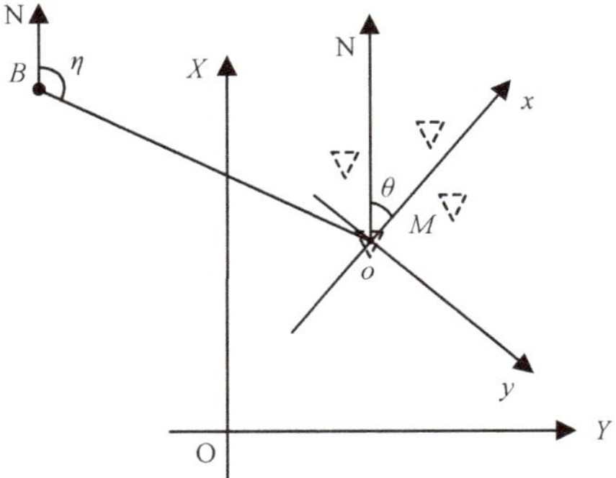

Figure 2. Global and local coordinate systems for positioning.

The distance between the reference shore station and the *i*th antenna on the vessel can be calculated using the following simple equation [28].

$$\overline{R}_i = c\left(T_{r_i} - T_t\right) \tag{1}$$

where *c* is the speed of light in the free space; T_t is the signal transmission time, which can be obtained according to the AIS VHF signal from the shore station; and T_{r_i} is the signal arrival time for the *i*th antenna of vessel *M*, which can be obtained by the vessel.

However, as time synchronization between a vessel and shore stations is a difficult task to achieve in reality, $\overline{R}_i$ is the measured distance between the *i*th antenna and the shore station *B*, not equal to the actual distance R_i. The positioning equation between the *i*th antenna of the vessel and the shore station is

$$\overline{R}_i = R_i(X_i, Y_i) + c\Delta T \tag{2}$$

where subscript *i* represents the *i*th antenna, and ΔT is the clock offset between *M* and *B*. The accurate distance R_i can be calculated by the Euclidean distance formula:

$$R_i(X_i, Y_i) = \left((X_i - X_B)^2 + (Y_i - Y_B)^2\right)^{\frac{1}{2}} \tag{3}$$

According to Equation (3), Equation (2) can be further written as

$$\overline{R}_i = \left((X_i - X_B)^2 + (Y_i - Y_B)^2\right)^{\frac{1}{2}} + c\Delta T \tag{4}$$

If one antenna is installed at the center of the vessel, the position coordinates of the vessel can be represented as (X_0, Y_0).

The heading angle θ of the vessel at any time can be obtained according to the outputs of shipborne equipment, including a magnetic compass, a gyrocompass, and so forth. In the global coordinate system, the coordinates of the *i*th antenna (X_i, Y_i) can be expressed as

$$\begin{bmatrix} X_i \\ Y_i \end{bmatrix} = \begin{bmatrix} X_0 \\ Y_0 \end{bmatrix} + T(\theta) \begin{bmatrix} x_i \\ y_i \end{bmatrix} \tag{5}$$

where $T(\theta)$ is called the rotation matrix and is given below:

$$T(\theta) = \begin{bmatrix} -\sin\theta & \cos\theta \\ \cos\theta & \sin\theta \end{bmatrix} \tag{6}$$

substituting Equation (5) into Equation (3), we have

$$R_i(X_i, Y_i) = R_i'(X_0, Y_0) \tag{7}$$

The positioning equation of (X_0, Y_0) can be written as

$$\overline{R}_i = R_i'(X_0, Y_0) + c\Delta T \tag{8}$$

As Equation (8) is a nonlinear equation, the first step of solving the positioning equation is linearization. A Taylor series is used, and the first-order terms are retained:

$$\overline{R}_i - R_i'(\hat{X}_0, \hat{Y}_0) - c\Delta\hat{T} = \left.\frac{\partial R_i'}{\partial X_0}\right|_{(\hat{X}_0,\hat{Y}_0)} \delta X_0 + \left.\frac{\partial R_i'}{\partial Y_0}\right|_{(\hat{X}_0,\hat{Y}_0)} \delta Y_0 + c\delta T \tag{9}$$

where $(\hat{X}_0, \hat{Y}_0)$ is the initial estimated coordinate of M, and $(\delta X_0, \delta Y_0, \delta T)$ are the corrections of the corresponding estimated values.

However, due to the limitation of the vessel size, the distances between different antennas to the shore station are approximately the same and not sufficiently independent. Therefore, the positioning matrix given by Equation (9) easily becomes a near-singular matrix or an ill-conditioned matrix, which is difficult to solve. Thus, the Taylor series expansion method [29,30] or the least-squares method [31,32] is widely used in position estimation. Both of these methods require good initial values; otherwise, it is difficult for the solution to achieve convergence [33]. Therefore, they are not suitable for solving the position estimation situation proposed in this paper. The Chan method is a TDOA-based localization algorithm, which can provide a closed-form solution for arbitrarily placed reference nodes [34]. Inspired by the Chan method, we solved the proposed positioning equation as follows:

According to Equations (7) and (8), we rewrote Equation (8) as

$$R_i(X_i, Y_i) = R_i'(X_0, Y_0) = \overline{R}_{i0} + R_0(X_0, Y_0) \tag{10}$$

where

$$\overline{R}_{i0} = \overline{R}_i - \overline{R}_0 \tag{11}$$

Substituting Equation (10) into Equation (3), we obtained

$$\overline{R}_{i0}^2 + 2\overline{R}_{i0}R_0 + R_0^2 = X_i^2 + Y_i^2 + X_B^2 + Y_B^2 - 2(X_i X_B + Y_i Y_B) \tag{12}$$

If $i = 0$ in Equation (3), it can be simplified as

$$R_0^2 = X_0^2 + Y_0^2 + X_B^2 + Y_B^2 - 2(X_0 X_B + Y_0 Y_B) \tag{13}$$

Then, subtracting Equation (13) from Equation (12), the result is

$$\overline{R}_{i0}^2 + 2\overline{R}_{i0}R_0 = X_i^2 + Y_i^2 - \left(X_0^2 + Y_0^2\right) - 2(X_{i0}X_B + Y_{i0}Y_B) \tag{14}$$

according to

$$X_i^2 + Y_i^2 = \begin{bmatrix} X_0 \\ Y_0 \end{bmatrix}^T \begin{bmatrix} X_0 \\ Y_0 \end{bmatrix} + 2\begin{bmatrix} x_i \\ y_i \end{bmatrix}^T T(\theta)\begin{bmatrix} X_0 \\ Y_0 \end{bmatrix} + \begin{bmatrix} x_i \\ y_i \end{bmatrix}^T \begin{bmatrix} x_i \\ y_i \end{bmatrix} \tag{15}$$

Substituting Equations (5) and (15) into Equation (14), we obtained

$$\overline{R}_{i0}^2 + 2\overline{R}_{i0}R_0 = 2\begin{bmatrix} x_i \\ y_i \end{bmatrix}^T T(\theta)\begin{bmatrix} X_0 \\ Y_0 \end{bmatrix} + \begin{bmatrix} x_i \\ y_i \end{bmatrix}^T\begin{bmatrix} x_i \\ y_i \end{bmatrix} - 2\begin{bmatrix} x_i \\ y_i \end{bmatrix}^T T(\theta)\begin{bmatrix} X_B \\ Y_B \end{bmatrix} \tag{16}$$

3.1. Condition of Three Antennas

If there are three antennas on the vessel, according to Equation (16), the positioning matrix can be written simply as

$$\begin{bmatrix} X_0 \\ Y_0 \end{bmatrix} = \left(\begin{bmatrix} x_1 & x_2 \\ y_1 & y_2 \end{bmatrix}^T T(\theta)\right)^{-1}\left\{\frac{1}{2}\left(\begin{bmatrix} \overline{R}_{10}^2 \\ \overline{R}_{20}^2 \end{bmatrix} - \begin{bmatrix} x_1 & y_1 \\ x_2 & y_2 \end{bmatrix}\begin{bmatrix} x_1 & y_1 \\ x_2 & y_2 \end{bmatrix}^T\right) + \begin{bmatrix} \overline{R}_{10} \\ \overline{R}_{20} \end{bmatrix}R_0\right\} + \begin{bmatrix} X_B \\ Y_B \end{bmatrix} \tag{17}$$

where R_0 can be expressed by (X_0, Y_0), given by

$$R_0^2 = \begin{bmatrix} X_0 - X_B \\ Y_0 - Y_B \end{bmatrix}^T\begin{bmatrix} X_0 - X_B \\ Y_0 - Y_B \end{bmatrix} \tag{18}$$

The equation can be solved by substituting Equation (17) into Equation (18), and we can calculate the vessel position coordinates (X_0, Y_0) according to Equation (17).

3.2. Condition of More Than Three Antennas

If there are M antennas on the vessel and $M > 3$, according to Equation (16), the positioning matrix can be written as

$$\begin{bmatrix} x_i \\ y_i \end{bmatrix}^T T(\theta)\begin{bmatrix} X_0 \\ Y_0 \end{bmatrix} - \overline{R}_{i0}R_0 = \frac{1}{2}\left(\overline{R}_{i0}^2 - \begin{bmatrix} x_i \\ y_i \end{bmatrix}^T\begin{bmatrix} x_i \\ y_i \end{bmatrix}\right) + \begin{bmatrix} x_i \\ y_i \end{bmatrix}^T T(\theta)\begin{bmatrix} X_B \\ Y_B \end{bmatrix} \tag{19}$$

Substituting Equation (6) into Equation (19), we obtained

$$\begin{bmatrix} \left(-\sin\theta \quad \cos\theta\right)\begin{bmatrix} x_i & y_i \end{bmatrix}^T \\ \left(\cos\theta \quad \sin\theta\right)\begin{bmatrix} x_i & y_i \end{bmatrix}^T \\ -\overline{R}_{i0} \end{bmatrix}^T\begin{bmatrix} X_0 \\ Y_0 \\ R_0 \end{bmatrix} = \frac{1}{2}\left(\overline{R}_{i0}^2 - \begin{bmatrix} x_i \\ y_i \end{bmatrix}^T\begin{bmatrix} x_i \\ y_i \end{bmatrix}\right) + \begin{bmatrix} x_i \\ y_i \end{bmatrix}^T T(\theta)\begin{bmatrix} X_B \\ Y_B \end{bmatrix} \tag{20}$$

The positioning matrix can be simplified as

$$\mathbf{HX} = \mathbf{b} \tag{21}$$

where

$$\mathbf{X} = \begin{bmatrix} X_0 & Y_0 & R_0 \end{bmatrix}^T \tag{22}$$

$$\mathbf{H} = \begin{bmatrix} \left(-\sin\theta \quad \cos\theta\right)\begin{bmatrix} x_1 \\ y_1 \end{bmatrix} & \left(\cos\theta \quad \sin\theta\right)\begin{bmatrix} x_1 \\ y_1 \end{bmatrix} & -\overline{R}_{10} \\ \left(-\sin\theta \quad \cos\theta\right)\begin{bmatrix} x_2 \\ y_2 \end{bmatrix} & \left(\cos\theta \quad \sin\theta\right)\begin{bmatrix} x_2 \\ y_2 \end{bmatrix} & -\overline{R}_{20} \\ \vdots & \vdots & \vdots \\ \left(-\sin\theta \quad \cos\theta\right)\begin{bmatrix} x_{M-1} \\ y_{M-1} \end{bmatrix} & \left(\cos\theta \quad \sin\theta\right)\begin{bmatrix} x_{M-1} \\ y_{M-1} \end{bmatrix} & -\overline{R}_{(M-1)0} \end{bmatrix} \tag{23}$$

$$
\mathbf{b} = \begin{bmatrix}
\frac{1}{2}\left(\overline{R}_{10}^2 - \begin{bmatrix} x_1 \\ y_1 \end{bmatrix}^T \begin{bmatrix} x_1 \\ y_1 \end{bmatrix}\right) + \begin{bmatrix} x_1 \\ y_1 \end{bmatrix}^T T(\theta) \begin{bmatrix} X_B \\ Y_B \end{bmatrix} \\[2ex]
\frac{1}{2}\left(\overline{R}_{20}^2 - \begin{bmatrix} x_2 \\ y_2 \end{bmatrix}^T \begin{bmatrix} x_2 \\ y_2 \end{bmatrix}\right) + \begin{bmatrix} x_2 \\ y_2 \end{bmatrix}^T T(\theta) \begin{bmatrix} X_B \\ Y_B \end{bmatrix} \\[2ex]
\vdots \\[2ex]
\frac{1}{2}\left(\overline{R}_{(M-1)0}^2 - \begin{bmatrix} x_{M-1} \\ y_{M-1} \end{bmatrix}^T \begin{bmatrix} x_{M-1} \\ y_{M-1} \end{bmatrix}\right) + \begin{bmatrix} x_{M-1} \\ y_{M-1} \end{bmatrix}^T T(\theta) \begin{bmatrix} X_B \\ Y_B \end{bmatrix}
\end{bmatrix}
\tag{24}
$$

As measurement errors exist in $\overline{R}_{i0}$, the positioning matrix of Equation (21) was rewritten as

$$
\psi = \mathbf{b} - \mathbf{HX}
\tag{25}
$$

where

$$
\psi_i = R_i c n_i + 0.5 c^2 n_i^2
\tag{26}
$$

where n_i is an error corresponding to $\overline{R}_{i0}$. Due to $R_i \approx R_0$, $\mathbf{R} \approx R_0 \mathbf{I}$. $\mathbf{I}$ is an identity matrix. Equation (25) can be written as

$$
\psi \approx c R_0 \mathbf{n} + 0.5 c^2 \mathbf{n} \odot \mathbf{n}
\tag{27}
$$

Therefore, solving the positioning matrix requires actually finding the value of $\mathbf{X}$ where $\|\mathbf{b} - \mathbf{HX}\|$ is the minimum. Using the weighted least-squares method, the solution of Equation (25) is

$$
\mathbf{X} = \left(\mathbf{H}^T \mathbf{Q}^{-1} \mathbf{H}\right)^{-1} \mathbf{H}^T \mathbf{Q}^{-1} \mathbf{b}
\tag{28}
$$

where $\mathbf{Q}$ is the covariance matrix of the measurement error. The covariance matrix of $\mathbf{X}$ can be expressed as

$$
\mathrm{cov}(\mathbf{X}) \approx c^2 R_0^2 \left(\mathbf{H}^T \mathbf{Q}^{-1} \mathbf{H}\right)^{-1}
\tag{29}
$$

In order to improve the accuracy, assuming that the elements of $\mathbf{X}$ are independent, we defined

$$
\mathbf{X} = \begin{bmatrix} \widetilde{X}_0 + e_X \\ \widetilde{Y}_0 + e_Y \\ \widetilde{R}_0 + e_R \end{bmatrix}
\tag{30}
$$

where $\mathbf{e}$ represents the estimation errors of $\mathbf{X}$. Subtracting the first two components of $\mathbf{X}$ by (X_B, Y_B), we obtained

$$
\varphi = \mathbf{h} - \mathbf{GZ}
\tag{31}
$$

where

$$
\mathbf{Z} = \begin{bmatrix} \left(\widetilde{X}_0 - X_B\right)^2 \\ \left(\widetilde{Y}_0 - Y_B\right)^2 \end{bmatrix}
\tag{32}
$$

$$
\mathbf{G} = \begin{bmatrix} 1 & 0 \\ 0 & 1 \\ 1 & 1 \end{bmatrix}
\tag{33}
$$

$$
\mathbf{h} = \begin{bmatrix} (X_0 - X_B)^2 \\ (Y_0 - Y_B)^2 \\ R_0^2 \end{bmatrix}
\tag{34}
$$

The solution of Equation (31) is

$$
\mathbf{Z} = \left(\mathbf{G}^T \mathbf{\Phi}^{-1} \mathbf{G}\right)^{-1} \mathbf{G}^T \mathbf{\Phi}^{-1} \mathbf{h}
\tag{35}
$$

where

$$\mathbf{\Phi} = E\big(\varphi^T \varphi\big) \tag{36}$$

we know that

$$\left.\begin{aligned}
\varphi_1 &= 2(X_0 - X_B)e_X + e_X^2 \approx 2(X_0 - X_B)e_X \\
\varphi_2 &= 2(Y_0 - Y_B)e_Y + e_Y^2 \approx 2(Y_0 - Y_B)e_Y \\
\varphi_3 &= 2R_0 e_R + e_R^2 \approx 2R_0 e_R
\end{aligned}\right\} \tag{37}$$

Equation (36) can be derived as

$$\mathbf{\Phi} = 4\mathbf{D}\mathrm{cov}(\mathbf{X})\mathbf{D} \tag{38}$$

where

$$\mathbf{D} = \mathrm{diag}\{(X_0 - X_B), (Y_0 - Y_B), R_0\} \tag{39}$$

Substituting Equation (28) into Equation (37),

$$\mathbf{\Phi} = E\big(\varphi^T \varphi\big) = 4\mathbf{D}\mathrm{cov}(\mathbf{X})\mathbf{D} = 4c^2 R_0{}^2 \mathbf{D}\big(\mathbf{H}^{0T}\mathbf{Q}^{-1}\mathbf{H}^0\big)^{-1}\mathbf{D} \tag{40}$$

Similarly, substituting Equation (38) into Equation (34),

$$\mathbf{Z} = \big(\mathbf{G}^T\mathbf{D}^{-1}\mathbf{H}^{0T}\mathbf{Q}^{-1}\mathbf{H}^0\mathbf{D}^{-1}\mathbf{G}\big)^{-1}\mathbf{G}^T\mathbf{D}^{-1}\mathbf{H}^{0T}\mathbf{Q}^{-1}\mathbf{H}^0\mathbf{D}^{-1}\mathbf{h} \tag{41}$$

The covariance of **Z** is

$$\mathrm{cov}(\mathbf{Z}) = \big(\mathbf{G}^T\mathbf{\Phi}^{-1}\mathbf{G}\big)^{-1} \tag{42}$$

Finally, the vessel position is estimated as

$$\mathbf{X}' = \pm\sqrt{\mathbf{Z}} + \begin{bmatrix} X_B \\ Y_B \end{bmatrix} \tag{43}$$

According to the approximate position of the vessel, the minus sign or the plus sign can be determined in Equation (43). The covariance matrix of position estimation **X'**, corresponding to positioning precision, is given by

$$\mathrm{cov}\big(\mathbf{X}'\big) = \frac{1}{4}\mathbf{B}^{-1}\mathrm{cov}(\mathbf{Z})\mathbf{B}^{-1} \approx c^2 R_0{}^2\big(\mathbf{B}\mathbf{G}^T\mathbf{D}^{-1}\big(\mathbf{H}^{0T}\mathbf{Q}^{-1}\mathbf{H}^0\big)\mathbf{D}^{-1}\mathbf{G}\mathbf{B}\big)^{-1} \tag{44}$$

where

$$\mathbf{B} = \mathrm{diag}\{(X_0 - X_B), (Y_0 - Y_B)\} \tag{45}$$

The position root-mean-square error (RMSE) is equal to the root of the sum of the diagonal elements of cov(**X'**). Furthermore, to further improve the accuracy of the estimated position, smooth filtering was also used based on the estimated position **X'** of Equation (43). We formulated this smooth problem in MATLAB by simply using the smoothdata function to smooth noisy position data.

4. Simulation Scenario

In order to verify the proposed position estimation method for a single shore station in AIS R-Mode, numerical simulation of positioning errors was performed in different scenarios using MATLAB. A real AIS shore station named Huangbaizui was used, the latitude of which is 38°54.2850′ N, and the longitude is 121°42.9500′ E, as shown in Figure 3a. The location of the vessel was very close to the Huangbaizui shore station, where the vessel could only receive signals from the Huangbaizui shore station. Its initial latitude was 38°40.1690′ N, and the longitude was 122°10.1020′ E. We used a Gauss–Krüger projection with six-degree zones [35]; the shore station coordinates (X_B, Y_B) were $[4.3087 \times 10^6, -1.1139 \times 10^5]$; and the vessel's initial position (X_0, Y_0) was $[4.3092 \times 10^6, -1.0938 \times 10^5]$. The unit of measurement was meters. Figure 3b shows the geometric relationship between the shore station and the antennas of the vessel when the heading angle of the vessel was 0°. The blue star denotes the position of the AIS shore station. The red asterisk is the position of the vessel's center. The green crosses denote the other three antennas on the vessel. As the antennas' positions were linked with the vessel's dimensions, we introduced a common engineering vessel, named Maochang 526, to set up the antennas' location parameters. As Maochang 526's cargo was 650 t, the length was 68 m, and the width was 14 m, the position coordinates of the four antennas were [0, 0], [15, 7], [30, 0], [15, −7] in the local coordinate system of the vessel, as shown in Figure 3b.

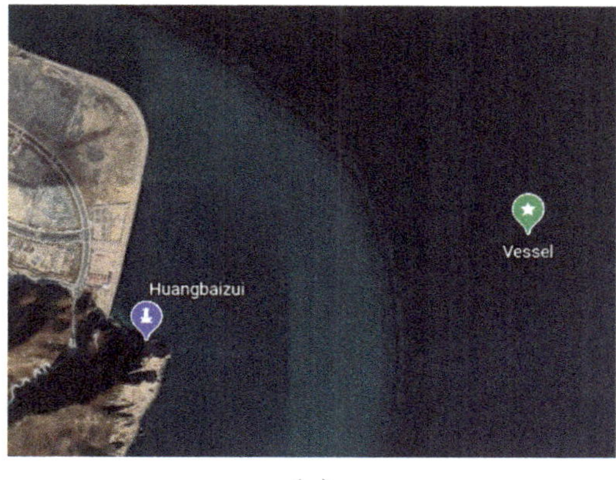

（**a**）

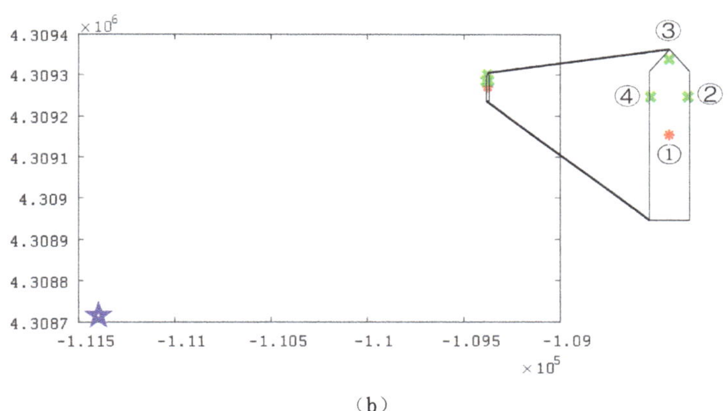

（**b**）

Figure 3. Distribution of single shore station and initial vessel location. (**a**) distribution on google map (**b**) distribution using a Gauss–Krüger projection.

In order to assess the performance of the proposed positioning method for a single shore station, positioning errors were evaluated in this study in the following four simulation scenarios.

- Scenario 1: The heading angle was 0°. In order to evaluate the effects of the heading angle of the vessel on the positioning errors, the location of the vessel was fixed and just the heading angle was changing.
- Scenario 2: The vessel moved along a circular trajectory, the center of which was the location of the shore station. During the movement, the heading angle of the vessel was constant.

- Scenario 3: The vessel moved along a circular trajectory, the center of which was the location of the shore station. During the movement, the heading angle of the vessel was continuously adjusted to maintain the relative position between the antennas and the shore station.
- Scenario 4: The vessel moved toward or away from the shore station with a constant heading angle. Only the distance between the shore station and the vessel was changing.

To bring simulation closer to reality, we used the Monte Carlo method to create measurement noises [36]. We ran this 1000 times at every vessel position in the above scenarios by adding Gaussian random noise with a mean root square of 0.001.

5. Simulation Results Analysis

5.1. Scenario 1: Positioning Errors Vary with Heading Angles

In Scenario 1, the position of the vessel was fixed. Further, the heading angle of the vessel was increased from 0° to 90°. Figure 4 shows the antenna locations of the vessel with different heading angles θ. The red cross in Figure 4 denotes the vessel's location—the center of the vessel. The green quadrilaterals represent the layout of the antennas in the vessel with different heading angles.

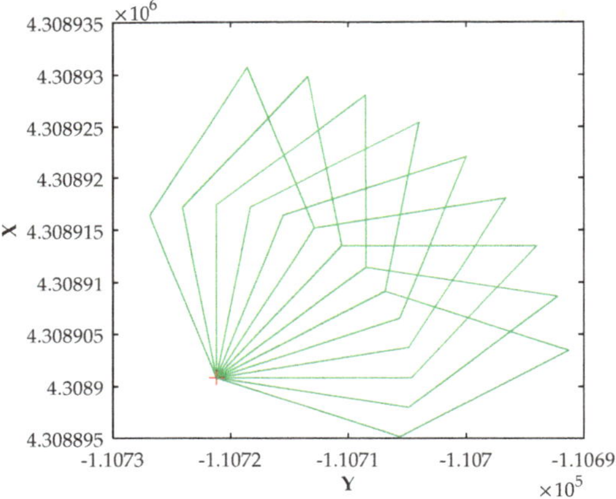

Figure 4. Antenna layout with different heading angles.

The theoretical RMSE of the proposed method in this study at different heading angles is given in Table 1. Moreover, the calculated RMSE and the average positioning error after smooth filtering according to the simulation results in Scenario 1 are also given.

Table 1. Positioning results of Scenario 1.

Heading Angle θ	5	15	25	35	45	55	65	75	85
Theoretical RMSE	1.609	2.849	4.439	6.381	8.682	11.324	14.334	17.685	21.350
Simulated RMSE	1.609	2.888	4.441	6.386	8.683	11.332	14.334	17.688	21.395
Filtered position error	0.019	0.057	0.165	0.240	0.298	1.916	2.354	3.412	7.312

From Table 1, it can be seen clearly that the positioning errors increased as the heading angles increased. The reason for the increase of the RMSE was that the correlation between distances from different antennas to the shore station was increasing with the relative position change between the antennas and the shore station. These results indicate that the calculated RMSE based on the simulation positioning results is consistent with the RMSE according to the theoretical derivation. In addition, positioning errors could be reduced effectively by using the smoothdata function in MATLAB.

5.2. Scenario 2: Positioning Errors Vary with Vessel's Position

In Scenario 2, the vessel navigated around the shore station. The relative azimuth angle η shown in Figures 2 and 5 is the angle between the north vector and the shore to the vessel vector. It started at 5° and stopped every 10° to stabilize its heading angle on 0° and get its position. The vessel continued to navigate until η became 85°. The shore station was the center of the motion curve. Figure 5 shows the positioning results of the vessel as its position changed. The red cross is the actual position of the vessel. The black asterisk is the estimated position using the proposed method. The vertexes of the green diamond represent the positions of antennas.

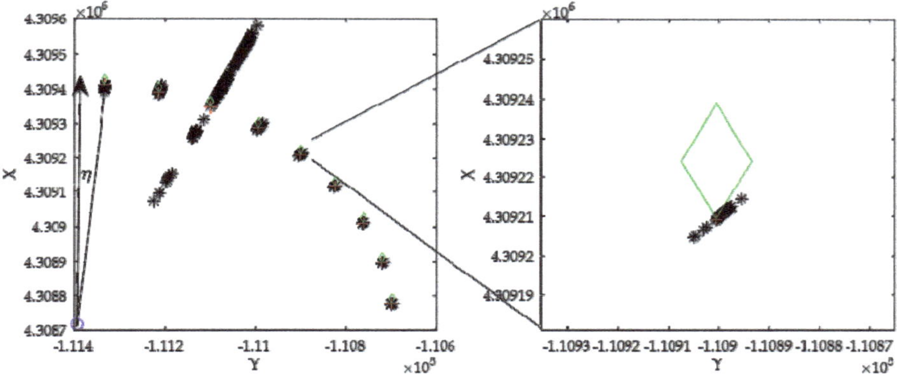

Figure 5. Positioning results at different locations.

η is the relative azimuth angle of the shore station shown in Figures 2 and 5. Table 2 shows the positioning results when the vessel stopped every 10° until η became 85° in Scenario 2.

Table 2. Positioning results of Scenario 2.

Relative Azimuth Angle η	85	75	65	55	45	35	25	15	5
Theoretical RMSE	3.117	3.418	4.077	5.412	8.451	18.688	271.811	22.850	15.274
Simulated RMSE	3.118	3.420	4.078	5.408	8.473	18.690	352.870	22.950	15.276
Filtered position error	0.092	0.102	0.092	0.231	1.464	3.797	588.233	7.592	3.400

It can be seen from Figure 5 and Table 2 that the RMSE was changing with the relative position between the antennas and the shore station. The RMSE was largest and the positioning result was the worst when the relative azimuth angle between the vessel and the shore station was 25°. In this situation, the distances between different antennas and the shore station were not sufficiently independent. Antennas ② and ④, and the shore station, were almost on the same line, shown in Figure 5. Therefore, the positioning matrix attained singularity in this situation. The calculated RMSE based on the simulation positioning results was much larger than the theoretical RMSE, as the noise had a significant impact on the positioning matrix. Even using a smoothing filter could not reduce the position errors.

5.3. Scenario 3: Positioning Errors Vary with Relative Position

In Scenario 3, the vessel navigated around the shore station, and the shore station was still the center of the motion curve. It stopped every 10° to adjust its heading angle to maintain the relative positional relationship between the antennas and the shore station, and get its position. The positioning results of the vessel are shown in Figure 6. We can see that the relative position between the antennas and the shore station remained the same. In addition, the position errors were similar, although the vessel was in different locations.

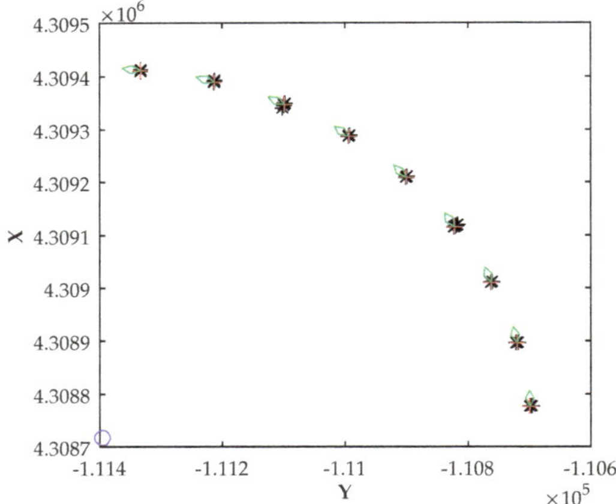

Figure 6. Positioning results with unchanged relative position.

The numerical positioning results in Scenario 3 are given in Table 3.

Table 3. Positioning results of Scenario 3.

Heading Angle θ	85	75	65	55	45	35	25	15	5
Theoretical RMSE	3.117	3.117	3.117	3.117	3.117	3.117	3.117	3.117	3.117
Simulated RMSE	3.119	3.115	3.119	3.118	3.113	3.117	3.118	3.116	3.117
Filtered position error	0.142	0.077	0.142	0.141	0.233	0.041	0.245	0.072	0.031

According to the simulation results shown in Table 3, the theoretical RMSE was the same. The reason for this was that the relative position between the antennas and the shore station was the same. Further, the calculated RMSE was almost the same. Positioning errors by the smoothing filter were reduced and were at the same level, substantially, when the vessel was in different positions. In Scenario 1, the heading angles of the vessel were changing. In Scenario 2, the position of the vessel was changing. However, in this scenario, the relative position was the same, although both the heading angle and the position were changing. Therefore, it can be concluded that, as long as the distance is constant and the relative positional relationship remains unchanged, the positioning performance is the same, regardless of whether the position and the heading angles of the vessel are changing.

5.4. Scenario 4: Positioning Errors Vary with Distances

In Scenario 4, the vessel moved toward or away from the shore station. The distance between the shore station and the vessel was changing from 100 to 1000 m. During the movement, the heading angle of the vessel was always 0°. Figure 7 shows the estimated positioning results as the distance changed. Position errors varied with the distance, as depicted in Figure 8.

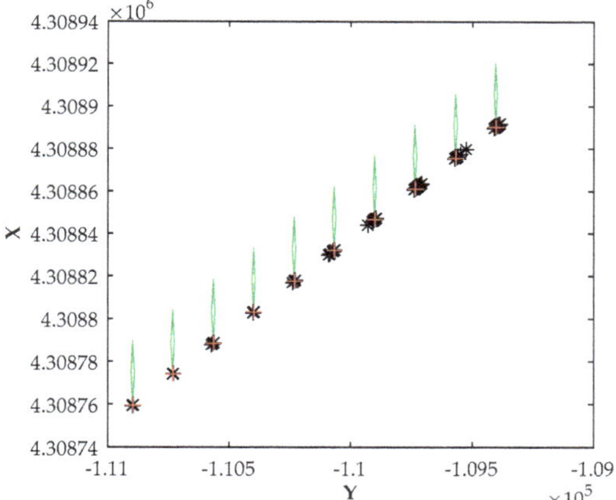

Figure 7. Positioning results with changing distance.

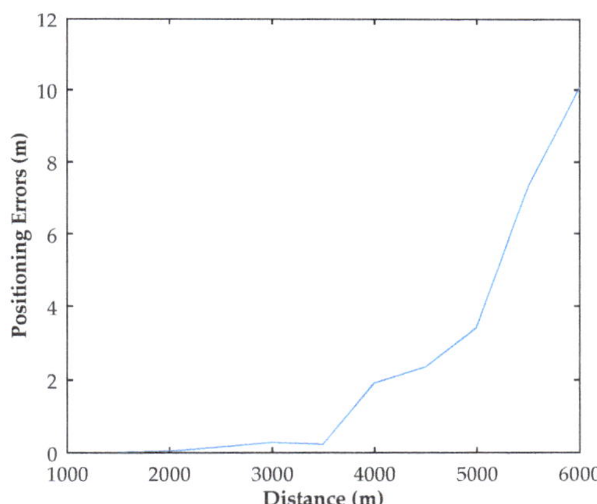

Figure 8. Positioning results at different distances.

Table 4 shows the positioning results in Scenario 4. It can be observed that the simulated RMSE and position errors increased as the distance increased. Therefore, the performance was better in the area closer to the shore station. Correspondingly, the proposed method is precisely used for positioning based on a single shore station, which is suitable for areas closer to the shore station.

Table 4. Positioning result of Scenario 4.

Distance	1500	2000	2500	3000	3500	4000	4500	5000	5500	6000
Theoretical RMSE	1.609	2.849	4.441	6.385	8.682	11.332	14.334	17.688	21.350	25.422
Simulated RMSE	1.609	2.849	4.449	6.386	8.683	11.334	14.334	17.689	21.395	25.454
Filtered position error	0.019	0.057	0.1654	0.298	0.4202	1.916	2.354	3.412	7.312	10.057

6. Conclusions

As the existing AIS is a communication system, when AIS shore stations are used as positioning reference stations in AIS R-Mode, a vessel can only receive signals from a single shore station in some areas, especially in an area close to the shore station. In this situation, traditional positioning methods are not suitable. A position estimation method using multiple antennas was proposed in this paper to solve this problem. The geometric relationship between the antennas in the local coordinate system is converted into the global positioning coordinate system using the heading angle. Then, the positioning matrix is obtained according to measurements from multiple antennas. However, due to the size of the vessel, the distances between different antennas to the shore station are not sufficiently independent. Therefore, the positioning matrix of this proposed method is easily near singularity or ill-conditioned. A novel method to solve the positioning matrix was presented here, rather than the Taylor series expansion method, which can effectively prevent singularity. Finally, we verified the validity of the proposed method in diverse scenarios by numerical simulations. The influencing factors of positioning performance were analyzed, such as heading angle, relative position, and distances. The positioning performance became worse as the distance increased. Fortunately, the proposed method was exactly suited for the area close to the shore station, where the vessel can only receive signals from this shore station. Further, the positioning errors were mainly affected by the relative positional relationship between the antennas and the shore station. As it is not necessary to establish new AIS shore stations, the proposed method can help the AIS R-Mode positioning system to expand its application scope. A possible area of future research would be to investigate the improved position estimation method based on a single shore station, which can reduce positioning errors in the far area of the shore station. This would help to further expand AIS R-Mode applications.

Author Contributions: Conceptualization, Y.J. and K.Z.; methodology, Y.J.; software, Y.J. and K.Z.; validation, Y.J. and K.Z.; investigation, Y.J. and K.Z.; writing—original draft preparation, Y.J.; writing—review and editing, Y.J. and K.Z.; funding acquisition, Y.J. All authors have read and agreed to the published version of the manuscript.

Funding: This research was funded by the Chinese National Science Foundation (numbers 61501079 and 61971083); the Remote Sensing Youth Science and Technology Innovative Research; the Dalian Technology Star Program (number 2017Q058); the Dalian Science and Technology Innovation Fund (number 2019J11CY015); and the Fundamental Research Funds for the Central Universities (number 017190327).

Conflicts of Interest: The authors declare no conflict of interest.

References

1. Grant, A.; Williams, P.; Shaw, G.; de Voy, M.; Ward, N. Understanding GNSS availability and how it impacts maritime safety. In Proceedings of the Institute of Navigation International Technical Meeting, San Diego, CA, USA, 24–26 January 2011; pp. 687–695.
2. Williams, P.; Grant, A.; Hargreaves, C. Resilient PNT—Making way through rough waters. In Proceedings of the 7th GNSS Vulnerabilities and Solutions Conference, Baška, Krk Island, Croatia, 18–20 April 2013.
3. Maritime Safety Committee. Adoption of amendments to the international convention to the safety of life at sea, 1974, as amended. *Resolut. MSC* **2006**, *216*, 82.
4. Available online: http://www.imo.org/en/about/conventions/listofconventions/pages/international-convention-for-the-safety-of-life-at-sea-(solas),-1974.aspx (accessed on 10 January 2020).
5. International Maritime Organization. *SOLAS 1974 Amendments*; International Maritime Organization: London, UK, 2000.
6. Creech, J.; Ryan, J. AIS the cornerstone of national security? *J. Navig.* **2003**, *56*, 31–44. [CrossRef]
7. International Association of Lighthouse Authorities. *IALA Worldwide Radio Navigation Plan, Version 2*; International Association of Lighthouse Authorities: St Germain en Laye, France, 2012.
8. Williams, P.; Grant, A.; Hargreaves, C.; Bransby, M.; Ward, N.; Last, D. Resilient PNT for e-Navigation. In Proceedings of the ION 2013 Pacific PNT Meeting, Honolulu, HI, USA, 23–25 April 2013; pp. 477–484.
9. Johnson, G.; Swaszek, P.; Alberding, J.; Hoppe, M.; Oltmann, J.H. The feasibility of r-mode to meet resilient PNT requirements for e-navigation. In Proceedings of the 27th International Technical Meeting of the Satellite Division of the Institute of Navigation, Tampa, FL, USA, 8–12 September 2014; pp. 3076–3100.
10. Ward, N.; Shaw, G.; Williams, P.; Grant, A. The Role of GNSS in E-Navigation and the Need for Resilience. Available online: http://www.forschungsinformationssystem.de/servlet/is/342313/ (accessed on 10 January 2020).
11. Hu, Q.; Jiang, Y.; Zhang, J.B.; Sun, X.W.; Zhang, S.F. Development of an automatic identification system autonomous positioning system. *Sensors* **2015**, *15*, 28574–28591. [CrossRef] [PubMed]

12. Johnson, G.; Swaszek, P. Feasibility Study of R-Mode Using AIS Transmissions. Available online: http://www.accseas.eu/publications/r-mode-feasibility-study/ (accessed on 10 January 2020).

13. Chang, S.J. Development and analysis of AIS applications as an efficient tool for vessel traffic service. In Proceedings of the MTTS/IEEE TECHNO-OCEAN '04, Kobe, Japan, 9–12 November 2004.

14. Johnson, G.; Swaszek, P.; Hoppe, M.; Grant, A.; Safar, J. Initial results of MF-DGNSS R-Mode as an alternative position navigation and timing service. In Proceedings of the 2017 International Technical Meeting of the Institute of Navigation, Monterey, CA, USA, 30 January–2 February 2017; pp. 1206–1226.

15. Julia, H.; Stefan, G. Launch of R-Mode Baltic Project—An Alternative Navigation System at Sea. Available online: http://www.dlr.de/dlr/en/desktopdefault.aspx/tabid-10260/370_read-24695#/gallery/28882/ (accessed on 10 January 2020).

16. Zhang, J.B.; Zhang, S.F.; Wang, J.P. Pseudorange measurement method based on AIS signals. *Sensors* **2017**, *17*, 1183. [CrossRef] [PubMed]

17. Han, Y.; Son, P.; Lee, S.; Park, S.G.; Fang, T.H.; Park, S. A measurement based accuracy prediction of terrestrial radio navigation system for maritime backup in South Korea. In Proceedings of the 32nd International Technical Meeting of the Satellite Division of the Institute of Navigation, Miami, FL, USA, 16–20 September 2019; pp. 1512–1523.

18. Dziewicki, M.; Gewies, S.; Hoppe, M. R-Mode Baltic—A user need driven testbed development for the Baltic Sea. In Proceedings of the 31st International Technical Meeting of the Satellite Division of The Institute of Navigation, Miami, FL, USA, 24–28 September 2018; pp. 1736–1764.

19. Nguyen, N.H.; Doğançay, K. Optimal geometry analysis for multistatic TOA localization. *IEEE Trans. Signal Process.* **2016**, *64*, 4180–4193. [CrossRef]

20. Shalaby, M.; Shokair, M.; Messiha, N.W. Performance enhancement of TOA localized wireless sensor networks. *Wirel. Pers. Commun.* **2017**, *95*, 4667–4679. [CrossRef]

21. Lee, H.K.; Kim, H.; Shim, J.; Heo, M.B. Analytic equivalence of iterated TOA and TDOA techniques under structured measurement characteristics. *Multidimens. Syst. Signal Process.* **2011**, *22*, 361–377. [CrossRef]

22. Ionel, R.; Ionel, S. The majority principle in TDOA estimation. In Proceedings of the International Conference on Communications, Bucharest, Romania, 10–12 June 2010; pp. 21–24.

23. Ma, S.; Liu, Z.; Jiang, W.L. Pulse sorting algorithm using TDOA in multiple sensors system. In Proceedings of the Advanced Materials Research, Shenyang, China, 27–29 July 2012; Volume 571, pp. 665–670.

24. Xu, J.; Ma, M.; Law, C.L. Position estimation using UWB TDOA measurements. In Proceedings of the IEEE 2006 International Conference on Ultra-Wideband, Waltham, MA, USA, 24–27 September 2007; pp. 605–610.

25. Zheng, K.; Hu, Q.; Zhang, J. Positioning error analysis of ranging-mode using AIS signals in China. *J. Sens.* **2016**, *2016*, 1–11. [CrossRef]

26. Jiang, Y.; Zhang, S.F.; Yang, D.K. A novel position estimation method based on displacement correction in AIS. *Sensors* **2014**, *14*, 17376–17389. [CrossRef] [PubMed]

27. Jiang, Y.; Wu, J.; Zhang, S. An improved positioning method for two base stations in AIS. *Sensors* **2018**, *18*, 991. [CrossRef] [PubMed]

28. Shi, G.; Ming, Y. Survey of indoor positioning systems based on ultra-wideband (UWB) technology. *Wirel. Commun. Netw. Appl.* **2016**, *348*, 1269–1278.

29. Ren, J.; Chen, J.; Feng, L. A novel positioning algorithm based on self-adaptive algorithm of RBF network. *Open Electr. Electron. Eng. J.* **2016**, *10*, 141–148. [CrossRef]

30. Zhang, X.; Huang, D.; Liao, H. New mathematical model for GNSS relative positioning resolving. *J. Southwest Jiaotong Univ.* **2015**, *50*, 485–489.

31. Pulford, G.W. Analysis of a nonlinear least squares procedure used in global positioning systems. *IEEE Trans. Signal Process.* **2010**, *58*, 4526–4534. [CrossRef]

32. He, Y.; Martin, R.; Bilgic, A.M. Approximate iterative Least Squares algorithms for GPS positioning. In Proceedings of the IEEE International Symposium on Signal Processing and Information Technology, Luxor, Egypt, 15–18 December 2010; pp. 231–236.

33. Liu, L.; Deng, P.; Fan, P. A cooperative location method based on Chan and Tayor Algorithms. *J. Electron. Inf. Technol.* **2004**, *26*, 41–46.

34. Chan, Y.T.; Ho, K.C. A simple and efficient estimator for hyperbolic location. *IEEE Trans. Signal Process.* **2002**, *42*, 1905–1915. [CrossRef]

35. Jiang, Y.; Hu, Q.; Yang, D.; Zheng, K. Map projection positioning method in ranging-mode of automatic identification system. *ICIC Express Lett.* **2016**, *10*, 1093–1100.
36. Binder, K.; Heermann, D.W. Guide to practical work with the Monte Carlo method. In *Monte Carlo Simulation in Statistical Physics*; Springer: Berlin/Heidelberg, Germany, 2002.

Article

Sensitivity of Safe Trajectory in a Game Environment on Inaccuracy of Radar Data in Autonomous Navigation

Józef Lisowski

Faculty of Marine Electrical Engineering, Gdynia Maritime University, 81-225 Gdynia, Poland; j.lisowski@we.umg.edu.pl; Tel.: +48-694-458-333

Received: 27 March 2019; Accepted: 15 April 2019; Published: 16 April 2019

Abstract: This article provides an analysis of the autonomous navigation of marine objects, such as ships, offshore vessels and unmanned vehicles, and an analysis of the accuracy of safe control in game conditions for the cooperation of objects during maneuvering decisions. A method for determining safe object strategies based on a cooperative multi-person positional modeling game is presented. The method was used to formulate a measure of the sensitivity of safe control in the form of a relative change in the payment of the final game; to determine the final deviation of the safe trajectory from the set trajectory of the autonomous vehicle movement; and to calculate the accuracy of information in terms of evaluating the state of the control process. The sensitivity of safe control was considered in terms of both the degree of the inaccuracy of radar information and changes in the kinematics and dynamics of the object itself. As a result of the simulation studies of the positional game algorithm, which used an example of a real situation at sea of passing one's own object with nine other encountered objects, the sensitivity characteristics of safe trajectories under conditions of both good and restricted visibility at sea are presented.

Keywords: autonomous navigation; automatic radar plotting aid; safe objects control; game theory; computer simulation

---

1. Introduction

The subject of this article directly concerns sensors, which are the basic part of vessel detection and navigation in the process of ensuring safe control of marine objects. Sensors, such as radars, logs and gyro-compasses, form a source of information for the Automatic Radar Plotting Aid ARPA system, which are mandatory pieces of equipment for each ship to prevent collisions. However, its functional scope is limited to the determination of the safe maneuver of the object to the most dangerous object encountered, followed by its simulation in an accelerated time scale [1–3]. Modern navigation systems aim to use computer decision support systems that take many factors into account [4–9].

First, we take into account the subjectivity of the navigator in the assessment of the situation as well as the kinematics and dynamics of the objects encountered [10]. According to Lloyd Register statistics, the human errors caused by subjectivity account for about 60% of the causes of maritime accidents [11,12].

Secondly, we take into account the game nature of the anti-collision process, which results from the imperfection of maritime law and the complexity of the actual navigational situation at sea [13].

The influence of the information accuracy from sensors on the current state of the transport process and on the quality of safe control, which can be determined through the analysis of the control sensitivity to information inaccuracy, becomes important [14–17]. Most of the scientific literature concerns the sensitivity analysis of deterministic systems [18–21]. Therefore, the purpose of this study is to conduct the sensitivity analysis of the game control system for safe moving objects.

The process of managing the autonomous marine vehicles as a complex dynamic control object depends both on the accuracy of the measurements determining the current navigational situation, which was measured using the devices of the automatic radar plotting aids (ARPA) anti-collision system, and on the mathematical model of the process used to synthesize the object control algorithm.

The ARPA system allows us to track the automatically encountered object jth by determining its motion parameters, including velocity V_j and course ψ_j, and approach elements to its own ship. Furthermore, we can also determine D_{jmin} (distance of the nearest approach point, DCPA$_j$) and T_{jmin} (time to the nearest approach point, TCPA$_j$) (Figure 1).

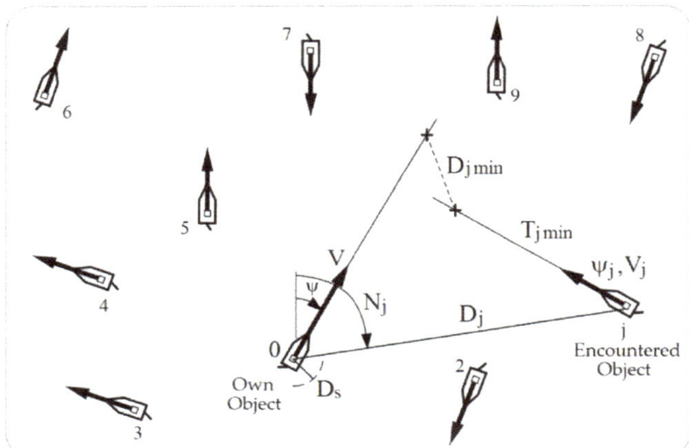

Figure 1. The navigational situation of the passage of the own object 0 moving at speed V and course ψ with the jth object encountered when moving at speed V_j and course ψ_j. In this figure, D_j is distance, N_j is bearing and D_s is safe distance.

In theory and practice, there are many methods for determining a safe maneuver or the safe trajectory of the own object while passing other objects. The simplest method is to determine the course change maneuver or the speed of the own object in relation to the most dangerous object encountered.

In one article [22], the "time to safe distance" upon the detection of dangerous objects was proposed as a potentially important parameter, which should be accompanied by a display of possible evasive maneuvers. The acceptable solutions for altering the course range should comply with the international regulations for preventing collisions at sea (COLREGs), as presented in [23].

The most important purpose of the control process is to determine a certain sequence of maneuvers in the form of a safe trajectory of the own object. The safe trajectory of the own object can be distinguished in deterministic terms without considering the maneuvering of other objects encountered. In the game approach, this is based on the use of a cooperative or non-cooperative game model of the control process [24,25].

The safety distance D_s of the passing objects, which was subjectively determined by the navigator in the current navigational situation, is important for safe navigation. This value depends on the current state of visibility at sea, which is classified by the international rules for preventing collisions of ships at sea (COLREGs) as either good or restricted visibility at sea.

Therefore, the aim of this study is to assess the sensitivity of the quality of security checks and games under the conditions of good and limited visibility at sea.

2. The Safe and Game Object Control in Autonomous Ship Navigation

The complexity of the situation when many dynamic objects are passed at sea provides the possibility of using a game model for the control process, given the fairly limited rules of international law of the sea route (COLREGs) in terms of good and limited visibility at sea and a large influence of the navigator's subjectivity in making the final maneuvering decision. It is possible to describe the process in the form of static positional and matrix games or in the form of dynamic differential games. This article proposes the use of the positional game, which is the most appropriate for this type of control process.

The basis of the positional game involves assigning the maneuver strategy of the own object to the current positions of $p(t_k)$ objects in the current step k. As such, the process model considers all possible changes in the course and speed of the encountered objects during the control [26–42] (Figure 2).

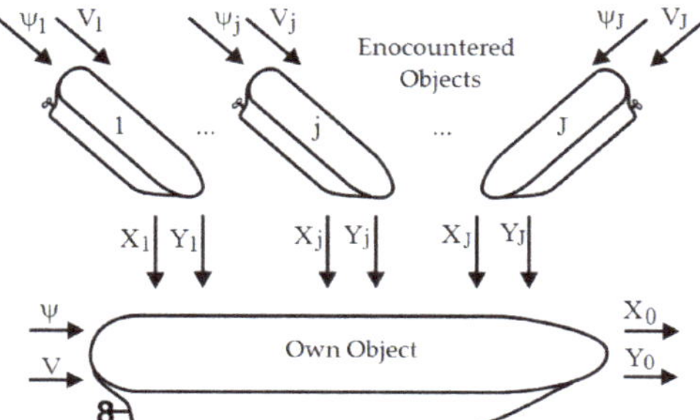

Figure 2. Block diagram of the positional game model in the situation with passing objects.

The process state is determined by the coordinates of the object's own position and the positions of the objects encountered as follows:

$$\left. \begin{array}{l} x_0 = (X_0,\ Y_0)\,,\ x_j = \left(X_j,\ Y_j\right) \\ j = 1,\ 2,\ ...,\ J \end{array} \right\}. \tag{1}$$

The control algorithm generates the own object's movement strategy at the present time t_k, based on information from the ARPA anti-collision system on the relative position of the objects that it meets:

$$p(t_k) = \left[\begin{array}{c} x_0(t_k) \\ x_j(t_k) \end{array} \right] \quad j = 1,\ 2,...,\ J \quad k = 1,\ 2,...,\ K. \tag{2}$$

Thus, in the multi-stage positional game model, at each discrete time t_k, the own object knows the positions of the objects encountered.

The following navigation restrictions are imposed on the components of the process state, which consist of the acceptable coordinates of the own position and the objects encountered:

$$\left\{ x_0(t),\ x_j(t) \right\} \ \in P. \tag{3}$$

The limits of the control values of the own and met objects are determined as:

$$u_0 \in U_0\,,\ u_j \in U_j \quad j = 1,\ 2,...,\ J. \tag{4}$$

which considers the ship movement kinematics, recommendations of the COLREGs rules and the conditions required to maintain a safe passing distance:

$$D_{j\min} = \min D_j(t) \geq D_s. \tag{5}$$

The sets of acceptable maneuvering strategies of the own object $U_{0j}(p)$ and the met objects $U_{j0}(p)$ are dependent, which means that the choice of control u_j by the jth object changes the sets of acceptable strategies of other objects.

Figure 3 shows the geometrical structure of the sets of acceptable safe strategies for the own object and for one met jth object. First, the value of the safe ships passing at a distance D_s is assumed. After this, they are positioned to be at a tangent to circles with a radius D_s, which cut off the areas of safe possible changes in the courses and speeds of the own and the encountered objects in the districts with the V and V_j rays.

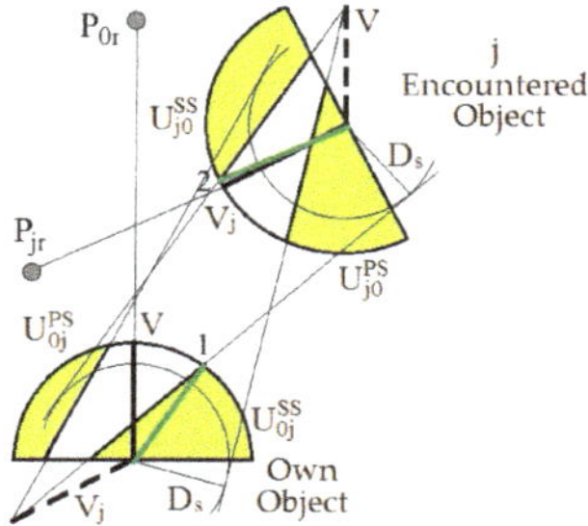

Figure 3. The method of determining the acceptable strategy sets of the own object $U_{0j} = U_{0j}{}^{PS} \cup U_{0j}{}^{SS}$ and the jth encountered object $U_{j0} = U_{j0}{}^{PS} \cup U_{j0}{}^{SS}$ for the port side (PS) and starboard side (SS). In this figure, P_{0r} is the turning point of the rhumb line of the own object, P_{jr} is the turning point of the rhumb line of the encountered object, 1 is a safe maneuver for changing the course of the own object and 2 is a safe maneuver for changing the course of the encountered object.

The method for determining the total sets of acceptable safe strategies of the own object, while passing with many objects encountered simultaneously, is shown in Figure 4, which utilizes the example of passing the own object with six objects encountered at a safe distance D_s.

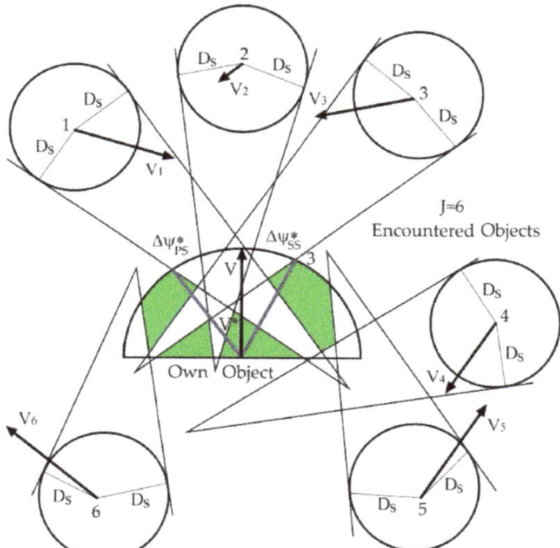

Figure 4. Areas of acceptable safe strategies of the own object in relation to the six objects encountered: 3 is the optimal maneuver for changing the course of the own object $u_0{}^* = \Delta\psi^*_{SS}$ when safely passing the six met objects.

The algorithm of the safe cooperative control of the own object $u_0^*(t_k)$ at each stage k is implemented using the following three tasks:

(1) Task min1: determine the safe admissible and optimal control of the own object u_{0j} in relation to the encountered jth object from the appropriate set of acceptable strategies U_{0j}^{PS} or U_{0j}^{SS}, which must comply with the 15th rule of COLREGs and ensure the lowest loss of the path for the safe passage of the encountered object (point 1 in Figure 3),

(2) Task min2: determine the safe admissible and optimal control of the encountered jth object u_{j0} in relation to the own object from the appropriate set of acceptable strategies U_{j0}^{PS} or U_{j0}^{SS}, which must comply with the 15th rule of COLREGs and ensure the lowest loss of the path for safe pasting of the own object (point 2 in Figure 3),

(3) Task min3: determine the safe and optimal control of the own object u_0 in relation to all J objects (point 3 in Figure 4) of the following dependence:

$$D^*(x_0) = \min_{u_0} 3 \ \min_{u_{j0}} 2 \ \min_{u_{0j}} 1 \ D[x_0(t_k)], \quad j = 1, 2, ..., J. \tag{6}$$

where D is the distance of the own object to the nearest point of return P_{0r} on the reference route.

The criterion for choosing the best trajectory of the own object is to calculate such course values and the speed, which would provide the smallest loss of the path for the safe passage of the encountered objects at a distance that is no less than the value of D_s previously accepted by the navigator.

Through the three-fold use of the *linprog* function, which is linear programming from the MATLAB Optimization Toolbox software, a cooperative multi-stage Positional Game (PG) algorithm was developed to determine the safe trajectory of the own object.

3. Control Sensitivity Analysis

The sensitivity analysis refers to the identification of the static and dynamic properties of control objects and to the synthesis of automatic control systems, particularly optimal, adaptive and game systems. A distinction is made between the sensitivity of the object model itself or the control process of changes in its operating parameters and the sensitivity of the optimal, adaptive or game control. This is both in terms of changes in parameters and the influence of disturbances, and impacts of other objects. Therefore, the s_x sensitivity functions of the optimal control u of the game process described by the state variables x can be represented as the following partial derivatives of quality control index Q:

$$s_x = \frac{\partial Q[x(u)]}{\partial x}. \tag{7}$$

The game control quality index Q acts as the form of payment for the game, which consists of integral payments and the final payment:

$$Q = \int_{t_0}^{t_K} [x(t)]^2 + r_j(t_K) + d(t_K). \tag{8}$$

The integral game payment represents the loss of the path through its own object when passing the objects encountered and the final payment determines the final collision risk $r_j(t_k)$ with respect to the jth object encountered and the final deviation of the trajectory of the object $d(t_k)$ from the reference trajectory.

Testing the sensitivity of game control will complete the sensitivity analysis of the final game payment $d(t_k)$:

$$s_i = \frac{\partial d(t_K)}{\partial x_i}. \tag{9}$$

Considering the practical application of the game control algorithm for the own object in a collision situation, it is recommended that the sensitivity analysis of a secure control should be conducted in terms of the information accuracy obtained from the ARPA anti-collision radar system in the current situation and in relation to changes in the kinematic parameters and dynamic control.

The permissible average errors that may be caused by an anti-collision system sensors may have the following values for:

- radar: bearing ±0.25° and distance ±0.05 nm,
- gyrocompass: ±0.5°,
- log: ±0.5 kn.

The algebraic sum of all errors affecting the image of the navigational situation cannot exceed ±5% for absolute values and ±3° for angular quantities.

3.1. Sensitivity of Safe Ship Control to Inaccuracy of Information from Sensors of ARPA System

SP represents such a set of information about the control of the State Process in a navigational situation:

$$SP = \left\{ V, \psi, V_j, \psi_j, D_j, N_j \right\}. \tag{10}$$

SP_e represents a set of information from the sensors of ARPA system, which contains errors in measurement and processing parameters:

$$SP_e = \left\{ V \pm \delta V, \psi \pm \delta\psi, V_j \pm \delta V_j, \psi_j \pm \delta\psi_j, D_j \pm \delta D_j, N_j \pm \delta N_j \right\}. \tag{11}$$

The relative sensitivity of the final payment in the s_x game as the final deviation of the safe trajectory of the ship d_k from the reference trajectory is expressed as follows:

$$s_x = \left| \frac{d_K(SP_e) - d_K(SP)}{d_K(SP)} \right| 100\%. \tag{12}$$

$$s_x = \left\{ s_V, s_\psi, s_{Vj}, s_{\psi j}, s_{Dj}, s_{Nj} \right\}. \tag{13}$$

3.2. Sensitivity of Safe Own Object Control to Autonomous Navigation Process Parameter Alterations

PP is a set of state Parameter Processes of control, which is expressed as follows:

$$PP = \{t_m, D_s, \Delta t_k, \Delta V\}. \tag{14}$$

PP_e represents a set of parameters containing errors in measurement and processing parameters:

$$PP_e = \{t_m \pm \delta t_m, D_s \pm \delta D_s, t_k \pm \delta t_k, \Delta V \pm \delta\Delta V\}. \tag{15}$$

The relative sensitivity of the final payment in the game s_p, which represents the final deviation of the safe trajectory of the ship d_K from the assumed trajectory, will be:

$$s_p = \left| \frac{d_K(PP_e) - d_K(PP)}{d_K(PP)} \right| 100\%. \tag{16}$$

$$s_p = \{s_{tm}, s_{Ds}, s_{Ts}, s_{\Delta tk}, s_{\Delta V}\}. \tag{17}$$

where t_m is the advance time of the maneuver with respect to the dynamic properties of the own ship, t_k is the duration of one stage of the ship's trajectory, D_s is the safe distance and T_s is the safe time of the approach.

4. Sensitivity Characteristics

The computer simulation of the PG algorithm, which represents the computer software supporting the navigator's maneuvering decision, was conducted using an example of a real navigational situation of the $J = 9$ objects encountered.

4.1. Sensitivity Characteristics of Game Own Object Control in Good Visibility at Sea

The safe trajectory of the own object and sensitivity characteristics, which were determined by the PG algorithm in the MATLAB/Simulink software, are presented in Figures 5 and 6.

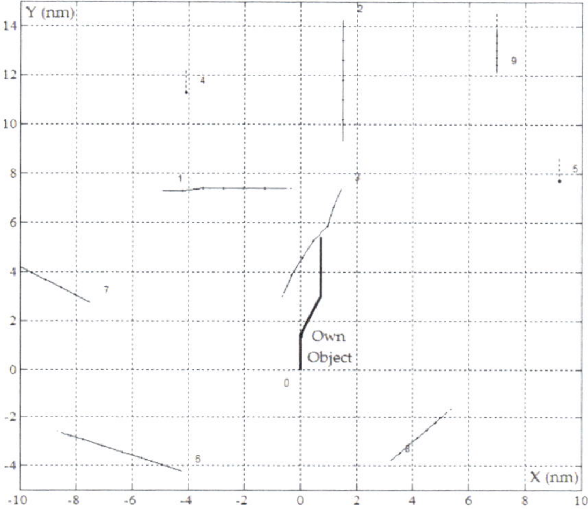

Figure 5. The safe trajectory of the own object for the positional game PG_gv algorithm in good visibility at sea where $D_s = 0.5$ nm in the situation of passing $J = 9$ encountered objects, $r(tK) = 0$ and $d(tK) = 0.71$ nm.

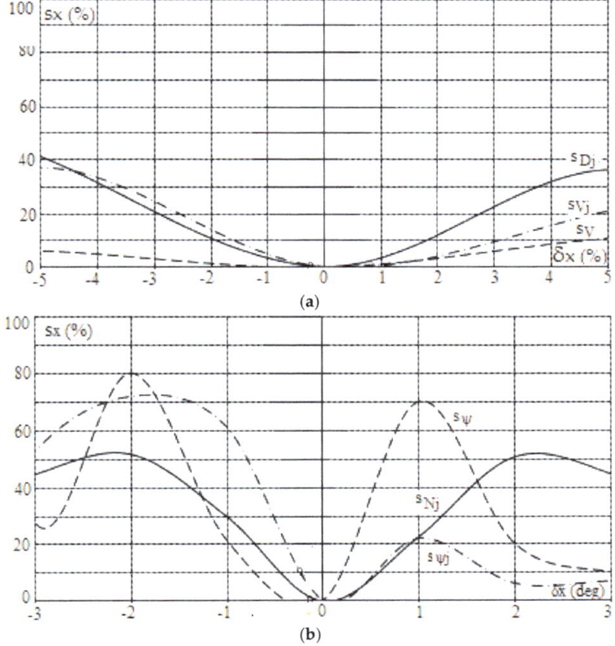

Figure 6. *Cont.*

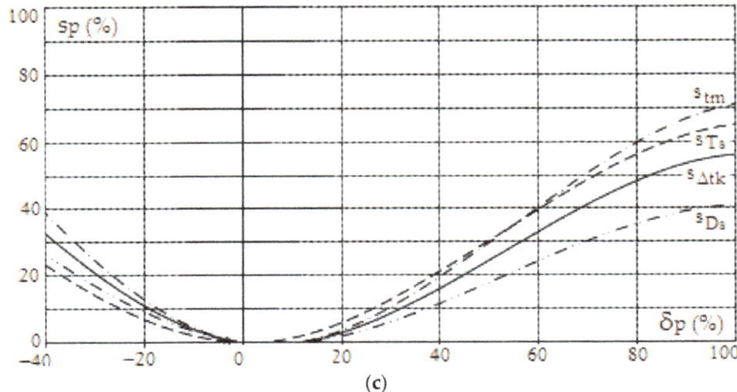

Figure 6. Sensitivity characteristics of the positional game control of the own object in good visibility at sea according to PG_gv algorithm as a function of: (**a**) absolute values of the information from sensors, (**b**) angular values of the information from sensors and (**c**) values of the control process parameters.

4.2. Sensitivity Characteristics of Game Own Object Control in Restricted Visibility at Sea

The safe trajectory of the own object and sensitivity characteristics, which were determined by the PG algorithm in the MATLAB/Simulink software, are presented in Figures 7 and 8.

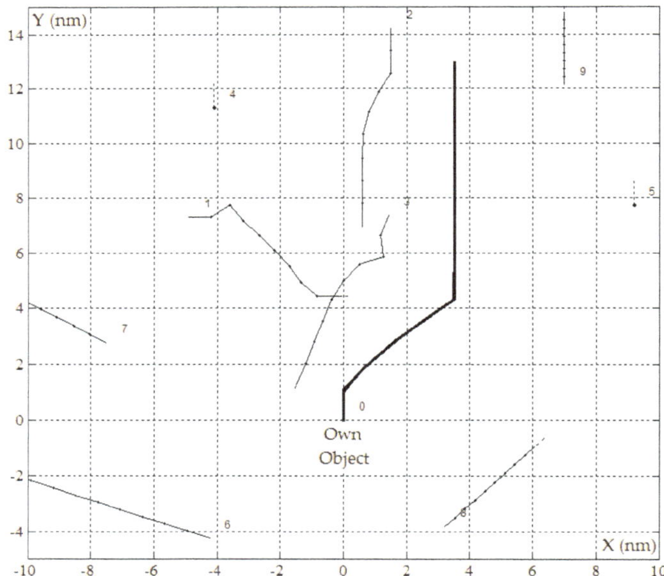

Figure 7. The safe trajectory of the own object for positional game PG_rv algorithm in restricted visibility at sea where D_s = 1.5 nm in the situation of passing J = 9 encountered objects, $r(tK)$ = 0 and $d(tK)$ = 3.47 nm.

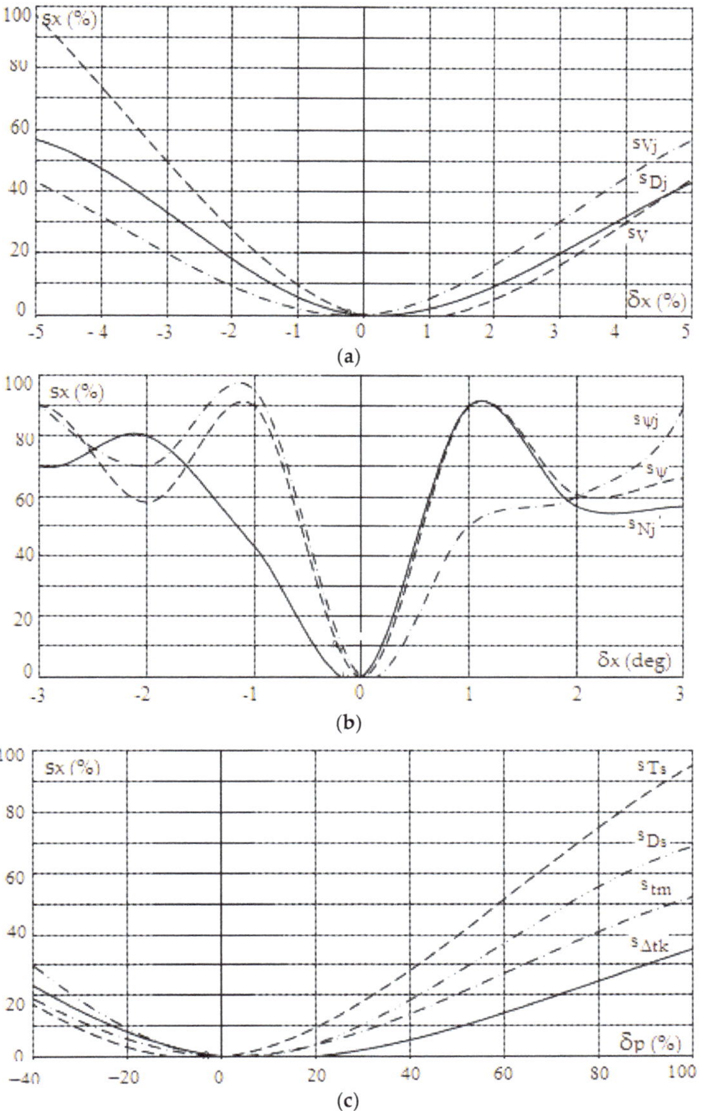

Figure 8. Sensitivity characteristics of the positional game control of the own object in restricted visibility at sea according to the PG_rv algorithm as a function of: (**a**) absolute values of the information from sensors, (**b**) angular values of the information from sensors and (**c**) values of the control process parameters.

5. Conclusions

The use of simplified models of a dynamic process game for the synthesis of optimal control allowed us to determine the safe trajectories of an own object in situations that involve passing a large number of encountered objects in a certain course sequence and speed maneuvers.

The developed algorithms also consider the COLREGs rules and maneuver advance time in addition to estimating the object's dynamic properties and assessing the final deviation of the actual trajectory from the reference value.

The following conclusions follow from the course of the sensitivity characteristics presented in Figures 5–8:

- the sensitivity characteristics are nonlinear, with most being angular values,
- in the range of the most important changes in input quantities (1% for absolute values, 1 degree for angular quantities and 20% for process parameters), the sensitivity usually does not exceed 20%,
- together with the deterioration of visibility at sea, the sensitivity to the inaccuracy of information about the object encountered and the change in the value of safe distance and time of approach increases.

The considered control algorithms are the formal models of the navigator's thinking process that controls the objects' movement and maneuvering decisions. Therefore, they can be used in the construction of a new model of the ARPA system containing a computer that supports the decision-making of the navigator.

Funding: The project was financed under the program of the Minister of Science and Higher Education under the name "Regional Initiative of Excellence" from 2019 to 2022, project number 006/RID/2018/19 and the amount of financing was 11 870 000 PLN.

Conflicts of Interest: The author declares no conflict of interest regarding the publication of this paper. The funders had no role in the design of the study; in the collection, analyses or interpretation of data; in the writing of the manuscript; or in the decision to publish the results.

References

1. Bole, A.; Dineley, B.; Wall, A. *Radar and ARPA Manual*; Elsevier: Amsterdam-Tokyo, The Netherlands, 2006; ISBN 978-0-08-048052-7.
2. Bist, D.S. *Safety and Security at Sea*; Butter Heinemann: Berlin, Germany, 2000; ISBN 0-75064-774-4.
3. Perera, L.P.; Carvalho, J.P.; Soares, C.G. Decision making system for the collision avoidance of marine vessel navigation based on COLREGs rules and regulations. In Proceedings of the 13th Congress of International Maritime Association of Mediterranean, Istanbul, Turkey, 12–15 October 2009.
4. Schuster, M.; Blaich, M.; Reuter, J. Collision avoidance for vessels using a low-cost radar sensor. *IFAC Proc.* **2014**, *47*, 9673–9678. [CrossRef]
5. Zhang, L.; Gao, W.; Li, Q.; Li, R.; Yao, Z.; Lu, S. A novel monitoring navigation method for cold atom interference gyroscope. *Sensors* **2019**, *19*, 222. [CrossRef] [PubMed]
6. Lin, H.; Chen, H.; Wang, H.; Yin, J.; Yang, J. Ship detection for PolSAR images via task-driven discriminative dictionary learning. *Remote Sens.* **2019**, *11*, 769. [CrossRef]
7. Wang, K.; Zhang, P.; Niu, J.; Sun, W.; Zhao, L.; Ji, Y. A performance evaluation scheme for multiple object tracking with HFSWR. *Sensors* **2019**, *19*, 1393. [CrossRef]
8. Huang, Y.; Shi, Y.; Song, T.L. An efficient multri-path multitarget tracking algorithm for Over-The-Horizon radar. *Sensors* **2019**, *19*, 1384. [CrossRef] [PubMed]
9. Chang, Y.-L.; Anagaw, A.; Chang, L.; Wang, Y.C.; Hsiao, C.-Y.; Lee, W-H. Ship detection bvased on YOLOv2 for SAR imagery. *Remote Sens.* **2019**, *11*, 786. [CrossRef]
10. Perez, T. *Ship Motion Control*; Springer: Berlin, Germany, 2005; ISBN 978-1-84628-157-0.
11. EMSA—European Maritime Safety Agency. *Annual Overview of Marine Casualties and Incidents 2018*; EMSA: Lisboa, Portugal, 2018.
12. Gluver, H.; Olsen, D. *Ship Collision Analysis*; Balkema: New York, NY, USA, 1998; ISBN 90-5410-962-9.
13. Lisowski, J. The safe control sensitivity functions in matrix game of ships. *Transnav Int. J. Mar. Navig. Saf. Sea Transp.* **2018**, *12*, 527–532. [CrossRef]
14. Eslami, M. *Theory of Sensitivity in Dynamic Systems*; Springer: Berlin, Germany, 1994; ISBN 978-3-662-01632-9.
15. Rosenwasser, E.; Yusupov, R. *Sensitivity of Automatic Control Systems*; CRC Press: Boca Raton, FL, USA, 2000; ISBN 978-0-849-3229293-8.

16. Wierzbicki, A. *Models and Sensitivity of Control Systems*; WNT: Warszawa, Poland, 1977; ISBN 0-444-996-20-6.
17. Cruz, J. *Feedback Systems*; Mc Graw-Hill Book Company: New York, NY, USA, 1972; ISBN 0-691-135-76-2.
18. Cao, J.; Sun, Y.; Kong, Y.; Qian, W. The sensitivity of grating-based SPR sensors with wavelength interrogation. *Sensors* **2019**, *19*, 405. [CrossRef] [PubMed]
19. Seok, G.; Kim, Y. Front-inner lens for high sensitivity of CMOS image sensors. *Sensors* **2019**, *19*, 1536. [CrossRef] [PubMed]
20. Ahsani, V.; Ahmed, F.; Jun, M.B.G.; Bradley, C. Tapered fiber-optic Mach-Zehnder interferometer for ultr-high sensitivity measurement of refractive index. *Sensors* **2019**, *19*, 1652. [CrossRef] [PubMed]
21. Kowal, D.; Statkiewicz-Barabach, G.; Bernas, M.; Napiorkowski, M.; Makara, M.; Czyzewska, L.; Mergo, P.; Urbanczyk, W. Polarimetric sensitivity to torsion in spun highly birefringent fibers. *Sensors* **2019**, *19*, 1639. [CrossRef] [PubMed]
22. Lenart, A.S. Analysis of Collision Threat Parameters and Criteria. *J. Navig.* **2015**, *68*, 887–896. [CrossRef]
23. Borkowski, P. Presentation algorithm of possible collision solutions in a navigational decision support system. *Sci. J. Marit. Univ. Szczec.* **2014**, *38*, 20–26.
24. Basar, T.; Olsder, G.J. *Dynamic Noncooperative Game Theory*; Siam: Philadelphia, PA, USA, 2013; ISBN 978-0-898-714-29-6.
25. Bressan, A.; Nguyen, K.T. Stability of feedback solutions for infinite horizon noncooperative differential games. *Dyn. Games Appl.* **2018**, *8*, 42–78. [CrossRef]
26. Breton, M.; Szajowski, K. *Advances in Dynamic Games: Theory, Applications and Numerical Methods for Differential and Stochastic Games*; Birkhauser: Boston, MA, USA, 2010; ISBN 978-0-8176-8089-3.
27. Broek, W.A.; Engwerda, J.C.; Schumacher, J.M. Robust equilibria in indefinite linear-quadratic differential games. *J. Optim. Theory Appl.* **2003**, *119*, 565–595. [CrossRef]
28. Dockner, E.; Feichtinger, G.; Mehlmann, A. Noncooperative solutions for a differential game model of fishery. *J. Econ. Dyn. Control* **1989**, *13*, 1–20. [CrossRef]
29. Engwerda, J.C. *LQ Dynamic Optimization and Differential Games*; John Wiley & Sons: West Sussex, UK, 2005; ISBN 978-0-470-01524-7.
30. Gromova, E.V.; Petrosyan, L.A. On an approach to constructing a characteristic function in cooperative differential games. *Autom. Remote Control* **2017**, *78*, 1680–1692. [CrossRef]
31. Isaacs, R. *Differential Games*; John Wiley and Sons: New York, NY, USA, 1965; ISBN 0-48640-682-2.
32. Krawczyk, P.; Zaccour, G. *Games and Dynamic Games*; World Scientific: New York, NY, USA, 2012; ISBN 9-78981-440-126-5.
33. Miloh, T. *Determination of Critical Manoeuvres for Collision Avoidance Using the Theory of Differential Games*; Institut Für Schiffbau: Hamburg, Germany, 1974.
34. Mesterton-Gibbons, M. *An Introduction to Game Theoretic Modeling*; American Mathematical Society: New York, NY, USA, 2001; ISBN 978-0-82-181929-6.
35. Millington, I.; Funge, J. *Artificial Intelligence for Games*; Elsevier: Amsterdam-Tokyo, The Netherlands, 2009; ISBN 978-0-12-374731-0.
36. Modarres, M. *Risk Analysis in Engineering*; Francis Group: London, UK, 2006; ISBN 1-57444-794-7.
37. Nisan, N.; Roughgarden, T.; Tardos, E.; Vazirani, V.V. *Algorithmic Game Theory*; Cambridge University Press: New York, NY, USA, 2007; ISBN 978-0-521-87282-9.
38. Olsder, G.J.; Walter, J.L. A differential game approach to collision avoidance of ships. *Proceed. VIII Symp. IFIP Optim. Techn.* **1977**, *6*, 264–271.
39. Osborne, M.J. *An Introduction to Game Theory*; Oxford University Press: New York, NY, USA, 2004; ISBN 978-0-19-512895-6.
40. Sadler, D.H. The mathematics of collision avoidance at sea. *J. Navig.* **1957**, *10*, 306–319. [CrossRef]
41. Sanchez-Soriano, J. An overview on game theory applications to engineering. *Int. Game Theory Rev.* **2013**, *15*, 1–18. [CrossRef]
42. Wells, D. *Games and Mathematics*; Cambridge University Press: London, UK, 2013; ISBN 978-1-78326-752-1.

Article

Vessel Detection and Tracking Method Based on Video Surveillance

Natalia Wawrzyniak [1,*], Tomasz Hyla [2] and Adrian Popik [3]

[1] Faculty of Navigation, Maritime University of Szczecin, 70-500 Szczecin, Poland

[2] Faculty of Computer Science and Information Technology, West Pomeranian University of Technology in Szczecin, 70-210 Szczecin, Poland; thyla@zut.edu.pl

[3] Marine Technology Ltd., 81-521 Gdynia, Poland; a.popik@marinetechnology.pl

* Correspondence: n.wawrzyniak@am.szczecin.pl

Received: 29 October 2019; Accepted: 26 November 2019; Published: 28 November 2019

Abstract: Ship detection and tracking is a basic task in any vessel traffic monitored area, whether marine or inland. It has a major impact on navigational safety and thus different systems and technologies are used to determine the best possible methods of detecting and identifying sailing units. Video monitoring is present in almost all of them, but it is usually operated manually and is used as a backup system. This is because of the difficulties in implementing an efficient and universal automatic detection method that would work in quickly alternating environmental conditions for all kind of sailing units—from kayaks to seagoing merchant vessels. This paper presents a method that allows the detection and tracking of ships using the video streams of existing monitoring systems for ports and rivers. The method and the results of experiments on three sets of data using cameras with different characteristics, settings, and scene locations are presented. The experiments were carried out in variable light and weather conditions, and a wide range of unit types were used as detection objectives. The results confirm the usability of the proposed solution; however, some minor issues were encountered in the presence of ships wakes or highly unfavourable weather conditions.

Keywords: vessel detection; video monitoring; inland waterway; real-time detection

---

1. Introduction

Video surveillance systems are typically used to monitor vessels' movement on coastal and inland waterways—especially those with heavy traffic, complicated organisation or which are in direct proximity to ports. However, the most typical way to track ships is using radar or Automatic Identifications System (AIS) infrastructure [1,2]. In most cases, waterside video surveillance works as a support for one of existing vessel traffic monitoring systems that uses wide collection of sensors and subsystems to acquire and distribute information on current traffic to systems users. (1) Vessel Traffic Services (VTS) is a marine system for ports and harbors that collects data on current traffic and assists ships in decision-making on area covered by the system. (2) River Information Services (RIS) are implemented in European waterways and serve as an information center for all systems' recipients (ships, port authorities, and ship-owners) [3]. The information on detected and identified vessels in RIS can be pushed to other interconnected systems of administration (customs, police, etc.). (3) Integrated waterside security systems are developed in sensitive areas (port, naval bases, power plants, etc.) [4]. These systems detect and identify intruders using and fusing information from both underwater and above water with the use of sonars, echosounders [5,6] and autonomous vehicles [7]. In all of these systems, video surveillance helps to visually confirm the identification of a vessel or to monitor non-conventional (according to the International Convention for the Safety of Life at Sea (SOLAS)) units that are not obliged to be equipped with AIS transponders. Also, video monitoring is often used as a backup system and is seen as more reliable as a passive sensor. Currently, in most cases, a systems operator must visually monitor vessels that are passing in front of the cameras, particularly when it is necessary to detect and identify a wide range of vessels, which range from large cargo vessels to motorboats. Implementing detection and tracking algorithms to analyse live video streams allows for later automatic classification and identification of ships that enter or leave ports or other areas that need traffic support.

A video monitoring system can be used to track the status of all vessels that are present in a monitored zone. Especially, when the zone borders are established on rivers or canals. Additionally, each camera, that has a view from one bank to another, can be used to update vessels' status or to count vessels passing a certain point on the waterway. The cameras can be placed in several different positions (Figure 1). The best view for the camera is usually from a bridge as passage under a bridge is often narrower than a waterway and zoom is not required to obtain a high-resolution image of a vessel. When a waterway is wide (hundreds of meters), a set of cameras is required to detect and identify passing vessels.

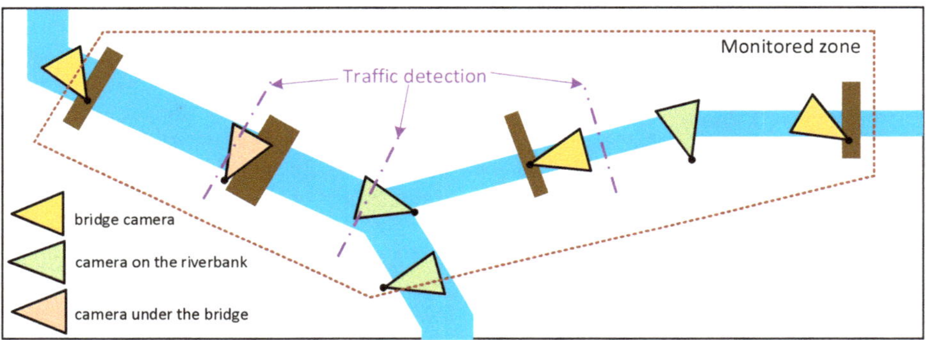

Figure 1. Vessel monitoring on a river based on video surveillance.

The automatic detection of vessels based on video stream analysis is difficult, mainly because the scene condition is constantly changing; e.g., the lighting conditions can be very dynamic due to sun reflections and the waves generated by wind and passing vessels. This causes difficulties in separating moving objects from the background. Generally, the object in a video stream can be detected using two basic solutions. The first solution is based on a pixel-based detection method that allows the detection of any moving object on a constant or slightly changing background. The second solution is object-based detection using a classifier; this second solution is usually used when it is possible to find a distinctive property of a class of objects—e.g., a specific type of prow.

Tracking vessels passing in front of the camera requires matching vessels that were detected in different video frames or using one of the standard tracking algorithms. The second approach is used when it is easy to specify the area of camera view in which objects enter and leave the scene. The tracking algorithms are generally faster than detection algorithms. Therefore, the choice of a tracking approach depends largely on performance requirements.

1.1. Related Works

Several researchers have proposed methods for object detection and tracking in video streams that did not assume a constant background, but were not specially dedicated for waterside systems. Their work was brought up and thoroughly compared in recent literature [8–10]. However, these solutions are usually too slow for use in real-time systems. Recent trend in object tracking and detection is to exploit the machine learning and pattern recognition methods; but again, the computational complexity of these solutions is high [11,12]. There are much fewer methods developed precisely for marine or costal environment with different approaches to detect ships.

Ferreira et al. [13] proposed a solution for the vessel plate number identification of fishing vessels that enter and leave the harbor with the use of two cameras. The filming camera, with a low resolution, detects movements, and the photographic camera, with a high resolution, takes a photo when a movement is detected. In this solution, the type of the vessel is determined using object-based detection that looks for the prow of the vessel. Object-based detection is mainly based on a Histogram of Oriented Gradients (HOG) classifier [14].

In contrary, a pixel-based detection was used by Hu et al. [15] in a video surveillance system designed to detect and track intruder vessels approaching a cage aquaculture. They used the median scheme to create a background image from previous N frames and used a two-stage procedure to remove wave ripples. In the first stage, they used brightness and chromatic distortion to select some wave candidates, and in the second stage, they used brightness variation to finally select waves.

In 2010, Szpak and Tapamo [16] also proposed a solution to the problem of vessel detection in the presence of waves. They used a background subtraction method and a real-time approximation of level-set-based curve evolution to distinguish moving vessels' outlines. Another approach to improve detection (tracking) quality, based on a fusion of Bayesian and Kalman filters and the adaptive tracking algorithm, was proposed by Kim et al. [17]. Also, Kaido et al. [18] in 2016 proposed a two-stage method for detecting and tracking vessels based on edge selection and the support vector machine in the detection stage, and on a particle filter based on a colour histogram in the tracking stage.

In 2014, Moreira et al. [19] reviewed state-of-the-art algorithms related to the detection and tracking of maritime vessels. They concluded, among other results, that detection and tracking algorithms do not produce efficient results when applied to a maritime environment without proper adjustments; especially algorithms that have problems in real situations such as with small vessels that are hard to distinguish from the background due to low contrast.

There is also dynamic ongoing research into the methods used for the satellite optical imagery that is used to detect vessels. In 2010, Corbane et al. [20] proposed an operational vessel detection algorithm using high spatial resolution optical imagery. The algorithm is based on statistical methods and signal processing techniques (wavelet analysis, Radon transform). Another method based on shape and texture features was presented by Zhu et al. [21]. Later in 2013, Yang et al. [22] proposed another detection method based on sea surface analysis. More detection methods based on satellite imagery were published in [23–25].

In recent years, significant progress has been made in background/foreground segmentation algorithms that allow the detection of a moving object in the presence of a changing background. The background subtraction algorithms were evaluated in [26] and compared in [27]. Several background subtraction algorithms are implemented in the OpenCV library [28]. To begin with, the Gaussian Mixture-based Background–Foreground Segmentation (MOG) algorithm uses a mixture of three to five Gaussian distributions to model each background picture. The probable values of background pixels are the ones that are more static and present in most of the previous frames [29]. Next, Gaussian Mixture-based Background–Foreground Segmentation Algorithm version 2 (MOG2) [30] is available, which is an improved version of MOG. Other popular algorithms are the Godbehere–Matsukawa–Goldberg (GMG) method [31], which uses per-pixel Bayesian segmentation; CouNT (CNT), designed by Zeevi [32], which is designed for variable outdoor lighting conditions; k Nearest Neighbours (KNN), which implements K-nearest neighbours background subtraction, as shown in [33]; the algorithm created during the Google Summer of Code (GSOC) [28]; and Background Subtraction using the Local SVD Binary Pattern (LSBP) [34].

1.2. Motivation and Contribution

This paper is a part of an ongoing research in the Vessel Recognition (SHREC) [35] project, which concerns the automatic recognition and identification of non-conventional vessels in areas covered either by River Information Services or Vessel Traffic Service systems. The detection method is a first step in the automatic vessel identification and classification process.

The main contribution of this research is a new vessel detection and tracking method. The method (1) detects all kind of moving vessels; (2) works in variable lightning conditions; (3) tracks vessels, i.e., it assigns a unique identifier to each vessel passing in front of the camera; and (4) is designed to be efficient, i.e., it can process Full High-Definition (FHD) and 4K video streams using economically acceptable amount of server resources. Based on several observations related to vessels movement and size characteristic, several rules were created that make it possible to distinguish between vessels and other moving objects. These rules were incorporated into several steps of the proposed method. Therefore, the proposed vessel detection method uses less image processing operations than existing ones. Additionally, a simple water area detection algorithm is used that detects water based on number and length of edges in a bounding box containing a moving object. The method was implemented, and the results from the experiments are described, which confirm that the proposed method is good to use in practice.

The preliminary version of the method was published in [36]. In contrast to that version, the proposed method contains a tracking function, a status update algorithm that includes additional filtering, a simple water detection algorithm, and several other minor improvements. The main difference compared to other vessel detection methods is the use of more logical processing (object filtering, tracking) than image processing techniques. Such an approach allows us to improve processing speed.

1.3. Paper Organisation

The rest of this paper is organised as follows. Section 2 contains the description of a novel Vessel Detection Method. Section 3 presents our test environment, test application, and experimental data sets. The final section discusses the experiment results and provides conclusions concerning the practical implementation of the proposed method.

2. Materials and Methods

2.1. Vessel Detection Method

The proposed vessel detection method is designed using the following approach. The method assumes that for each camera view, there is a determined detection zone that eliminates areas of the scene where either ships cannot appear (e.g., on land) or they are too far for the detection process to make sense (Figure 2). The background subtraction algorithm is used for each frame from a video stream to obtain foreground objects, find their contours, and to obtain bounding boxes for these contours. These boxes are then used in analysis throughout the whole method. First, the bounding boxes outside of the detection zone (Figure 2i), boxes too small to be a vessel (Figure 2e,f), or with improper width to height ratio (Figure 2d) are removed. Next, the bounding boxes that contain water areas (e.g., waves, sun reflexes) are removed (Figure 2h). Then, the method matches the boxes from the current frame to boxes from previous frames based on several properties such as location or overlapping ratio and stores them in a buffer. Finally, once per every five frames, the boxes stored in the buffer that have movement features similar to passing vessels are returned as vessels (Figure 2a,b), while others are discarded (Figure 2c).

The vessel detection method consists of the Moving Vessel Detection Algorithm (MVDA), Status Update Algorithm (SUA), and Temporary Buffer (TB). The MVDA is responsible for returning bounding boxes with probable vessel locations for a given input video stream. The results are returned every 1 s and are stored in TB (Figure 3). At the end of every round (every 5 s), the SUA algorithm is run. The SUA inspects the content of TB, filters out probable artefacts, and returns 0 or more series of pictures containing the detected moving vessel. The example MVDA output for one passing vessel is presented in Figure 4.

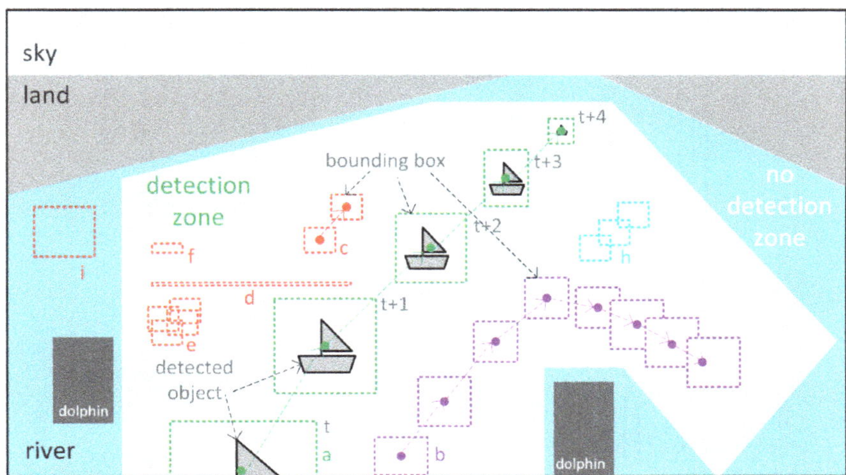

Figure 2. Schema of an exemplary video scene from a bridge camera with distinguished detection zone and different cases of bounding boxes being analysed in the method.

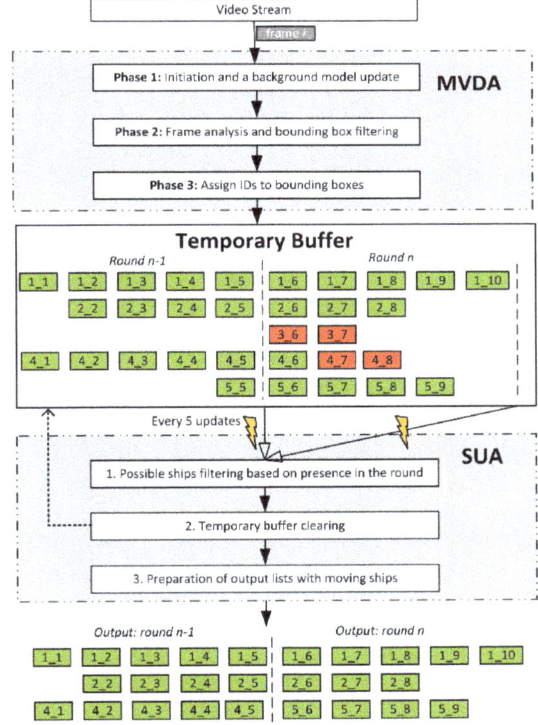

Figure 3. Vessel detection method. MVDA: Moving Vessel Detection Algorithm, SUA: Status Update Algorithm.

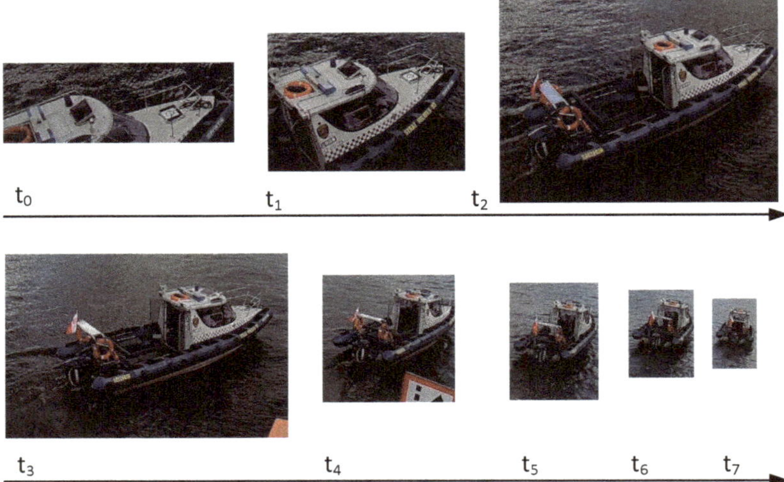

Figure 4. Example of MVDA output (a bridge camera—size corresponds to the size of the vessel in the video frames).

2.2. Moving Vessel Detection Algorithm

The MVDA takes as an input a frame from a video stream and returns one or more bounding boxes with possible vessel locations. The algorithm works in three phases. In the first phase, the background model is initiated. In the second phase, a frame from a video stream is analysed, which results in a set of bounding boxes with probable locations of vessels in the frame. In the third and final phase, the algorithm assigns an identification number to each bounding box.

The detailed Algorithm 1 is as follows:

Algorithm 1 Moving Vessel Detection Algorithm

1:    **Phase 1**—Initiate a background subtractor (GSOC algorithm from OpenCV (Open Source BSD Licence):
2:      For the first 250 frames, convert each frame to 720 p resolution, convert to grayscale and blur;
3:      Update the background model with the result from step 1.1.
4:    **Phase 2**—Frame analysis of the video stream:
5:      Convert the frame to 720 p resolution, convert to grayscale and blur;
6:      If the frame is a multiple of 10, update the background model using it (for a stream of 30 frames per second; when streams have a different framerate, update the model using 3 frames per second).
7:      For 1 frame per second, continue; otherwise, go to the analysis of the next frame;
8:      Download the current background mask from the background model;
9:      Search for edges in the background mask;
10:    For each edge:
11:      Calculate the bounding box;
12:      Check the following conditions; if any of them is not met, proceed to the next edge analysis:
13:        The bounding box is outside the negative area (the defined area on the scene for which no motion is detected);

14:             The bounding box height is greater than the defined minimum value;

15:             The bounding box width is greater than the defined minimum value;

16:             The ratio of height to width of the envelope is less than the defined value;

17:             The bounding box area is greater than the defined minimum value;

18:             Check that the bounding box does not contain water using the Water Detection Algorithm (this algorithm detects the water area by checking the number and length of the edges).

19:       Merge overlapping bounding boxes.

20:    **Phase 3**—Write the bounding box list into the Temporary Buffer:

21:       For each bounding box:

22:          Calculate the percentage of intersection of the bounding box with the bounding box detected in the previous frame;

23:          Calculate the distance between the centre of the box and the boxes from the previous round;

24:          Calculate the angle between the centre of the box and the boxes from the previous round;

25:          Assign the current bounding box the ID number of the bounding box from the previous round with which the intersection area is the largest—if there is no such area, select the bounding box at a distance not greater than 1.5 from the current bounding box and with an angle difference not greater than the maximum value;

26:          If the previous step failed to obtain the ID number, then create a new one;

27:          Write the bounding box in the Temporary Buffer with the ID number.

2.3. Status Update Algorithm

The SUA's goal is to eliminate any reaming artefacts. The algorithm is run every 5 s. It uses as its input data from the *Temporary Buffer*. It outputs 0 or more moving vessels; i.e., their pictures and movement direction.

The detailed Algorithm 2 is as follows:

Algorithm 2 Status Update Algorithm

1:    For each set of bounding boxes with a given ID number that was present in the last 5 updates of the buffer (in the same round):

2:       Verify if the set contains at least 3 bounding boxes in the round;

3:       Verify if the average overlapping coefficient of the boxes' areas for the set is greater than 0.2;

4:       Verify if the average change of angle between the bounding boxes in the set is less than 60 degrees;

5:       Verify if the centre of the bounding boxes is moving;

6:       If one or more of the above conditions are not met, discard the set in this round;

7:       Find the direction of movement of the vessel using the camera position information;

8:       Cut the vessels' pictures from the original video stream using a set of bounding boxes (requires bounding box scaling to the original stream resolution);

9:       Create the output list with the pictures from the previous step—add the ID number and direction of movement to the output list; and

10:       Add the output list to the output list set.

11:    If the vessel with the ID number has not been updated by the last 5 updates of the buffer, remove it from the buffer.

12:    Return a set of output lists.

2.4. Water Detection Algorithm

The Water Detection Algorithm is based on an observation that all moving vessels have some kind of edges in contrast to water areas. The only situation when this algorithm is not able to detect water are waves behind moving vessels that have a lot of edges. In such a case, the algorithm is not able to differentiate them from vessels characteristics.

The Algorithm 3 is as follows:

Algorithm 3 Water Detection Algorithm

1:    Convert an image to grayscale, use bilateral filter, and blur the image;
2:    Detect edges using Canny edge detector and find contours;
3:    For each contour:
4:      Calculate length of the contour (using curve approximation);
5:      Calculate the contour area;
6:      When the length is less than 100 or the area is less than 30, discard the contour;
7:    Calculate average length of the remaining contours and find the longest contour;
8:    Return water when number of contours is less than 3, the maximum length is less than 250, and average length is less than 40.

---

3. Results

3.1. Test Environment

The method was implemented using C# (Microsoft Corporation, Redmond, Washington, DC, USA) and Emgu CV version 4.0.1 (EMGU Corporation, Markham, Ontario, Canada) (C# wrapper for OpenCV). The application (Figure 5) allows the user to select different parameters and see how they affect the detection quality. It is possible to observe the detection performed by two methods simultaneously to allow for better comparison. Some of the intermediate results are visualised to help us to better understand the impact of each element of the method on the final result. Additionally, it is possible to run a batch test for a given set of video files and store the detection results into the files.

Figure 5. A screenshot from the test application.

Three data sets were used to test the quality of the proposed detection method:

1. Data Set A—the preliminary set, containing 15 different video samples (Full High Definition 1920 × 1080, 30 fps, bitrate: 20 Mb/s, AVC Baseline@L4, duration: between 30 s and 120 s), which shows different types of vessels from different angles, recorded in different places using GoPro Hero 6 cameras (GoPro Inc., San Mateo, CA, USA) The set was also used to test the first version of our method [24].
2. Data Set B—the set containing 20 video samples (4000 × 3000, 20 fps, bitrate 8 Mb/s, AVC High@L5.1, duration: between 27 s and 434 s) recorded using an AXIS IP camera Q1765-LE 1080 p motozoom AXIS (Axis Communications AB, Lund, Sweden)
3. Data Set C—the set containing 20 video samples (Full High Definition 1920 × 1080, 25 fps, bitrate: 2–7 Mb/s, AVC Main@L4.1, duration: between 48 s and 224 s) recorded using a Dahua IP camera IPC-HFW81230E-ZEH 12Mpx (Dahua Technology Co., Ltd., Hangzhou, China).

Figure 6 presents sample (not all) camera views from sets A–C. View (a) is the view from the center of the bridge, view (b) is the view from the river bank, view (c) is the view from below a high suspended bridge, and view (d) is a canal bank.

Figure 6. A screenshot from different video samples from sets A, B, and C: (**a**) bridge view; (**b**) bank view–close perspective; (**c**) under bridge view; (**d**) bank view–distant perspective.

Two sets of settings were used in the tests:

-   Standards settings: blur_size = 21, use_red_channel = false, min_width = 50, min_height = 40, max_ratio = 9, min_area = 2500, water_sensitivity = normal;
-   High threshold settings: blur_size = 21, use_red_channel = false, min_width = 65, min_height = 50, max_ratio = 10, min_area = 5000, water_sensitivity = high;

The method was tested using a test computer (Core i7-8700K (Intel Corporation, Santa Clara, CA, USA)., 32GB RAM., SSD 1TB NVIDIA Quadro P4000 (Nvidia, Santa Clara, CA, USA)). Each test outputted a set of image files that were categorised to each catgory of detection events by two experts.

3.2. Experiments

The detection events were divided into the following categories:

1. Correct detection—a vessel was detected once per passage in front of the camera view, i.e., a series of pictures of the vessel with one identification number was returned;
2. Semi-correct detection Type I—a vessel was detected more than once per passage in front of the camera view, i.e., two or more series of pictures of one vessel with different idetification numbers was returned;
3. Semi-correct detection Type II—a vessel was detected once per passage in front of the camera view, but some pictures in the series contained small artefacts;
4. Incorrect detection Type I—a vessel was not detected;
5. Incorrect detection Type IIa—a series of artefacts were returned incorrectly as a vessel; in this category, all artefacts exluding water are included;
6. Incorrect detection Type IIb—a series of artefacts were returned incorrectly as a vessel; in this category, all water artefacts are included.

The semi-correct results can be later corrected by the final vessel identification algorithm, which is not a part of the detection method. For example, when a series containing 20 vessel images contains a water artefact in the end, this is not a problem, as the identification algorithm does not use boundary images.

The results for datasets A, B, and C and standard settings are presented in Figure 7. The method returned around 75% of correct detection events for sets A and B and 60% for set C. The semi-correct detection events accounted for 16% of the total events for set A, 24% for set B, and 27% for set C. The results from set A contain more Type II semi-correct events, and those from set B and C exhibit more Type I semi-correct events. For all sets, there were no inncorrect detection Type I and IIa events; i.e., all vessels were detected, and series containing only artefacts were not returned. The only incorrect results are Type IIb events, which are series of pictures with water-only artefacts.

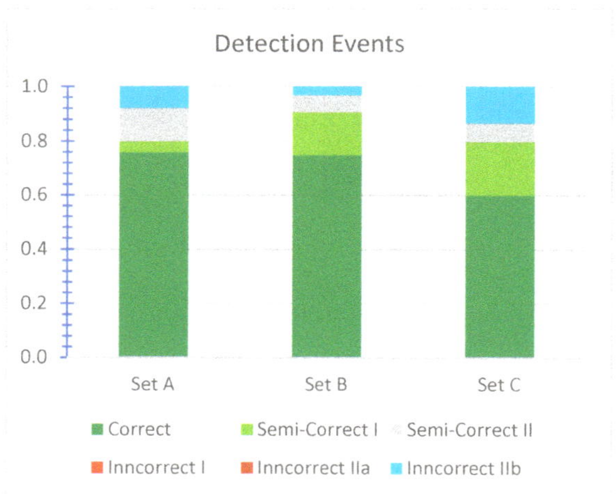

Figure 7. Detection events for datasets A, B, and C and standard settings.

The results for high-threshold settings are presented in Figure 8. In contrast to standard settings, the incorrect detection events of type I (the vessel was not detected) are presents for sets A and B. Also, water artefacts are not present in sets A and C. The method with high settings returned more correct results for set C (85%) and fewer for sets A (70%) and B (64%), mainly because these sets contain more small vessels that were filtered out.

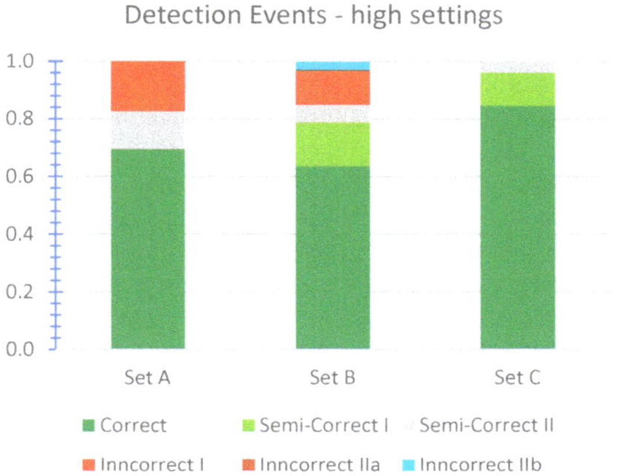

Figure 8. Detection events for datasets A, B, and C and high-threshold settings.

4. Discussion

The incorrect detection events mainly arose from a few video samples with unfavourable lightning conditions. The video samples that come from a camera facing the sun have less colour saturation, and sun reflexes are present. This causes the background subtraction algorithm to provide worse results. Another difficult case is of a camera placed below a high suspended bridge (Figure 6c) that has a large shadow in the middle of the camera view. In practical deployment, such situations can be avoided in most cases by carefully placing surveillance cameras; for example, the cameras can be placed on the north side of the bridge (on northern hemisphere) or before it to avoid such cases.

The method for the input resizes the input frames to 1280×720 resolution. This resolution is a trade-off between processing speed and accuracy. Set B, despite the fact that it has higher resolution, is more compressed and has visible compression artefacts. This means that, after downsizing the frame, the picture has fewer sharp edges. The method uses a blurred frame for background–foreground separation, but a non-blurred frame is used for water artefact filtering. Because of that, water artefacts were better filtered in set B, as water detection is based mostly on counting edges in the image.

It is worthwhile to note that the professional camera used to obtain datasets B and C produced lower quality images than our GoPro camera. This is mainly because video samples were recorded locally on an memory card. The video samples from the other two cameras were recorded from the video streams coming from these cameras with the highest possible quality settings.

One of the main problems in vessel detection are waves. However, the waves in inland waters caused by wind are significantly smaller than waves at sea. In the evaluated datasets, the proposed method removes all these waves. The few water artefacts that are returned (Type IIb incorrect detection events) are caused by the wakes created by a moving vessel. These water artefacts have more small sharp edges than small boats and therefore are not eliminated. This can be changed by setting the sensitivity to a higher threshold in the water detection algorithm. The output from the proposed method is further used in the vessel classification method based on deep neural networks (DNN). One of the defined classes is water, so that it can be later eliminated. The only reason that we do not use DNN in the MVDA is efficiency, as our tests showed that using DNN in the detection method takes too much time.

The results are not uniformly distributed among samples from different camera views. Samples from camera views with sunlight from behind and which are placed on the centre of the bridge or placed perpendicular to a narrow waterway provide practically no incorrect detection results.

The main limitation of the study is that the method is designed to work in daytime without large atmospheric precipitation. One of the other known limitations is the problem of distinguishing vessels in a situation where one big vessel is in front of the camera and small ones enter the camera view when the other vessel is in the background. In such cases, the method returns a frame with two vessels. In further steps in the final identification algorithm, it will be possible to detect hull inscriptions from both vessels, and in this way, the vessel can be distinguished.

Future works include improving the method by adjusting it to work with video streams obtained from infrared cameras at night and detecting vessels passing in proximity. Future improvements also include adding the ability to work in low-light conditions; for example, when the camera is placed on a bridge in a city during the night, when there is residual light from street lamps.

Author Contributions: Conceptualization, N.W. and T.H.; methodology, T.H.; software, T.H.; validation, N.W. and T.H.; investigation, N.W., T.H. and A.P.; resources, A.P.; data curation, A.P.; writing—Original draft preparation, N.W. and T.H.; writing—Review and editing, N.W. and T.H.

Funding: This scientific research work was supported by the National Centre for Research and Development (NCBR) of Poland under grant No. LIDER/17/0098/L-8/16/NCBR/2017.

Conflicts of Interest: The authors declare no conflict of interest.

References

1. Kazimierski, W.; Stateczny, A. Radar and automatic identification system track fusion in an electronic chart display and information systems. *J. Navig.* **2015**, *68*, 1141–1154. [CrossRef]
2. Borkowski, P. Data fusion in a navigational decision support system on a sea-going vessel. *Polish Marit. Res.* **2012**, *19*, 78–85. [CrossRef]
3. Bodus-Olkowska, I.; Uriasz, J. The Integration of Image and Nonimage Data to Obtain Underwater Situation Refinement. In Proceedings of the 2017 BGC Geomatics, Gdansk, Poland, 22–25 June 2017. [CrossRef]
4. Stateczny, A.; Lubczonek, J. Radar Sensors Implementation in River Information Services in Poland. In Proceedings of the 15th International Radar Symposium (IRS), Gdańsk, Poland, 16–18 June 2014.
5. Wawrzyniak, N.; Zaniewicz, G. Detecting small moving underwater objects using scanning sonar in waterside surveillance and complex security solutions. In Proceedings of the 17th International Radar Symposium (IRS), Krakow, Poland, 10–12 May 2016. [CrossRef]
6. Maleika, W.; Koziarski, M.; Forczmanski, P. A Multiresolution Grid Structure Applied to Seafloor Shape Modeling. *ISPRS Int. J. Geo-Inf.* **2018**, *7*, 119. [CrossRef]
7. Stateczny, A.; Kazimierski, W.; Gronska-Sledz, D.; Motyl, W. The Empirical Application of Automotive 3D Radar Sensor for Target Detection for an Autonomous Surface Vehicle's Navigation. *Remote Sens.* **2019**, *11*, 1156. [CrossRef]
8. Singh, S.; Shekhar, C.; Vohra, A. Real-Time FPGA-Based Object Tracker with Automatic Pan-Tilt Features for Smart Video Surveillance Systems. *J. Imaging* **2017**, *3*, 18. [CrossRef]
9. Zingoni, A.; Diani, M.; Corsini, G. A Flexible Algorithm for Detecting Challenging Moving Objects in Real-Time within IR Video Sequences. *Remote Sens.* **2017**, *9*, 1128. [CrossRef]
10. Lee, G.-B.; Lee, M.-J.; Lee, W.-K.; Park, J.-H.; Kim, T.-H. Shadow Detection Based on Regions of Light Sources for Object Extraction in Nighttime Video. *Sensors* **2017**, *17*, 659. [CrossRef] [PubMed]
11. Smeulders, A.W.M.; Cucchiara, R.; Dehghan, A. Visual Tracking: An Experimental Survey. *IEEE Trans. Pattern Anal. Mach. Intell.* **2014**, *36*, 1422–1468.
12. He, Z.; He, H. Unsupervised Multi-Object Detection for Video Surveillance Using Memory-Based Recurrent Attention Networks. *Symmetry* **2018**, *10*, 375. [CrossRef]

13. Ferreira, J.C.; Branquinho, J.; Ferreira, P.C.; Piedade, F. Computer vision algorithms fishing vessel monitoring—identification of vessel plate number. In *Ambient Intelligence—Software and Applications—8th International Symposium on Ambient Intelligence (ISAmI 2017); Advances in Intelligent Systems and Computing*; De Paz, J., Julián, V., Villarrubia, G., Marreiros, G., Novais, P., Eds.; Springer International Publishing: New York, NY, USA, 2017; Volume 615, pp. 9–17.

14. McConnell, R.K. Method of and Apparatus for Pattern Recognition. U.S. Patent 4,567,610, 28 June 1986.

15. Hu, W.-C.; Yang, C.-Y.; Huang, D.-Y. Robust real-time ship detection and tracking for visual surveillance of cage aquaculture. *J. Vis. Commun. Image Represent.* **2011**, *22*, 543–556. [CrossRef]

16. Szpak, Z.L.; Tapamo, J.R. Maritime surveillance: Tracking ships inside a dynamic background using a fast level-set. *Expert Syst. Appl.* **2011**, *38*, 6669–6680. [CrossRef]

17. Kim, Y.J.; Chung, Y.K.; Lee, B.G. Vessel tracking vision system using a combination of Kaiman filter, Bayesian classification, and adaptive tracking algorithm. In Proceedings of the 16th International Conference on Advanced Communication Technology, Pyeong Chang, Korea, 16–19 February 2014; pp. 196–201.

18. Kaido, N.; Yamamoto, S.; Hashimoto, T. Examination of automatic detection and tracking of ships on camera image in marine environment. In *2016 Techno Ocean*; IEEE: Piscataway, NJ, USA, 2016; pp. 58–63.

19. da Silva Moreira, R.; Ebecken, N.F.F.; Alves, A.S.; Livernet, F.; Campillo-Navetti, A. A survey on video detection and tracking of maritime vessels. *Int. J. Res. Rev. App. Sci.* **2014**, *20*, 37–50.

20. Corbane, C.; Najman, L.; Pecoul, E.; Demagistri, L.; Petit, M. A complete processing chain for ship detection using optical satellite imagery. *Int. J. Remote Sens.* **2010**, *31*, 5837–5854. [CrossRef]

21. Zhu, C.R.; Zhou, H.; Wang, R.S.; Guo, J. A Novel Hierarchical Method of Ship Detection from Spaceborne Optical Image Based on Shape and Texture Features. *IEEE Trans. Geosci. Remote Sens.* **2010**, *48*, 3446–3456. [CrossRef]

22. Yang, G.; Li, B.; Ji, S.F.; Gao, F.; Xu, Q.Z. Ship Detection from Optical Satellite Images Based on Sea Surface Analysis. *IEEE Geosci. Remote Sens. Lett.* **2014**, *11*, 641–645. [CrossRef]

23. Zou, Z.X.; Shi, Z.W. Ship Detection in Spaceborne Optical Image with SVD Networks. *IEEE Trans. Geosci. Remote Sens.* **2016**, *54*, 5832–5845. [CrossRef]

24. Liu, Z.; Hu, J.; Weng, L.; Yang, Y. Rotated region based CNN for ship detection. In Proceedings of the 2017 IEEE International Conference on Image Processing (ICIP), Beijing, China, 17–20 September 2017; pp. 1–5.

25. Kang, M.; Leng, X.-G.; Liu, Z. A modified faster R-CNN based on CFAR algorithm for SAR ship detection. In Proceedings of the 2017 International Workshop on Remote Sensing with Intelligent Processing (RSIP), Shanghai, China, 18–21 May 2017; pp. 1–4.

26. Brutzer, S.; Höferlin, B.; Heidemann, G. Evaluation of background subtraction techniques for video surveillance. In Proceedings of the CVPR 2011, Colorado Springs, CO, USA, 20–25 June 2011; pp. 1937–1944.

27. Benezeth, Y.; Jodoin, P.-M.; Emile, B.; Laurent, H.; Rosenberger, C. Comparative study of background subtraction algorithms. *J. Electron. Imaging* **2010**, *19*, 1–30.

28. Emgu CV Library Documentation Version 4.0.1. Available online: http://www.emgu.com/wiki/files/4.0.1/document (accessed on 15 October 2019).

29. KaewTraKulPong, P.; Bowden, R. An Improved Adaptive Background Mixture Model for Real-time Tracking with Shadow Detection. In *Video-Based Surveillance Systems*; Remagnino, P., Jones, G.A., Paragios, N., Regazzoni, C.S., Eds.; Springer: Boston, MA, USA, 2002.

30. Zivkovic, Z. Improved adaptive gaussian mixture model for background subtraction. In Proceedings of the 17th International Conference on Pattern Recognition (ICPR'04), Cambridge, UK, 23–26 August 2004; IEEE Computer Society: Washington, DC, USA, 2004; Volume 2, pp. 28–31.

31. Godbehere, A.B.; Matsukawa, A.; Goldberg, K. Visual tracking of human visitors under variable-lighting conditions for a responsive audio art installation. In Proceedings of the 2012 American Control Conference (ACC), Montreal, QC, Canada, 27–29 June 2012; pp. 4305–4312. [CrossRef]

32. Zeevi, S. Background Subtractor CNT Project. Available online: https://sagi-.github.io/BackgroundSubtractorCNT (accessed on 15 October 2019).

33. Zivkovic, Z.; van der Heijden, F. Efficient adaptive density estimation per image pixel for the task of background subtraction. *Pattern Recogn. Lett.* **2006**, *27*, 773–780. [CrossRef]

34. Guo, L.; Xu, D.; Qiang, Z. Background Subtraction Using Local SVD Binary Pattern. In Proceedings of the 2016 IEEE Conference on Computer Vision and Pattern Recognition Workshops, Las Vegas, NV, USA, 26 June–1 July 2016; pp. 1159–1167.

35. Wawrzyniak, N.; Stateczny, A. Automatic watercraft recognition and identification on water areas covered by video monitoring as extension for sea and river traffic supervision systems. *Polish Marit. Res.* **2018**, *25*, 5–13. [CrossRef]
36. Hyla, T.; Wawrzyniak, N. Ships Detection on Inland Waters Using Video Surveillance System. In *Computer Information Systems and Industrial Management*; Springer: Berlin/Heidelberg, Germany, 2018; Volume 11703, pp. 39–49.

Article

Classification of Non-Conventional Ships Using a Neural Bag-Of-Words Mechanism

Dawid Polap[1] **and Marta Wlodarczyk-Sielicka**[2,*]

[1] Marine Technology Ltd., 81-521 Gdynia, Poland; d.polap@marinetechnology.pl
[2] Department of Navigation, Maritime University of Szczecin, Waly Chrobrego 1-2, 70-500 Szczecin, Poland
[*] Correspondence: m.wlodarczyk@am.szczecin.pl; Tel.: +48-513-846-391

Received: 14 February 2020; Accepted: 12 March 2020; Published: 13 March 2020

Abstract: The existing methods for monitoring vessels are mainly based on radar and automatic identification systems. Additional sensors that are used include video cameras. Such systems feature cameras that capture images and software that analyzes the selected video frames. Methods for the classification of non-conventional vessels are not widely known. These methods, based on image samples, can be considered difficult. This paper is intended to show an alternative way to approach image classification problems; not by classifying the entire input data, but smaller parts. The described solution is based on splitting the image of a ship into smaller parts and classifying them into vectors that can be identified as features using a convolutional neural network (CNN). This idea is a representation of a bag-of-words mechanism, where created feature vectors might be called words, and by using them a solution can assign images a specific class. As part of the experiment, the authors performed two tests. In the first, two classes were analyzed and the results obtained show great potential for application. In the second, the authors used much larger sets of images belonging to five vessel types. The proposed method indeed improved the results of classic approaches by 5%. The paper shows an alternative approach for the classification of non-conventional vessels to increase accuracy.

Keywords: bag-of-words mechanism; machine learning; image analysis; ship classification; marine system; river monitoring system; feature extraction

---

1. Introduction

Ship classification is an important process in practical applications in different places. In coastal cities, ships enter from the mouth of a river or moor at ports. This type of activity is quite often reported and recorded. However, for measurement, statistical, or even analytical purposes, it is often necessary to record vessels that arrive but do not report anywhere. To this end, the simplest solution is to create a monitoring system and analyze acquired images. This type of system architecture is based primarily on three main components: video recording, image processing, and classifying possible water vehicles.

While the solution itself seems simple, each component has its disadvantages, which also affect the others. First, the video recorder may be a simple camera, but often one needs to take good-quality photos for easier analysis. The second component is image processing. Image processing should consider the location of a possible ship on an image, or even perform some extraction of features. It is particularly important to remove unnecessary areas such as the background, houses, and even water. The third element is classifying these images, i.e., based on the obtained images, the algorithm should determine with some probability the type of ship.

In this paper, we considered the third aspect of such a system to model a solution enabling the most accurate classification of a given type of ship based on a photo entered into the system. In the analyzed system [1] an important element was the recording of information about passing vessels in water bodies. Unfortunately, this task is not easy due to the similarities between ships and the many factors that can be mistaken for a ship.

2. Related Works

In the last decade, the number of methods for classifying images has increased, and the main contribution is the description of convolutional neural networks (CNNs). These mathematical structures can be modeled for specific classification problems, as can be seen in [2]. The classification problem might also be improved by extracting some important objects. For this purpose, segmentation can be used [3,4]. In this paper, the authors proposed a convolutional network architecture based on three dimensions of the incoming image. Moreover, CNN was used for classifying objects from different points of view, which is very practical using drones [5], or even for sleep stage scoring based on electroencephalogram (EEG) signals [6]. Moreover, CNN has been used for recognition of vehicle driving behavior [7]. In addition to classic architectures, there are others, such as U-net, which are used as segmenting elements and in inverse problems [8]. Many applications of CNN can be found, primarily in monitoring systems and medicine.

In general, a database for training such structures has the biggest impact on the classifier. A very common problem is the lack of enough samples, which results in low efficiency or even overfitting. Data augmentation, i.e., generating new samples based on existing samples using image processing techniques, is a popular solution [9]. In [10], the authors discuss the effect of augmented data on CNN based on images from chest radiographs. Similarly, in [11], the authors use augmentation to increase the dataset by adding some distortions, such as changing the brightness, rotating, or adding some mirroring effects. Moreover, in recent years learning transfer has been enabled, i.e., the use of trained network architectures to minimize training time. The main idea was to create architectures and train them on huge databases. Trained classifiers are those whose coefficients are specialized in searching for features and classifying, so the learning transfer consists of using the finished model, modifying only selected layers, and overtraining only selected values to meet the needs of new bases. Not modifying any of the layers is called freezing. One of the first architectures for that was AlexNet [12]. Another was VGG16, modeled by a group from Oxford who primarily reduced the size of the filter in a convolution layer [13]. Another popular model is Inception [14], which drastically reduced the number of architecture parameters.

The ship classification problem depends on using images. Commonly used are synthetic aperture radar (SAR) images, by which ships can be classified based on their shape [15]. Similar research was described in [16,17], where superstructure scattering features were analyzed in the process of classification. Similarly, in [18], the idea of ship classification was solved by analyzing sound signals and removing the background sound of the sea. Other input data are aerial images that present a top view of the scenery and the ship. In all of these solutions, CNN was used for faster feature extraction and classification. An interesting approach was presented in [19], where the authors described the impact of simulated data on the training process of neural classifiers in the problem of ship classification. Moreover, in [20], a neural approach for ship type analysis with sea traffic was presented as an automatic identification system. All these studies used neural classifiers for image processing and classification.

In this paper, we propose a solution for the classification of different non-conventional ships using images made from the side and not from the top, like SAR images. This problem is hard, because images can be created using different light, from a different distance, or even from different sides of the object. The described solution was based on splitting the image of the ship into smaller parts (using keypoint algorithms with clustering) and classifying them by CNN into vectors that can be identified as features. This idea is a representation of a bag-of-words mechanism, where created feature vectors might be called words, and using these words, a solution can assign them a specific class. The main contribution of this article is the use of a bag-of-words mechanism to classify non-conventional ships, which in the future could be used in an innovative system for automatic recognition and identification in video surveillance areas. The solution proposed in the paper has not been applied anywhere and is a new approach to the subject.

3. Bag-Of-Words

Bag-of-words is an abstract model used in the processing of text or graphics. It is the representation of data described in words, i.e., linguistic values. In the case of two-dimensional images, with a word we can describe a feature or fragment of an object. The idea of using a bag can help classify processes, because the input image will be decomposed into smaller fragments and classified according to certain linguistic values. These values can help in the classification of larger objects. This is especially desirable when analyzing the same objects that differ in their small features.

The proposed idea consists of extracting small fragments of the image with certain features. All points are divided against a certain metric into smaller images containing a fragment of the object. Such images can represent everything, because the object can be on any background; for instance, a ship can be captured in the port or against a background of trees; in that case, smaller images can even show some trees. Thus, the use of a classical approach, that is, the creation of a bag-of-words using an algorithm such as k-nearest neighbors, is not very effective. The reason is the lack of connection between the features (in smaller parts), because it should be considered that the objects can be on different scales or turned at a certain angle or even have some noise, such as bad weather or additional objects. That is why we propose a bag-of-words model based on more complex structures, such as neural networks.

3.1. Feature Extraction

The main idea of this study was to extract features using one of the classic algorithms for obtaining keypoints, such as scale-invariant feature transform (SIFT) [21], speeded up robust features (SURF) [22], features from accelerated segment test (FAST) [23], or binary robust invariant scalable keypoints (BRISK) [24], and then create samples with found features. It should be noted that if these algorithms processed the original image, the found points would probably cover the entire image; in the case of a simple image where a ship is at sea, all points could be placed on this object or water or waves, but there may be an image with some additional background with many possible points. To remedy this, in the first step, the image must be processed, which means using graphic filters to minimize elements such as edges or points. We used only two filters, such as gamma correction and blur.

3.2. Feature Extraction Based on Keypoints

Using the described algorithms, we obtained a set of keypoints, which we can describe as $A = \{(x_0,y_0), (x_1,y_1), \dots , (x_{n-1},y_{n-1})\}$. To minimize the number of points (because unnecessary elements of the image can be indicated), all points were checked against their neighbors. If the point had a neighbor within a certain distance α, it remained in the set. Otherwise, the point was removed, and the cardinality was reduced by one. The distance between two points $p_i = (x_i, y_i)$ and $p_j = (x_j, y_j)$ was checked using one of the two classic metrics, Euclidean or river. The best known is the Euclidean, modeled as

$$d_E = \left((x_i,\ y_i),\left(x_j,\ y_j\right)\right) = \sqrt{\left(y_i - x_i\right)^2 + \left(y_j - x_j\right)^2} \tag{1}$$

A river metric is the distance between points but counted relative to a certain straight line between the points. For both points, a perpendicular projection is made, as a result of which an additional two points are obtained, (x_o, y_o) and (x_p, y_p). The distance in this metric will be calculated as the sum of the distance of a given point to the straight line, the distance between these two points on the straight line, and the transition from the straight line to the second point. Formally, it can be stated as

$$d_R \left((x_i,\ y_i),\left(x_j,\ y_j\right)\right) = d_E((x_i,\ y_i),(x_o,\ y_o)) + d_E\left((x_o,\ y_o),\left(x_p,\ y_p\right)\right) + d_E\left((x_p,\ y_p),\left(x_j,\ y_j\right)\right) \tag{2}$$

Depending on the given metric, all points are checked to see if the distance is smaller, and if so, the point is removed. The next step is to divide the points into subsets B_q, where q is the number of objects. It is not possible to adjust the value of q without empirically checking and testing the data in the database. With this value, it is worth using existing algorithms to divide these points (for example, using the k-nearest neighbors algorithm). However, this value is unknown, so another approach to the topic should be taken. For this purpose, one of the previously described metrics can be used.

For all points in a given set A, the average distance value is calculated as

$$\xi(A) = \frac{1}{5 \cdot n^2} \sum_{i=0}^{n-1} \sum_{j=i}^{n-1} d_{metric}\big((x_i, y_i), (x_j, y_j)\big) \tag{3}$$

With the average distance, the points are divided concerning this value. The first subset is created by adding the first point to it, i.e., $(x_0, y_0) \in B_0$. Then, for each point $(x_r, y_r) \in A$, we check to see if the distance between this point and any other in a given subset $(x_0, y_0) \in B_0$ is less than the average distance of the set, i.e.,

$$d_{metric}((x_r, y_r), (x_0, y_0)) < \xi(A) \tag{4}$$

If the above equality is met for a point (x_r, y_r), it is added to subset B_0 and removed from A. In the case where none of the points is added to a given subset, another subset, B_1, is created. Then, the first point from A is added to subset B_1 and removed from A. In this way, the action is repeated to meet the stop condition, which is the emptiness of the set, $A = \varnothing$.

As a result, subsets B are generated, with each representing one feature. For each set, an image is created whose dimensions will depend on the subset. To find the dimensions, we look for the maximum and minimum values of both coordinates in a subset that we can mark as $x_{max}, y_{max}, x_{min}$, and y_{min}. Hence the image size will be $(x_{max}-x_{min}) \times (y_{max}-y_{min})$. Then, the images are saved and each one represents a part of the image. The left part of Figure 1 shows this process of extracting smaller parts of the image. Figure 1 shows a graphic visualization of the proposed model.

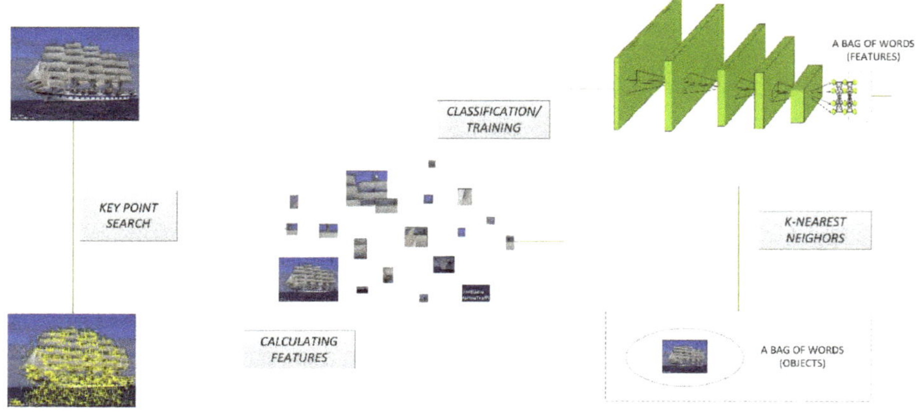

Figure 1. Graphic visualization of the proposed model.

4. Classification with Bag-Of-Words

Unfortunately, there was no unambiguous method to assign attributes to specific groups automatically. Therefore, we suggested creating groups at the initial stage of modeling the solution with the help of empirical division. In this way, the basic database of features were created, which will include a later bag-of-words.

4.1. Convolutional Neural Network

One of the most important branches of artificial intelligence methods is neural networks, which have been modeled for the needs of graphic image classification. Convolutional neural networks are models inspired by image processing by the cerebral cortex of cats. It is a mathematical structure built of three types of layers, where the layers between them are connected by synapses burdened with a certain weight. The weight is given randomly while creating the structure. Then, in the training process, the weights are modified to best match the training database.

One of the key layers of the network is the convolutional layer, which takes the image of the input with dimensions $w \times h \times d$, where w and h are the width and height of the image and d is understood as depth and depends on the model. For color images saved in the red–green–blue (RGB) model, the depth will be 3 due to the number of components. Formally, each image is saved as a set of matrices, each of which describes the image values for a given component. The convolutional layer works on a principle of image filter f of size $k \times k$. This filter is a matrix with k^2 coefficient defined randomly and modified during the training process. This filter is moved over the image and changes the value in pixel p on image I at position (i, j), which can be defined as

$$I[i, \, j] = \frac{1}{K} \sum_{t=-\left[\frac{k}{2}\right]}^{\left[\frac{k}{2}\right]} \sum_{r=-\left[\frac{k}{2}\right]}^{\left[\frac{k}{2}\right]} I[i+t, \, j+r]\cdot f[t, \, r] \tag{5}$$

where matrix f is located over an image and the central point of the matrix is over a pixel at position (i, j), and K is the sum of all weights of filter f. The main purpose of this layer is feature extraction and reduction of data redundancy on the image. Applying some filter on the image will change it; depending on the coefficient of filters, some objects might be deleted or highlighted.

The second type of layer is called pooling, which has only one purpose: to reduce the size of matrices. Reducing depends on some function $g(\cdot)$, which selects one pixel from each square $m \times m$. The most commonly used function is $max(\cdot)$ or $min(\cdot)$.

These two layers can be used alternately many times. In the end, there is the last layer, the fully connected layer, which is understood as a classical neural network. Each pixel from the last layer (pooling or convolutional) is input as a numerical value. This layer is composed of columns of neurons connected by synapses, which are burdened with some weight. Each neuron gets a numerical value that is processed and sent to the next column. This operation can be described as

$$x_m^t = f\left(\sum_{i=1}^{n} x_i^{t-1} \cdot \omega_i\right) \tag{6}$$

where x^t is the output from neuron m in layer t, and ω_i is a weight on the connection between x_m in layer t and x_i in layer $t-1$. The number of columns and neurons depends on the modeled architecture. In the last column, there should be k neurons (when a classification process is described as a k-classes problem). The final calculation of an image in such a structure gives a probability distribution that can be normalized by some function like softmax. These values are understood as the probability of belonging to this class.

Unfortunately, all weights in this model are generated randomly at the beginning. To change these values, the training algorithm must be used. The main idea is to minimize loss function during two iterations. One such algorithm commonly used in convolutional networks is adaptive moment estimation [25]. The modification of weights is based on a basic statistical coefficient like the correlation of mean $\widehat{m}$ or variation $\widehat{v}$:

$$\widehat{m}_t = \frac{m_t}{1 - \beta_1^t} \tag{7}$$

$$\widehat{v_t} = \frac{v_t}{1 - \beta_2^t} \tag{8}$$

where m_t and v_t are the mean and variation values in the tth iteration. The formulas for calculating this can be presented as

$$m_t = \beta_1 m_{t-1} + (1 - \beta_1) g_t \tag{9}$$

$$v_t = \beta_2 v_{t-1} + (1 - \beta_2) g_t^2 \tag{10}$$

where β_1, β_2 are distribution values.

Those two statistical coefficients are used in the modification of weight as

$$\theta_{t+1} = \theta_t - \frac{\eta}{\sqrt{\widehat{v_t}} + \epsilon} \widehat{m_t} \tag{11}$$

where η is the learning coefficient and $\epsilon \neq 0$, which prevents division by 0.

4.2. Bag-Of-Words

A trained classifier can be used as an element dividing incoming images into selected elements in a bag. For each image, smaller images representing features are created. Each of these features is classified using the pretrained convolutional neural network. As a result, the network will return the probability of belonging for each word in the set (each single output from the network is interpreted as a word). Based on a certain probability and features, it is possible to assign these attributes to an object. The selection of features for an object works on the principle of determining conditional affiliation to another word in the bag. To make it impossible to save the whole object to its characteristics, it is worth introducing division of the bag into two sets (or even two bags). The first bag will contain only features and the second full objects. For a better understanding of this idea, let us assume that the image presents a motorboat. The biggest bag will contain a class of ships, like motorboat, yacht, etc. The smallest bag (in the biggest one) will describe one ship. For motorboat, these words would be, for example "a man", "waves", and "no sails".

Each of these objects is defined as a numerical vector consisting of zeros and ones (ones as belonging to this class). Each item in the vector is assigned to one feature from the bag-of-words, so its creation consists of using the result returned by the classifier. It should be noted that for many smaller segments from basic images, there will be many classification results. These results are averaged by all returned decision from classifiers.

The evaluation of the feature vector to an object occurs by comparing these vectors. The simplest method is to approximate the values returned by the network to integers and compare them with the words in a bag. However, there may be a situation where the vector will be different in one position compared to the patterns. To prevent this, we suggest using the k-nearest neighbors algorithm, which will allow assigning to a given object. The full display of this process is shown in Figure 1.

The k-nearest neighbors algorithm consists of analyzing and assigning the sample to neighboring samples [26,27]. Suppose that the value x_i has an assigned class μ_i. In the case of the analyzed problem, x_i will correspond to 1 and values of μ_i are the appropriate values representing the objects. The algorithm finds the nearest neighbors (values) $x_n' \in \{x_0, x_1, \ldots, x_{n-1}\}$ for the given value x according to the following equation:

$$\min(d_{metric}(x_i, x)) = d_{metric}(x_n', x), \quad i = 0, 1, 2, \ldots, n-1. \tag{12}$$

5. Experiments

In our experiments, we tested two databases. The first one had two classes, sailing ship and others, and was used to create the first set of features and find the best combination of algorithms. The second database contained more classes and the biggest number of samples to show the potential application of such an approach.

5.1. Classification for Two Classes of Ships

In these experiments, we tested the proposed solution to find the best combination for our proposition. For this purpose, the database we used was very small. It contained two classes, sailing ship and others. A sailing ship should have sails, although they do not always have to be spread. Such an observation allows the creation of two features describing this object, i.e., masts and sails. In this way, a vector describing these two classes will be created:

$$\begin{cases} [1,\ 1] & sailing\ ship \\ [0,\ 0] & other \end{cases} \tag{13}$$

where individual values are understood as appropriate features, masts and sails. In these tests, a CNN architecture as described in Table 1 was used.

Table 1. Convolutional neural network (CNN) architecture. ReLU, rectified linear unit.

Type of Layer	Shape
Convolutional 5×5	(None, 96, 96, 20)
Activation: ReLU function	(None, 100, 100, 20)
MaxPooling 2×2	(None, 50, 50, 20)
Convolutional 5×5	(None, 50, 50, 50)
Activation: ReLU function	(None, 50, 50, 50)
MaxPooling 2×2	(None, 25, 25, 50)
Flatten	(None, 31, 250)
Dense 500	(None, 500)
Activation: ReLU function	(None, 500)
Dense 2	(None, 2)
Activation: softmax function	(None, 2)

In the experiments, we used a database contained 800 images (600 with sailing ships, and 200 with other ships). In the training process, 75% of the samples selected randomly from each class were used, and the remaining 25% was used for the validation process, which were 150 and 50 images.

For each sample, one of the keypoint algorithms was used, which allowed us to create a few smaller segments. We tested the algorithm for each segment, and the results of two selected metrics, Euclidean and river, are presented in Table 2. In the table, for each algorithm, there are two columns labeled "Object features" and "Background", which means that the extracted segment describes an important feature of a ship or not. Quite a common problem was to find the background, i.e., an insignificant fragment of the ship, and a large amount of sky or sea. The results shown are averaged over the entire base. It is easy to see that using the Euclidean metric generates many more features compared to the river metric. In both cases the ratio of images depicting features of the background exceeded 50%; however, that is not that big for the classic Euclidean metric.

Table 2. Average number of created objects using a key-search algorithm with the connection with Euclidean or river metrics. SIFT, scale-invariant feature transform; SURF, speeded up robust features; FAST, features from accelerated segment test; BRISK, binary robust invariant scalable keypoints.

	SIFT		SURF		FAST		BRISK	
	Object features	Background	Object features	Background	Object features	Background	Object features	Background
Euclidean metric	5	3	10	4	8	4	9	5
River metric	3	2	4	3	4	4	6	4

In our tests, we used the SIFT, SURF, BRISK, and FAST algorithms to find keypoints. After that, all found segments were resized to one size and calculated using CNN. The results obtained for each image were averaged and classified using the k-nearest neighbors algorithm (in this experiment, k = 2) and are presented in Tables 3 and 4 (the results in the second column represent classification of the whole image). Some examples of keypoint clustering are presented in Figure 2.

Table 3. Statistical coefficients for classification measurements using the selected keypoint search algorithm with Euclidean metric and CNN.

	CNN	CNN + SIFT	CNN + SURF	CNN + BRISK	CNN + FAST
Accuracy	0.78	0.814261	0.843831	0.832976	0.832347
Sensitivity	0.815476	0.903403	0.931257	0.909825	0.91635
Specificity	0.59375	0.453333	0.486842	0.534031	0.537778
Precision	0.91333	0.869979	0.881098	0.88366	0.874244
Negative predictive value	0.38	0.536842	0.634286	0.60355	0.647059
Miss rate	0.184524	0.096597	0.068743	0.090175	0.08365
Fallout	0.40625	0.546667	0.513158	0.465969	0.462222
False discovery rate	0.08667	0.130021	0.118902	0.11634	0.125756
False omission rate	0.62	0.463158	0.365714	0.39645	0.352941
F1 score	0.861635	0.886376	0.905483	0.896552	0.894802

Table 4. Statistical coefficients for classification measurements using the selected keypoint search algorithm with river metric and CNN.

	CNN	CNN + SIFT	CNN + SURF	CNN + BRISK	CNN + FAST
Accuracy	0.78	0.795932	0.825556	0.823859	0.820397
Sensitivity	0.815476	0.901408	0.916123	0.915133	0.923821
Specificity	0.59375	0.487113	0.521875	0.494465	0.516014
Precision	0.91333	0.837285	0.865317	0.867248	0.848889
Negative predictive value	0.38	0.627907	0.649805	0.617512	0.697115
Miss rate	0.184524	0.098592	0.083877	0.084867	0.076179
Fallout	0.40625	0.512887	0.478125	0.505535	0.483986
False discovery rate	0.08667	0.162715	0.134683	0.132752	0.151111
False omission rate	0.62	0.372093	0.350195	0.382488	0.302885
F1 score	0.861635	0.868164	0.889995	0.890547	0.884771

Figure 2. Examples of clustered keypoints for motorboat images.

The highest efficiency was obtained with the Euclidean metric using the SURF algorithm. For this combination, the results of classification compared to those without using the bag-of-words mechanism was nearly 6% higher than that with the convolutional network alone. However, it is worth noting that the significant difference between the results obtained indicates the negative predictive value, whose value was almost twice as high when using the bag mechanism. This factor determines the probability of assigning a false sample to the correct class; in this case, not a sailing ship. The situation is similar to other hybrids, where this value is always higher than 50%. A similar situation occurred with the F1 score, which is the harmonic average of the precision and recall coefficients. This factor allows us to evaluate the classification if its components have different values. In each case, the statistical coefficients indicated a more accurate process taking into account the proposed mechanism.

For a more detailed analysis, time measurements were also made for the image processing and training of a given architecture, as shown in Figure 3.

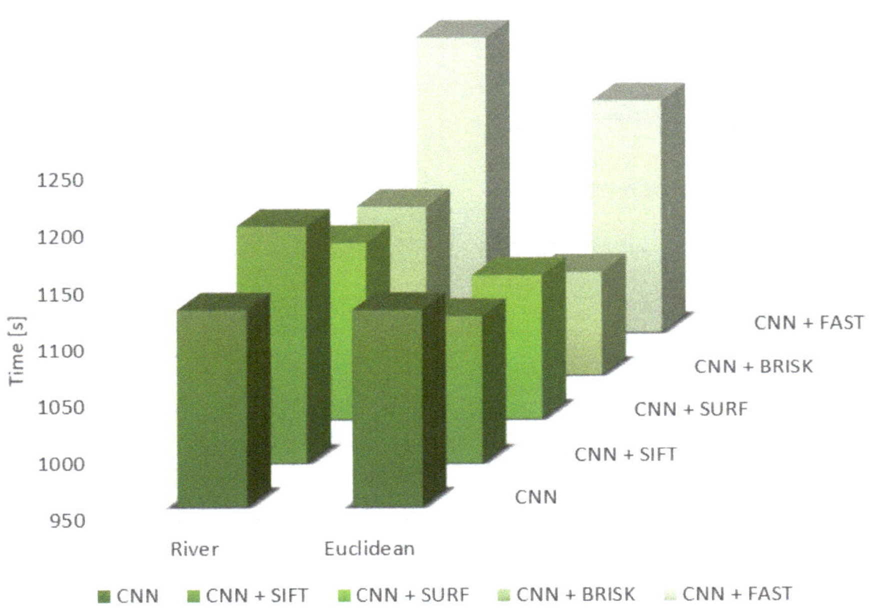

Figure 3. Average time for image processing and the training process.

The presented results are averaged data from 10 tests. In general, using the Euclidean metric saves approximately 10% more time than using the river metric. The tests showed that the longest processing time occurred using the FAST algorithm and the shortest with BRISK. As for the SIFT and SURF algorithms, the time measurement was at a similar level and was classified as in the middle.

5.2. Classification for Five Classes of Ships

Based on the previous results, the best accuracy was achieved with a combination of the SURF algorithm and CNN. We used this combination for classification of five classes: cargo (2120 images), military (1167 images), tanker (1217 images), yacht (688 images), and motorboat (512 images). For the first three classes, images were downloaded from a publicly available dataset from Deep Learning Hackathon organized by Analytics Vidhya. Each class was divided randomly into two sets in a 75%:25% (training/validation) ratio, and for the training process the data were split in the same proportion. Using the training set, the SURF algorithm was used to create smaller parts, and based on the created sets, these samples were put into features which can be described as the following vector:

$$[mast, sail, people, color, simplyShape], \tag{14}$$

where *people* means that on deck some people can be found, *color* means that a boat can have different colors (for a military ship, it is mainly gray), and *simplyShape* means that the ship can be recognized as a simple geometric figure, such as a rectangle. These features were chosen according to the database used and their possible location.

Using these features, words describing ship type were defined as follows:

$$\left\{ \begin{array}{ll} [0;\ 0;\ 0;\ 1;\ 1] & cargo \\ [1;\ 0;\ 0;\ 0;\ 0] & military \\ [0;\ 0;\ 0;\ 1;\ 0] & tanker \\ [1;\ 1;\ 1;\ 1;\ 0] & yacht \\ [0;\ 0;\ 1;\ 1;\ 1] & motorboat \end{array} \right. \tag{15}$$

The training database contained 4278 images, which resulted in almost 26,000 smaller segments. Data were split into features based on color clustering using the k-means algorithm [28] and corrected empirically (especially for shape). We trained the classifiers with the architecture described in Table 1, but in the end there were five because of the five classes of features. The classifier was trained for five different numbers of iterations, $t \in \{20,\ 40,\ \ldots,\ 100\}$, and the accuracy is presented in Table 5. The best accuracy was reached using 80 iterations; accuracy did not improve with more iterations.

Table 5. Average classification accuracy and number of iterations in the training process.

Iterations	20	40	60	80	100
Accuracy	29%	35%	56%	61%	61%

The obtained accuracy is not very promising in such a classification problem. The main cause of this is the selection of features and creating sets for them. In the experiments, the dataset was so big that whether the sample belonged to the set was determined by the algorithm. Moreover, a feature such as shape is not the best choice for ships.

Despite these drawbacks, we conducted an additional experiment to check the classification result for this database in terms of hybrids. We classically trained a CNN to classify full images. Next, we checked the effectiveness of the validation base. Then we combined the obtained results from this classification with the proposed solution. Our approach classified into a given class out of the bag, so we understand the assignment to this class as adding a constant value equal to 0.2 to the probability of assignment according to the classic classifier. This approach will allow one probability distribution to be changed by 20%. The results of such action are shown in Table 6. The table shows the exact numbers of correctly classified images from the validation set and the accuracy.

Table 6. Comparison of classic CNN usage and extended usage with the proposed approach.

	Cargo	Military	Tanker	Yacht	Motorboat	Accuracy
Classic CNN	478	195	234	78	57	0.731228
Classic CNN + proposed approach	479	236	236	105	78	0.795789

These data show that our proposition can be used as an additional component and increase the classification accuracy by nearly 5.5%. This result is better, but there is a problem with more time to train nets and classify samples because of much more operation. It is worth noting that values increased mainly for the military class, yachts, and motorboats. This is due to the good definitions of features such as the ones we have for military, or people and colors for the other two classes. The main conclusion is that the most difficult task is to initially declare a bag-of-words describing these features. This solution can be used in practice, but there are some additional tasks during the modeling of this solution, such as overseeing the creation of small images representing features and assigning them to individual groups. Also, the declaration of characteristics involves allocating image segments to these classes and analyzing them before training the classifier.

We used other CNN architectures, including VGG16, Inception, and AlexNet, and compared the results with and without our approach, as shown in Figure 4.

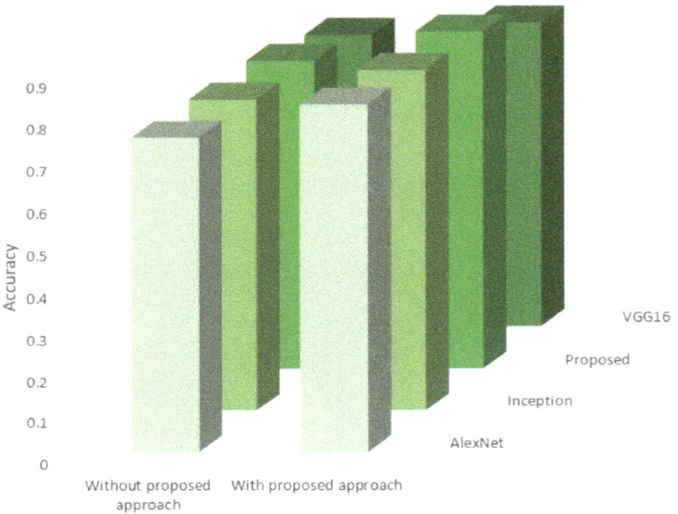

Figure 4. Comparison of the classic approach to image classification and hybrids.

The obtained results show that all tested architectures increased classification accuracy. The average value for all architectures was around 6%, which was a good result based on small datasets (neural networks are data-hungry algorithms). However, it was noted that apart from VGG16, where the increase was close to 3%, the other architectures achieved an increase of 7%. This was a good result, which could be significant for more extensive classes in the analyzed database.

6. Conclusions

Image classification is a problem for which solutions are being developed all the time. In recent years, revolutionary neural networks have been developed that have enabled a huge leap forward. Unfortunately, this solution also has its problems, such as requiring a large number of samples in the database, or architecture modeling. In this paper, we focused on analyzing images of selected ships. As part of the research, we proposed a classification mechanism based on sample segments that was determined based on algorithms searching for keypoints and subsequent classification.

As part of our experiments, we performed two tests. In the first, we analyzed two classes and the results obtained showed great potential for practical applications. In the second, we used much larger sets of images of five types of ships. The proposed solution in itself showed many disadvantages, especially at the stage of determining features and assigning samples to them to train the classifier. However, we used this solution as an additional element of classification after using the classic approach, including learning transfer. As a result, we noticed that the average efficiency increased by approximately 5% in almost all cases compared to the currently used convolutional network architectures.

An analysis of the database using a feature vector, which can be treated as a bag of words, shows potential practical application, especially if the features of the objects are well described. In future research, we plan to focus on how to automatically analyze images to extract features from them, as well as automatically assign classes as an unsupervised technique.

Author Contributions: Conceptualization, D.P. and M.W.-S.; methodology, D.P.; software, D.P.; validation, D.P.; data curation, D.P. and M.W.-S.; writing—original draft preparation, D.P.; writing—review and editing, D.P. and M.W.-S. All authors have read and agreed to the published version of the manuscript.

Funding: This work was supported by the National Centre for Research and Development (NCBR) of Poland under grant no. LIDER/17/0098/L-8/16/NCBR/2017 and grant no. 2/S/KG/20, financed by a subsidy from the Ministry of Science and Higher Education for statutory activities.

Conflicts of Interest: The authors declare no conflict of interest.

References

1. Wawrzyniak, N.; Stateczny, A. Automatic watercraft recognition and identification on water areas covered by video monitoring as extension for sea and river traffic supervision systems. *Pol. Marit. Res.* **2018**, *25*, 5–13. [CrossRef]
2. Anthimopoulos, M.; Christodoulidis, S.; Ebner, L.; Christe, A.; Mougiakakou, S. Lung pattern classification for interstitial lung diseases using a deep convolutional neural network. *IEEE Trans. Med. Imaging* **2016**, *35*, 1207–1216. [CrossRef] [PubMed]
3. Moeskops, P.; Viergever, M.A.; Mendrik, A.M.; de Vries, L.S.; Benders, M.J.; Išgum, I. Automatic segmentation of mr brain images with a convolutional neural network. *IEEE Trans. Med. Imaging* **2016**, *35*, 1252–1261. [CrossRef] [PubMed]
4. Zhang, Y.-D.; Dong, Z.; Chen, X.; Jia, W.; Du, S.; Muhammad, K.; Wang, S.-H. Image based fruit category classification by 13-layer deep convolutional neural network and data augmentation. *Multimed. Tools Appl.* **2019**, *78*, 3613–3632. [CrossRef]
5. Gray, P.C.; Fleishman, A.B.; Klein, D.J.; McKown, M.W.; Bézy, V.S.; Lohmann, K.J.; Johnston, D.W. A convolutional neural network for detecting sea turtles in drone imagery. *Methods Ecol. Evol.* **2019**, *10*, 345–355. [CrossRef]
6. Fernandez-Blanc, E.; Rivero, D.; Pazos, A. Convolutional neural networks for sleep stage scoring on a two-channel EEG signal. *Soft Comput.* **2020**, *24*, 4067–4079. [CrossRef]
7. Zhang, Y.; Li, J.; Guo, Y.; Xu, C.; Bao, J.; Song, Y. Vehicle driving behavior recognition based on multi-view convolutional neural network with joint data augmentation. *IEEE Trans. Veh. Technol.* **2019**, *68*, 4223–4234. [CrossRef]
8. Jin, K.H.; McCann, M.T.; Froustey, E.; Unser, M. Deep convolutional neural network for inverse problems in imaging. *IEEE Trans. Image Process.* **2017**, *26*, 4509–4522. [CrossRef] [PubMed]
9. Ding, J.; Chen, B.; Liu, H.; Huang, M. Convolutional neural network with data augmentation for sar target recognition. *IEEE Geosci. Remote Sens. Lett.* **2016**, *13*, 364–368. [CrossRef]
10. Ogawa, R.; Kido, T.; Mochizuki, T. Effect of augmented datasets on deep convolutional neural networks applied to chest radiographs. *Clin. Radiol.* **2019**, *74*, 697–701. [CrossRef] [PubMed]
11. Gómez-Ríos, A.; Tabik, S.; Luengo, J.; Shihavuddin, A.; Krawczyk, B.; Herrera, F. Towards highly accurate coral texture images classification using deep convolutional neural networks and data augmentation. *Expert Syst. Appl.* **2019**, *118*, 315–328. [CrossRef]
12. Krizhevsky, A.; Sutskever, I.; Hinton, G.E. ImageNet classification with deep convolutional neural networks. In Proceedings of the 26th Neural Information Processing Systems Conference, South Lake Tahoe, NV, USA, 2–8 December 2012; pp. 1097–1105.
13. Simonyan, K.; Zisserman, A. Very deep convolutional networks for large-scale image recognition. *arXiv* **2014**, arXiv:1409.1556.
14. Szegedy, C.; Vanhoucke, V.; Ioffe, S.; Shlens, J.; Wojna, Z. Rethinking the inception architecture for computer vision. In Proceedings of the IEEE Conference on Computer Vision and Pattern Recognition, Las Vegas, NV, USA, 26 June–1 July 2016; pp. 2818–2826.
15. Benot, C.; Domenico, V.; Bjorn, T. Ship classification interrasar-x images with convolutional neural networks. *IEEE J. Ocean. Eng.* **2017**, *43*, 258–266.
16. Jiang, M.; Yang, X.; Dong, Z.; Fang, S.; Meng, J. Ship classification based on superstructure scattering features in SAR images. *IEEE Geosci. Remote Sens. Lett.* **2016**, *13*, 616–620. [CrossRef]
17. Shen, S.; Yang, H.; Yao, X.; Li, J.; Xu, G.; Sheng, M. Ship type classification by convolutional neural networks with auditory-like mechanisms. *Sensors* **2020**, *20*, 253. [CrossRef] [PubMed]

18. Gallego, A.J.; Pertusa, A.; Gil, P. Automatic ship classification from optical aerial images with convolutional neural networks. *Remote Sens.* **2018**, *10*, 511. [CrossRef]

19. Ødegaard, N.; Knapskog, A.O.; Cochin, C.; Louvigne, J.C. Classification of ships using real and simulated data in a convolutional neural network. In Proceedings of the IEEE Radar Conference, Philadelphia, PA, USA, 2–6 May 2016; pp. 1–6.

20. Ichimura, S.; Zhao, Q. Route-based ship classification. In Proceedings of the IEEE 10th International Conference on Awareness Science and Technology, Morioka, Japan, 23–25 October 2019; pp. 1–6.

21. Cheung, W.; Hamarneh, G. N-sift: N-dimensional scale invariant feature transform. *IEEE Trans. Image Process.* **2009**, *18*, 2012–2021. [CrossRef] [PubMed]

22. Bay, H.; Tuytelaars, T.; Van Gool, L. Surf: Speeded up robust features. In *European Conference on Computer Vision*; Springer: Berlin, Germy, 2006; pp. 404–417.

23. Guo, L.; Li, J.; Zhu, Y.; Tang, Z. A novel features from accelerated segment test algorithm based on lbp on image matching. In Proceedings of the 2011 IEEE 3rd International Conference on Communication Software and Networks, Xi'an, China, 27–29 May 2011; pp. 355–358.

24. Leutenegger, S.; Chli, M.; Siegwart, R.Y. Brisk: Binary robust invariant scalable keypoints. In Proceedings of the 2011 International Conference on Computer Vision, Barcelona, Spain, 6–13 November 2011; pp. 2548–2555.

25. Kingma, D.P.; Ba, J. Adam: A method for stochastic optimization. *arXiv* **2014**, arXiv:1412.6980.

26. Tan, S. Neighbor-weighted k-nearest neighbor for unbalanced text corpus. *Expert Syst. Appl.* **2005**, *28*, 667–671. [CrossRef]

27. Zhang, M.-L.; Zhou, Z.-H. A k-nearest neighbor based algorithm for multi-label classification. In Proceedings of the 2005 IEEE International Conference on Granular Computing, Beijing, China, 25–27 July 2005; Volume 2, pp. 718–721.

28. Pena, J.M.; Lozano, J.A.; Larranaga, P. An empirical comparison of four initialization methods for k-means algorithm. *Pattern Recognit. Lett.* **1999**, *20*, 1027–1040. [CrossRef]

Article

A Detection and Tracking Algorithm for Resolvable Group with Structural and Formation Changes Using the Gibbs-GLMB Filter

Xinfeng Ru, Yudong Chi and Weifeng Liu *

College of Automation, Hangzhou Dianzi University, Hangzhou 310018, China; a840064210@hdu.edu.cn (X.R.); cyd0418@163.com (Y.C.)
* Correspondence: liuwf@hdu.edu.cn

Received: 25 May 2020; Accepted: 12 June 2020; Published: 15 June 2020

Abstract: In the field of resolvable group target tracking, further study on the structure and formation of group targets is helpful to reduce the tracking error of group bluetargets. In this paper, we propose an algorithm to detect whether the structure or formation state of group targets changes. In this paper, a Gibbs Generalized Labeled Multi-Bernoulli (GLMB) filter is used to obtain the estimation of discernible group target bluestates. After obtaining the state estimation of the group target, we extract relevant information based on the estimation data to judge whether the structure or formation state has changed. Finally, several experiments are carried out to verify the algorithm.

Keywords: target tracking; group targets; GLMB; structure; formation

---

1. Introduction

In many fields, the application of target tracking technology can be seen. For example, target tracking is an important part in the field of self-driving cars. Self-driving cars technology is one of the research focuses at present. In the process of driving a car, its control system needs to detect and discriminate the surrounding environment of the vehicle, and target tracking technology plays an important role in this part [1]. Target tracking technology can also be used in the field of national defense such as aircraft tracking in the air, ships tracking at sea, and vehicles tracking on land. Usually, they coordinate their movements in a certain way. Scholars have made many excellent achievements in the field of target tracking, among which the research on multi-target tracking based on random finite set (RFS) is also very fruitful; such as machine learning [2], computer vision [3], autonomous vehicle [4], sensor scheduling [5–12], sensor network [13–15], blue, in particular, a fast RFS based distributed tracking algorithm is presented for a sensor network in [15] and track-before-detect, tracking of merged measurements, and target tracking[16–20].

Mahler is the first one to apply the RFS theory to the field of target tracking [21–24] and gives the probability hypothesis density(PHD) filter [22,25], the cardinalized PHD (CPHD) filter [26,27], and the multi-target multi-Bernoulli (MeMBer) [28]. The MeMBer filter is different from the PHD filter and the CPHD filter. The MeMBer filter propagates the parameters of a multi-Bernoulli distribution that approximate the posterior multi-target density and the others propagate the moments and cardinality distributions. Subsequently, the CBMeMBer filter [29] is proposed to solve the problem of cardinal deviation of MeMBer filter. References [30,31] introduce label into RFS and propose a Generalized Labeled Multi-Bernoulli (GLMB) filter. Reference [30] gives the implementation method δ-Generalized Labeled Multi-Bernoulli (δ-GLMB) filter of GLMB, and the truncation method is used to refine the δ-GLMB in Reference [31]. In Reference [32,33], Vo et al. simplified the prediction step and update step of GLMB filter into a single step. This new method to implementation is known as Gibbs GLMB. The Gibbs-GLMB filter greatly reduces the computational complexity of GLMB filter and improves the computational efficiency. Based on these filters, some people have done some work on tracking hybrid targets. Hybrid targets is a collection of multiple target, group target and extended target. The problem of tracking multi-measurement targets is concerned in [34]. In Reference [35], group target tracking in the case of uncertain number of targets and groups. Reference [19] focuses on dynamic modeling and tracking estimation for multiple resolvable group targets. Reference [36] considers group state estimation. Under the framework of RFS, the work of modeling and tracking estimation for multi-extended targets is contributed in the Reference [37]. In the Reference [33], we apply a Gibbs-GLMB filter to estimate the state of resolvable group targets and track them.

In this paper, based on the collaborative relationships between targets and the state information of each independent target, we propose a method to determine the structure and formation change of resolvable group targets. We call the dependency relationship between group targets as structure and the shape formed by group targets with fixed spatial distance as formation.

The content of this paper is arranged as follows: Section 2 introduces relevant background knowledge including RFS, LRFS, and graph theory; Section 3 describes a state estimation method for resolvable group targets; Section 4 discusses how to determine the structure and formation of resolvable group targets in continuous and discrete environments. Then we carried out two simulations to verify our proposed method in Section 5. Finally, Section 6 is the summary of the paper.

2. Backgrounds

2.1. Labeled Random Finite Set (Labeled RFS)

In [30], Vo combines RFS with labels. RFS is essentially a set with random number of members, random state of members, and no fixed sorting rules among members. The random set is used to represent the target state of resolvable group targets. It can be expressed as Equation (1) at time k.

$$X_k = \{x_{k,1}, \cdots, x_{k,N_k}\}. \tag{1}$$

In group target tracking, the situation of the target is very complex and changeable. For example, a certain target may suddenly disappear, or a new target may appear at a certain moment, or a target may be decomposed into multiple targets, and multiple targets may be combined into one target set. Therefore, Equation (1) can be expressed as Equation (2).

$$X_k = [\bigcup_{x \in X_{k-1}} S_{k|k-1}(x)] \bigcup [\bigcup_{x \in X_{k-1}} B_{k|k-1}(x)] \bigcup \Gamma_k, \tag{2}$$

where $S_{k|k-1}(x)$, $B_{k|k-1}(x)$, Γ_k stand for the surviving targets, the spawned targets, and the birth targets, respectively. Similar to the state set, the measurement set of group targets is shown in the Equation (3).

$$Z_k = \{z_{k,1}, \cdots, z_{k,M_k}\}. \tag{3}$$

RFS improved to labeled RFS is the addition of a unique tag $\ell \in \mathbb{L} = \{\alpha_i : i \in \mathbb{N}\}$ for each target $x_{k,i}$. $\mathbb{N}$ is the set of positive integers. We use the constraint function Equation (4) to guarantee the uniqueness of the label.

$$\Delta(X) = \begin{cases} 1, |\mathcal{L}(X)| = |X| \\ 0, |\mathcal{L}(X)| \neq |X| \end{cases} \tag{4}$$

Reference [30] not only proposes the labeled RFS but also gives the densities of an LMB RFS and a labeled Poisson RFS. The LMB RFS's density is described as:

$$\pi(\{(x_1, l_1), \cdots, (x_n, l_n)\}) = \delta_n(|\{l_1, \cdots, l_n\}|) \times \prod_{\zeta \in \Psi} (1 - r^\zeta) \prod_{j=1}^{n} \frac{1_{\alpha(\Psi)}(l_j) r^{(\alpha^{-1}(l_j))}(x_j)}{1 + r^{(\alpha^{-1}(l_j))}}. \tag{5}$$

2.2. Graph Theory

Graph theory has a wide range of applications. In this paper, we use directed graph to describe the structure of group. We take independent targets as vertices V of the graph G and relationships between targets as edges E. At the same time, we are describing this graph as a asymmetric adjacency matrix, which is the structure of the resolvable group target. The matrix is shown by the Equation (6).

$$A_d = \begin{bmatrix} 0 & a(1,2) & \cdots & a(1,n) \\ a(2,1) & 0 & \cdots & a(2,n) \\ \vdots & \vdots & \ddots & \vdots \\ a(n,1) & a(n,2) & \cdots & 0 \end{bmatrix}, \tag{6}$$

Let target i be the parent node of target j, $a(i,j) = 1$. If target i is target j's child node, or target i has no relationship with target j, $a(i,j) = 0$.

2.3. Graph Theory Model of Labeled RFS

For the relationship between any two vertex v_i, we can express it in terms of Equation (7).

$$e_{i,j} : (x_i, x_j) \rightarrow \{1,0\}. \tag{7}$$

Since each vertex has an unique label, we can simplify Equation (7) to Equation (8).

$$e_{i,j} : (l_i, l_j) \rightarrow \{1,0\}. \tag{8}$$

Equation (8) shows that the structure of the group is encapsulated by the graph defined on the target labels.

2.4. Revolvable Group Tracking with Maneuver and the Efficient Implementation for the GLMB Filter

Let target state be $x_{k,i}$ given as follows:

$$x_{k,i} = [p_{k,x}(i), \dot{p}_{k,x}(i), p_{k,y}(i), \dot{p}_{k,y}(i)] \in X_k. \tag{9}$$

where $p_{k,x}(i)$ and $\dot{p}_{k,x}(i)$ are the position and velocity of target i on the x-axis, $p_{k,y}(i)$ and $\dot{p}_{k,y}(i)$ indicate the position and velocity of target i on the y-axis.

Suppose a target has a single parent node. We introduce the displacement $b_k(l,i)$ vector to describe this relationship and the resolvable group targets dynamic model is given as follows:

$$x_{k+1,i} = F_{k,l} x_{k,l} + b_k(l,i) + \Gamma_{k,i} \omega_{k,i} \tag{10}$$

where $b_k(l,i)$, l contains the direction and distance information between the parent node l and child nodes i.

Therefore, for group targets, the displacement vector $b_k(l,i)$ can be represented as:

$$b_k(l,i) = [R_k(l,i) \times \cos(\theta_k(l,i) - \beta_k(l,i)), 0, R_k(l,i) \times \sin(\theta_k(l,i) - \beta_k(l,i)), 0]^T, \tag{11}$$

where $R_k(l,i)$ denotes the designed distance between parent node l and child node i. $\beta_k(l,i)$ is the designed angle between nodes l and i. $\theta_k(l,i)$ is the motion direction for the parent target.

Under a maneuvering motion model, we assume that the formation of the group is stable within a certain time interval, so Equation (11) can be transferred to Equation (14).

$$b_k(l,i) = c_k(l,i)C_a(l,i)x_{k,l} \tag{12}$$

$$c_k(l,i) = \frac{R(l,i)}{\sqrt{\dot{p}^2_{k,x}(l) + \dot{p}^2_{k,y}(l)}} \tag{13}$$

$$C_a(l,i) = \begin{bmatrix} 0 & a_{\beta,2}(l,i) & 0 & a_{\beta,1}(l,i) \\ 0 & 0 & 0 & 0 \\ 0 & a_{\beta,1}(l,i) & 0 & -a_{\beta,2}(l,i) \\ 0 & 0 & 0 & 0 \end{bmatrix}. \tag{14}$$

The derivation process has been described in detail in Reference [33].

Another point is the relation between covariance and adjacency matrix. To calculate the means and covariance, the parent vertex l should be first known. This is dependent on adjacency matrix Equation (6). In this paper, the adjacency matrix is defined on the label space and known in prior. That is, in the predicted stage, the adjacency matrix can be gotten and adopted according to the predicted labels. In contrast, if the adacency matrix is unknown, it needs to be estimated according to the predicted states. In general, the adacency relation is based on the target states, or the motion information. A detailed discussion can be found in Reference [36].

In the original GLMB filter [30,31], both the prediction step and the update step have their own independent truncation. This makes the computational complexity of GLMB filter very high. In order to optimize this problem, Vo et al. proposed the Gibbs GLMB in Reference [32]. The Gibbs GLMB algorithm combines the prediction and update steps of the GLMB filter so that there is only one truncation in each iteration. The density of the joint step of prediction and update step at time k is shown as:

$$\pi_k(X) \propto \Delta(X) \sum_{J,\xi,\theta} \left(\sum_I \omega_{k-1}^{(I,\xi)} \omega_k^{(I,\xi,J,\theta)}(Z_k) \right) \delta_J(\mathcal{L}(X)) \left[p_k^{(\xi,\theta)}(\cdot|Z_k) \right]^X, \tag{15}$$

where

$$\omega_k^{(I,\xi,J,\theta)}(Z_k) = 1_{\Theta_k(J)}(\theta)[1 - r_{B,k}]^{B_k-J}[r_{B,k}]^{B_k \cap J}[1 - \overline{P}_{S,k}^{(\xi)}]^{I-J}[\overline{P}_{S,k}^{(\xi)}]^{I \cap J}[\overline{\Psi}_{Z_k,k}^{(\xi,\theta)}]^J$$

$$\overline{\Psi}_{Z_k,k}^{(\xi,\theta)}(l) = \left\langle \Psi_{Z_k,k}^{(\xi,\theta(l))}(\cdot,l), p_{k|k-1}^{(\xi)}(\cdot,l) \right\rangle$$

$$\overline{P}_{S,k}^{(\xi)}(l) = \left\langle P_{S,k}(\cdot,l), p_{k-1}^{(\xi)}(\cdot,l) \right\rangle$$

$$p_k^{(\xi,\theta)}(x,\cdot|Z_k) \propto \Psi_{Z_k,k}^{(\xi,\theta(l))}(x,l)p_{k|k-1}^{(\xi)}(x,l)$$

$$p_{k|k-1}^{(\xi)}(x,l) = 1_{B_k}(l)p(B,k)(x,l) + 1_{\mathbb{L}_{k-1}}(l)\frac{\langle P_{S,k}(\cdot,l)f_{k|k-1}(x|\cdot,l), p_{k-1}^{(\xi)}(\cdot,l) \rangle}{\overline{P}_{S,k}^{(\xi)}(l)},$$

Truncation by sampling $\{(I^{(i)}, \xi^{(i)}, J^{(i)}, \theta^{(i)})\}_{i=1}^{H_{k,max}}$ from some distribution π. It should be noted in the predicted density $p_{k|k-1}^{(\xi)}(\cdot,l)$ that the collaboration noise $\omega_{k,i}^o$ is adopted, instead of process noise $\omega_{k,i}$.

3. Analysis of the Structure and Formation of Discernible Group Targets

Structures and formations exist in every aspect of our lives. For example, every bridge needs a suitable structure to carry its weight. The motorcade of the bride and groom is organized into a beautiful pattern, that is, a formation. In the military field, structures and formations are easier to spot. The design of structure and formation is a very important part of the air force, ground fighting vehicle force, and sea ship force, which affects the combat effectiveness of troops. This section describes in detail how to identify structures and formations of resolvable group targets.

3.1. The Structure and Formation of Resolvable Group Targets

3.1.1. Structure

In this paper, the structure describes the collaborative relationship between the targets, we regard the collaboration between parent and child targets as the framework of the structure, while parent and child targets can be regarded as the important nodes of the structure. There is a detailed introduction on how to extract the structure of group targets in the Reference [35]. This paper focuses on how to determine whether the structure and formation of group targets have changed. First of all, let us talk about group structure in detail, focusing on when we think the group structure has not changed and when the group structure has changed. Since structure is the embodiment of collaboration, we can think of it this way: as long as the collaboration between group targets remains unchanged, the structure will remain unchanged. For example, in Figure 1, there are two independent subgroup targets, and each subgroup target has four subtargets. In these two subgroup targets, although the distance between its subtargets and the included angle of its relationship are different, their cooperative relationship is the same, so their structure is the same.

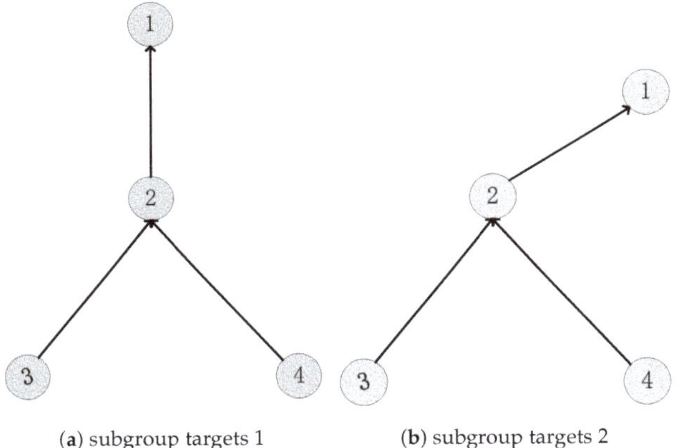

(**a**) subgroup targets 1   (**b**) subgroup targets 2

Figure 1. Two group targets with the same structure.

3.1.2. Formation

The formation and structure of group objects are very similar. First of all, formation and structure are dependent on the collaborative relationship between sub-targets. If the cooperative relationship changes, formation and structure will inevitably change. The difference between the two is that formation has more strict standards than structure. For example, in a military parade in the queue, keep formation is in the process of the whole travel distance between the two fighters to almost keep constant, each consisting of three soldiers geometry angles that are almost the same. In the group of target formation, too, so we can regard it as a group of target formation and have a child between target space distance between the fixed and the geometry of fixed structure. Still here in Figure 1 as an example, two group of target Figure 1a and Figure 1b the same known structure, but target x_1 and x_2 in the Figure 1a the space distance and Figure 1b in the target space distance differences between x_1, x_2, and Figure 1a in the target consisting of x_1, x_2, x_4 geometrical angle and Figure 1b the angle of the corresponding target in also have bigger difference, so although the structure of the two group of consistent, but its formation is not consistent. In the process of judging, as long as there is a big difference between spatial distance and geometric angle, we can determine that the formation of the two groups is not consistent.

3.2. The Determination of Distinguishable Group Target Structure and Formation

In this paper, it is assumed that the relationship between father and son nodes in the group target will not be reversed, that is, the parent node will always be the parent node, will die, will separate, but will not become the child of its original child node.

3.2.1. Determination of Group Target Structure in Continuous State

In this section we introduce the determination of group target structure in a continuous state. Based on the above description, we can clearly determine whether the elements of the structure are collaboration or not. The key to determine whether there is a collaboration relationship is the motion state of each subtarget and the spatial position between them. If the structure of a group is to remain in state, then the motion state of its members should be similar. If the motion state of a member differs greatly from that of other members, it is obvious that it must be separated from the group, and the structure of the group will also change accordingly. According to this principle, we can carry out a comparative analysis of its elements one by one. Firstly, in the discernible group target, the speed keeping approximation of all members is one of the important factors to judge whether the parent–child target cooperation relationship in the group target changes. We took the group target in Figure 2 as an example to demonstrate. At time k, the state of each target is $x_{k,i}$, and the velocity is $v_{k,i}$, then the state of target x_1, x_2, and x_3 at time $k+1$ can be expressed as:

$$\begin{aligned}
x_{k+1,1} &= x_{k,1} + v_{k,1} \\
x_{k+1,2} &= x_{k,2} + v_{k,2} \\
x_{k+1,3} &= x_{k,3} + v_{k,3}
\end{aligned} \tag{16}$$

Therefore, the position relationship $\mathcal{L}_k(i,j)$ between x_1 and x_2, x_1 and x_3, and x_2 and x_3 at time k and $k+1$ can be expressed as:

$$\begin{aligned}
\mathcal{L}_k(1,2) &= x_{k,2} - x_{k,1} \\
\mathcal{L}_k(1,3) &= x_{k,3} - x_{k,1} \\
\mathcal{L}_k(2,3) &= x_{k,3} - x_{k,2}
\end{aligned} \tag{17}$$

$$\begin{aligned}
\mathcal{L}_{k+1}(1,2) &= x_{k+1,2} - x_{k+1,1} \\
\mathcal{L}_{k+1}(1,3) &= x_{k+1,3} - x_{k+1,1} \\
\mathcal{L}_{k+1}(2,3) &= x_{k+1,3} - x_{k+1,2}
\end{aligned} \tag{18}$$

Substitute Equation (16) into Equation (18):

$$\begin{aligned}
\mathcal{L}_{k+1}(1,2) &= x_{k,2} - x_{k,1} + (v_{k,2} - v_{k,1}) \\
\mathcal{L}_{k+1}(1,3) &= x_{k,3} - x_{k,1} + (v_{k,3} - v_{k,1}) \\
\mathcal{L}_{k+1}(2,3) &= x_{k,3} - x_{k,2} + (v_{k,3} - v_{k,2})
\end{aligned} \tag{19}$$

If the formation of the group is guaranteed to remain unchanged, the following conditions must be satisfied:

$$\mathcal{L}_{k+1}(i,j) = \mathcal{L}_k(i,j) \tag{20}$$

That is:

$$v_{k,2} - v_{k,1} = 0$$
$$v_{k,3} - v_{k,1} = 0 \tag{21}$$
$$v_{k,3} - v_{k,2} = 0$$

Therefore, we can obtain the following conditions for the formation to remain stable:

$$v_{k,1} = v_{k,2} = v_{k,3} \tag{22}$$

Therefore, we can draw a conclusion that, in the continuous state, the resolvable group targets can maintain stable structure and formation by judging whether the velocity of each subtarget in the group is consistent

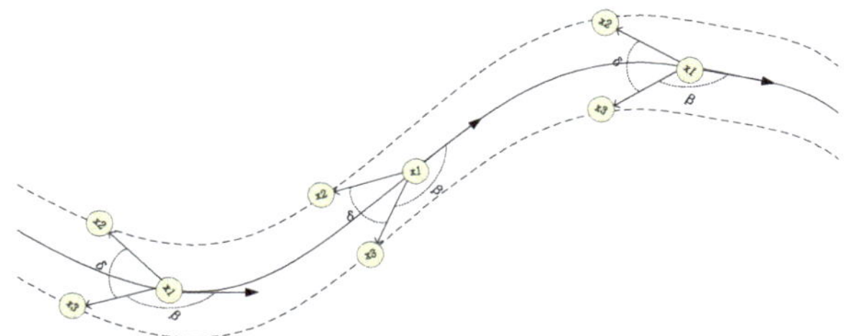

Figure 2. The trajectories of group targets under continuous conditions.

The degree of approximation, we can use the speed δ_{vel} and speed-direction difference threshold δ_{dir} to limit speed. The deviation of velocity has cumulative, namely with the accumulation of time, while its speed is still approximate, but its space relative position deviation of the group members will be accumulated, so make sure that the collaboration will remain with the original, with its father–child node distance also needing to be tested, the distance threshold δ_{dis} is as shown in Equation (23) :

$$\left\| \begin{bmatrix} p_x^1 \\ p_y^1 \end{bmatrix} - \begin{bmatrix} p_x^2 \\ p_y^2 \end{bmatrix} \right\| \prec \delta_{dis} \tag{23}$$

The constraint of the difference of velocity size is:

$$\left\| \begin{bmatrix} \dot{p}_x^1 \\ \dot{p}_y^1 \end{bmatrix} - \begin{bmatrix} \dot{p}_x^2 \\ \dot{p}_y^2 \end{bmatrix} \right\| \prec \delta_{vel} \tag{24}$$

The difference of velocity direction β_{vdif} deviation is:

$$\beta_{vdif} = \frac{\dot{p}_y^1}{\dot{p}_x^1} - \frac{\dot{p}_y^2}{\dot{p}_x^2} < \delta_{dir} \tag{25}$$

Through the above three constraints, we can determine the structure state of the group target, and the three elements of the structure are shown in Figure 3.

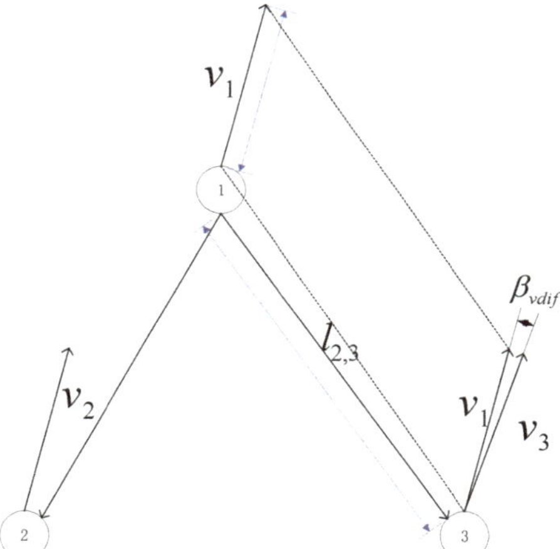

Figure 3. Schematic diagram of velocity in group target v, distance between father and son target l_{ij}, and velocity deviation β_{vdif}.

3.3. Determination of Group Target Formation in a Continuous State

The formation of the group target is a special case of the structure of the group target. Therefore, when analyzing the formation, we assume that the structure of the group target is unchanged. In other words, the speed factor of the group target member does not need to be considered, and we only need to judge whether the formation changes through its position state. In essence, it is to determine whether the relative position of two objects in a cooperative relationship has changed in two consecutive moments. The first step is to obtain the position deviation matrix A_P of all members of the group target. the acquisition method is shown in the Equation (26).

$$A_p = \begin{bmatrix} 0 & \|\eta_{12}\| & \cdots & \|\eta_{1n}\| \\ \|\eta_{21}\| & 0 & \cdots & \|\eta_{2n}\| \\ \vdots & \vdots & \ddots & \vdots \\ \|\eta_{n1}\| & \|\eta_{n2}\| & \cdots & 0 \end{bmatrix}_{n \times n} \tag{26}$$

where n is the number of group members, and η_{ij} is the deviation vectors of the relative position vectors of the two targets in the group at time k and time $k-1$. It can be seen from Figure 4 that the position deviation vector η_{ij} can be obtained from the displacement vector $rpos$ of the target and the target at time k and time $k-1$. The calculation method is as follows:

$$\eta_{k,ij} = rpos_{k,ij} - rpos_{k-1,ij} \tag{27}$$

$$rpos_{k,ij} = \begin{bmatrix} 1 & 0 & 1 & 0 \end{bmatrix} * (x_{k,i} - x_{k,j}) \tag{28}$$

After obtaining position deviation matrix A_P, we can obtain the formation change coefficient of group target through Equation (29).

$$\varphi = \|A_k A_P\|_2 \tag{29}$$

A_k is the adjacency matrix describing the group structure. Through A_k, the data with collaboration relationship targets in the group are screened out from A_p. According to the observation, we set a threshold of formation coefficient σ_{dis}. If the coefficient Φ is greater than the threshold σ_{dis}, the formation of group targets is considered to be fixed; if not, it is considered to have not changed, and the relationship is shown in the formula.

$$\begin{cases} 1, & \Phi > \sigma_{dis} \\ 0, & \Phi \leq \sigma_{dis} \end{cases} \tag{30}$$

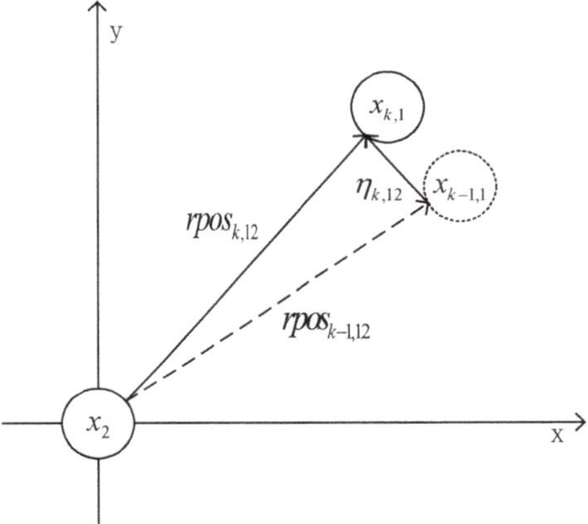

Figure 4. The schematic diagram of position deviation vector $\eta_{k,12}$.

4. Simulations

4.1. Experiment 1

4.1.1. Configuration Parameters for Experiment 1

In this experiment, we used a Gibbs-GLMB filter to get the state estimation of group targets. There are three subgroups of group target, including 4, 4, and 6 members, respectively. Their structure is shown in Figure 5.

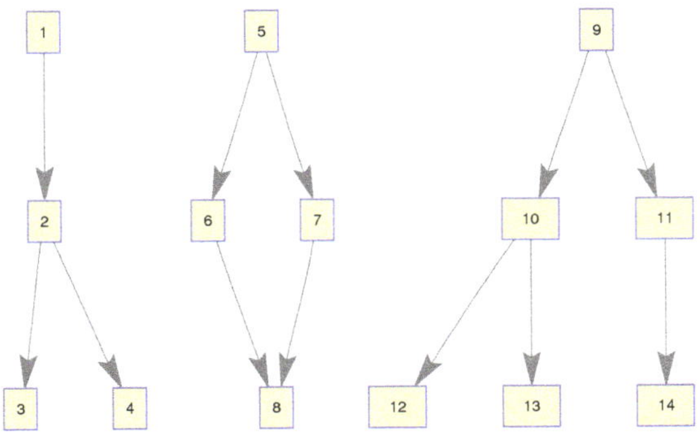

Figure 5. The structure of subgroup targets.

The initialized state of the distance between any parent and child targets is 100 meters. $\{(x_1, \ell_1), \cdots, (x_4, \ell_4)\}$ is subgroup 1, $\{(x_5, \ell_5), \cdots, (x_8, \ell_8)\}$ is subgroup 2, and $\{(x_9, \ell_9), \cdots, (x_{14}, \ell_{14})\}$ is subgroub 3. The three subgroups are independent of each other. In the initial state, the adjacency matrix of the three subgroups is set as known in this paper, which can be expressed as:

$$A_1(\ell_1, \cdots, \ell_4) = \begin{bmatrix} 0 & 0 & 0 & 0 \\ 1 & 0 & 0 & 0 \\ 0 & 1 & 0 & 0 \\ 0 & 1 & 0 & 0 \end{bmatrix} \tag{31}$$

$$A_2(\ell_5, \cdots, \ell_8) = \begin{bmatrix} 0 & 0 & 0 & 0 \\ 1 & 0 & 0 & 0 \\ 1 & 0 & 0 & 0 \\ 0 & 1 & 1 & 0 \end{bmatrix} \tag{32}$$

$$A_2(\ell_9, \cdots, \ell_{14}) = \begin{bmatrix} 0 & 0 & 0 & 0 & 0 & 0 \\ 1 & 0 & 0 & 0 & 0 & 0 \\ 1 & 0 & 0 & 0 & 0 & 0 \\ 0 & 1 & 0 & 0 & 0 & 0 \\ 0 & 1 & 0 & 0 & 0 & 0 \\ 0 & 0 & 1 & 0 & 0 & 0 \end{bmatrix} \tag{33}$$

The monitoring scope of experiment 1 is $[-\pi/2, \pi/2; 0m, 3000m]$. The duration of the experiment was 100 s: subgroup 1 was born at $k = 0$ s, subgroup 2 and subgroup 3 were both born at $k = 20$ s, target x_1, the head node in subgroup 1, died at $k = 30$ s, target x_8 in subgroup 2 and target x_{11} in subgroup 3 died at $k = 70$ s. In terms of structure, subgroup 3 is decomposed into two subgroup targets at $k = 30$ s. Target x_{11} and its child target x_{14} are separated from subgroup 3 into subgroup 4. At this time, subgroub 4 only has target x_{11} and target x_{14} and the dependency between target x_{11} and target x_{14} is the same as before. At $k = 70$ s, target x_2, x_3, and x_4 of subgroup 1 are completely separated into three independent targets, and target x_{11} of subgroup 4 becomes an independent target. The structure between $k = 30$ and $k = 70$ s is shown in Figure 6, and the structure of the group at $k > 70$ s is shown in Figure 7. The covariance of the observed noise is $R = \text{diag}[\begin{array}{cc} 0.0012 & 100 \end{array}]$. The covariance of process noise is $Q = \text{diag}[\begin{array}{ccc} 0.04 & 0.04 & 0.04 \end{array}]$. The real trajectory of the target is shown in Figure 8. The curve represents the trajectory, the circle represents the starting point, and the triangle represents the ending point.

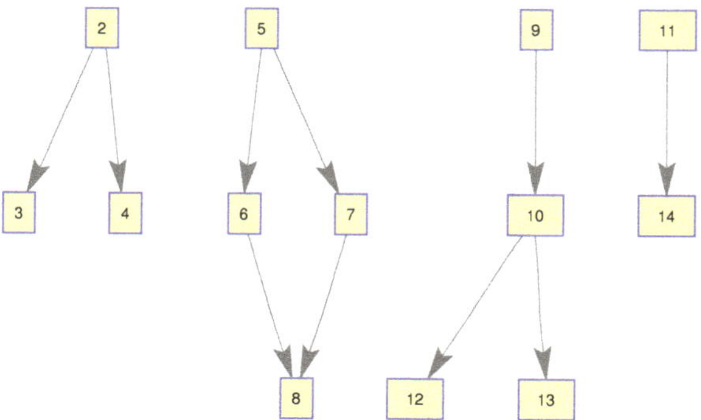

Figure 6. The structure of subgroup targets during $k = 30$ and $k = 70$.

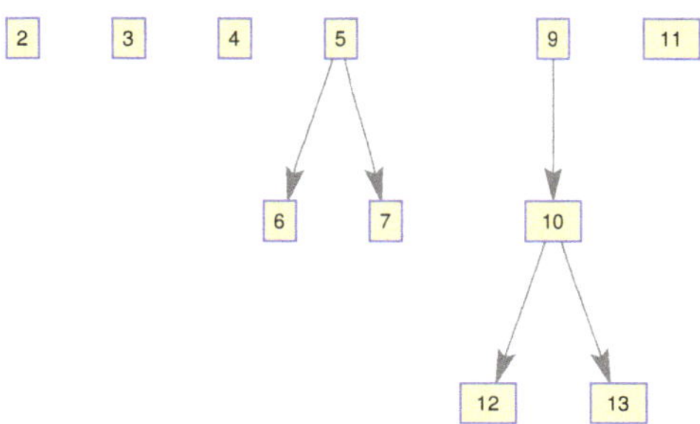

Figure 7. The structure of subgroup targets during $k > 70$.

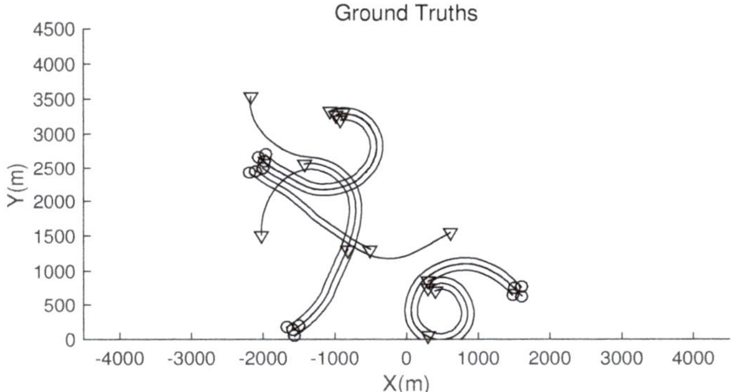

Figure 8. The structure of subgroup targets during $k > 70$.

4.1.2. The Result of the Experiment 1

In this experiment, we set that when the structure of the group target is detected to change, the estimated point of the group target and the structure of the group target at this time are immediately output. In the experimental environment parameter configuration, we can get that the group structure will change significantly when $k = 20$, $k = 30$, and $k = 70$. At $k = 20$, 10 targets will be born. At $k = 30$, a target dies, and subgroup 3 splits into two subgroups. At $k = 70$, there is both the death of the target and the change in the target structure. Therefore, we focus on the detection results of these three moments from the results. The following are the experimental result figures of the three moments in this experiment. When $k = 20$, the position of the target and the structure of the group target are shown in Figures 9 and 10. In Figure 9, we can see that Gibbs-GLMB tracks 13 points at this point, compared to $k = 19$, there are 9 target regenerations. Although the collaboration relationship between the targets has not changed compared with the previous moment, the overall structure of the group is bound to change with the increase of the members of the group, so the tracked target is traced and its structure is output. When $k = 30$, the head node of a subgroup dies, and a subgroup is separated into two subgroups, and the change is also detected successfully, as shown in Figures 11 and 12. When $k = 70$, the detection results are shown in Figures 13 and 14.

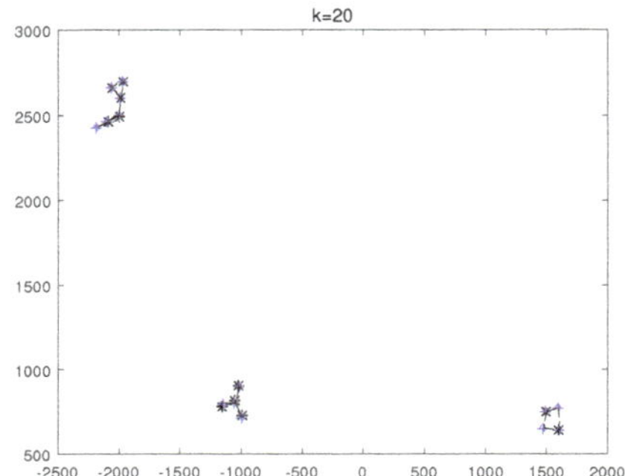

Figure 9. The state estimation of group targets at $k = 20$.

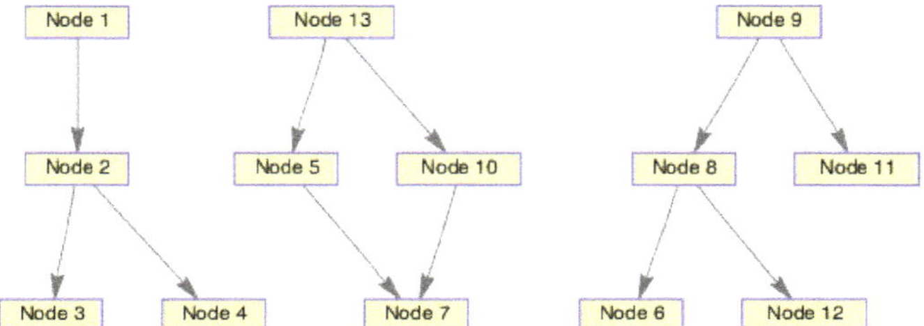

Figure 10. The structure estimation of group targets at $k = 20$.

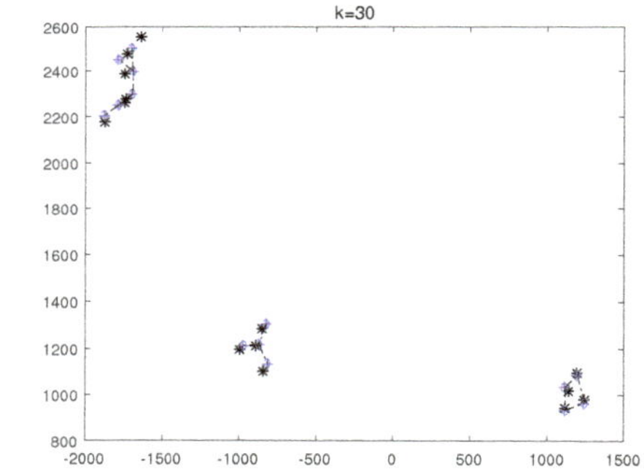

Figure 11. The state estimation of group targets at $k = 30$.

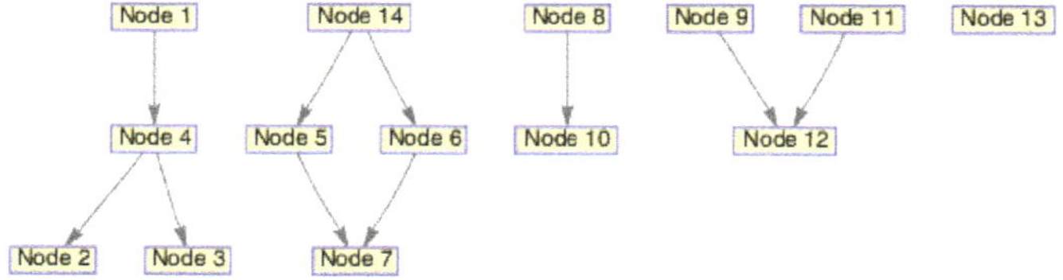

Figure 12. The structure estimation of group targets at $k = 30$.

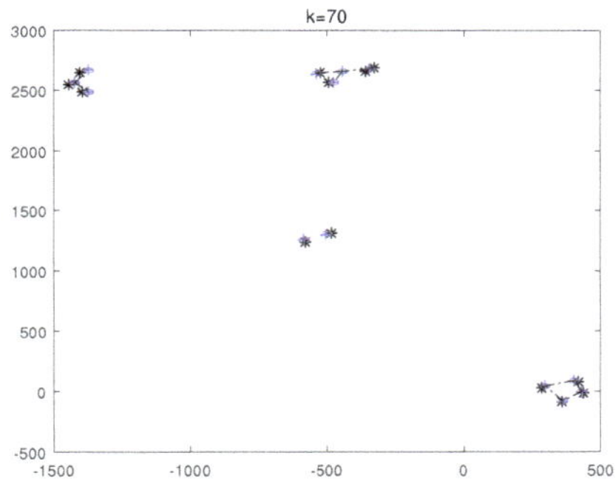

Figure 13. The state estimation of group targets at $k = 70$.

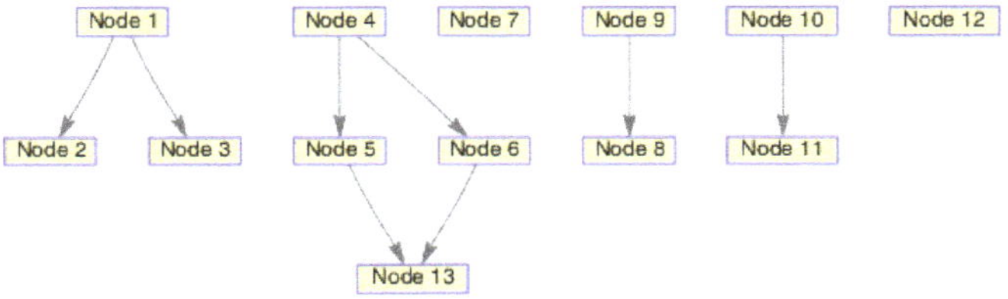

Figure 14. The structure estimation of group targets at $k = 70$.

The position state estimation of Gibbs-GLMB filter, OSPA distance, and target number estimation are shown in Figures 15–17.

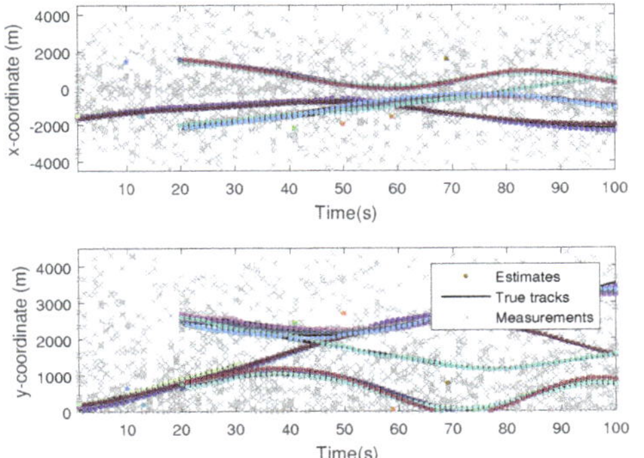

Figure 15. The state estimation by Gibbs Generalized Labeled Multi-Bernoulli (GLMB) filter.

The results show that the proposed method can effectively identify the change of group target structure, but there are some problems, such as false alarm. As a result that the method relies on Gibbs GLMB state estimation results, when the state estimation error occurs, the structure estimation error will also occur. In addition, the threshold setting too large or too small will cause error, the rationality of the threshold setting is one of the problems worth in-depth study.

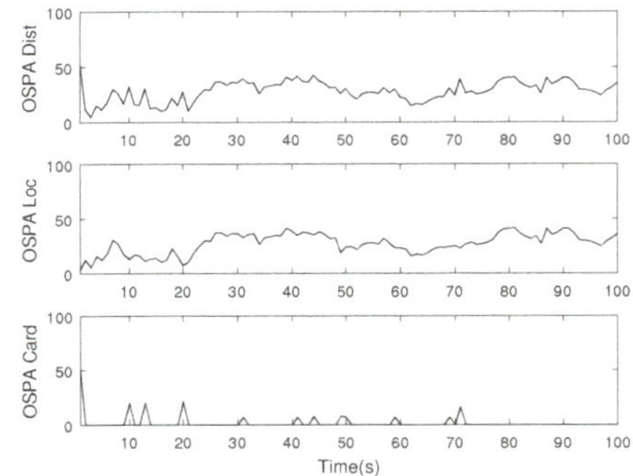

Figure 16. The OSPA distance by Gibbs-GLMB filter.

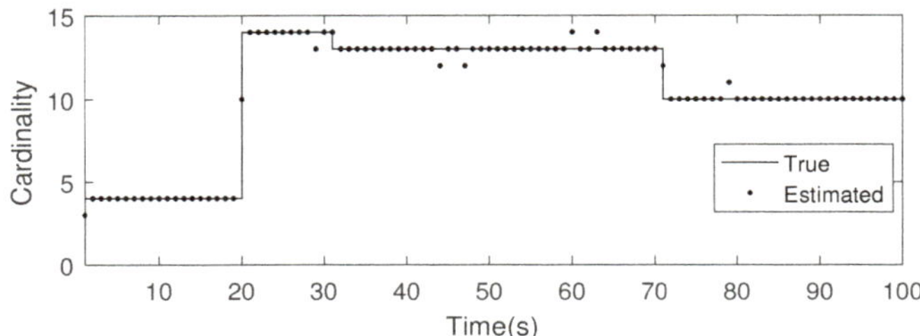

Figure 17. The estimated number of targets by GLMB filter.

4.2. Experiment 2

4.2.1. Configuration Parameters for Experiment 2

In this experiment, we only selected a single group target with four members for the experiment, whose structure is shown in the Figure 18:

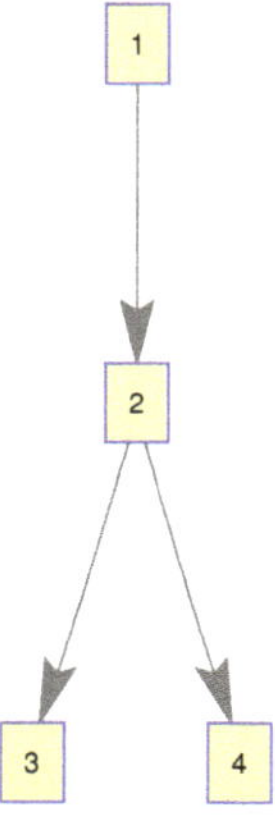

Figure 18. The structure of the single group targets.

The distance between the parent and child targets set in the initial state is 100 m, the same as in experiment 1. The adjacency matrix of the group $\{(x_1, \ell_1), \cdots, (x_4, \ell_4)\}$ is:

$$A_1(\ell_1, \cdots, \ell_4) = \begin{bmatrix} 0 & 0 & 0 & 0 \\ 1 & 0 & 0 & 0 \\ 0 & 1 & 0 & 0 \\ 0 & 1 & 0 & 0 \end{bmatrix} \tag{34}$$

The other parameters are basically the same as those in experiment 1, except that there are no birth, death and ovulation of the target in this experiment. As a result that this experiment studies the change of formation, we assume that its structure is stable and unchanged, which can simplify our modeling. In the model, we increase the distance between target x_1 and target x_2 in the group by $10m$ per second between $k = 30$ and $k = 40$ s. In other words, the formation of the group will continue to change during this period, and its track is as Figure 19:

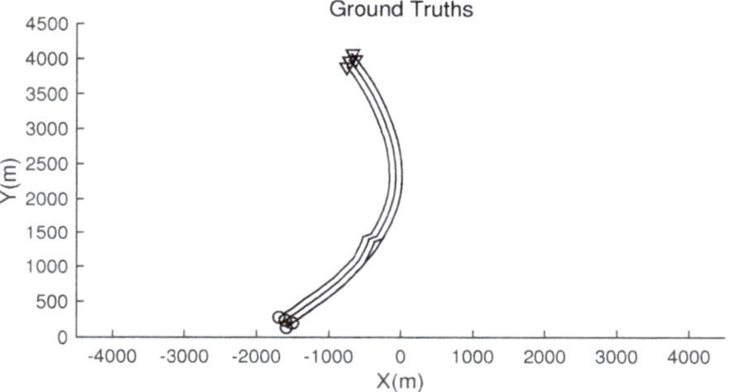

Figure 19. The track of the single group targets.

4.2.2. The Result of the Experiment 2

When a change in formation is detected, the current and previous positions of the group target are plotted immediately.The results show that all the preset change nodes can be detected successfully, and the detected position diagram at the time $k = 30$ to $k = 41$ is shown in Figure 20. The blue cross represents the current target position and the red circle represents the previous target position.

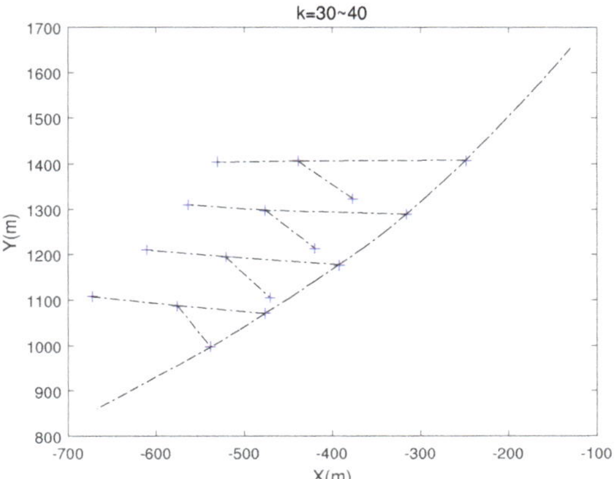

Figure 20. The position of the target at $k = 30$ to $k = 40$.

The experiment was carried out for ten times, and the detection accuracy of ten times was shown in Table 1. According to the data in the Table 1, we can know that the average recognition accuracy of the ten experiments is 74.70%.

Table 1. The judgment precision of ten experiments and the average value of ten experiments.

No.	1	2	3	4	5	6	7	8	9	10
precision	78.57%	64.71%	84.62%	68.75%	64.71%	81.82%	73.33%	78.57%	73.33%	78.57%
average	74.70%									

5. Conclusions

In this paper, we analyze the structure and formation of group targets based on the information of target position and velocity, and then propose a method to determine whether they have state changes. First, we use the Gibbs-GLMB filter to estimate the state of group target. Then, by analyzing and comparing the state estimation data of the target, we can determine whether its structure or formation has changed. The structure problem is mainly determined by the distance between the targets, the velocity difference between the two targets, and the angle difference between the velocity direction and the position vector. For the formation problem, it is mainly determined by the distance between the targets and the offset angle difference of the position vector. Finally, experiments show that the method proposed in this paper can effectively identify the structure and formation changes of group targets.

Author Contributions: For conceptualization, W.L.; methodology, X.R.; software, Y.C.; validation, W.L. and X.R.; formal analysis, W.L. and X.R.; investigation, W.L.; writing—original draft preparation, W.L. and X.R.; writing—review and editing, W.L. All authors have read and agreed to the published version of the manuscript.

Funding: This work is supported by the National Natural Science Foundation of China (61771177,61333011), and the Natural Science Foundation for Young Scientists of Jiangsu Province (BK20160148).

Conflicts of Interest: The authors declare no conflict of interest.

References

1. Clausen, U.; Klingner, M. Automated Driving: Computers take the wheel. In *Digital Transformation*; SpringVieweg: Berlin, Germany, 2019.
2. Vo, B.N.; Dam, N.; Phung, D.; Tran, Q.N.; Vo, B.T. Model-based learning for point pattern data. *Pattern Recognit.* **2018**, *84*, 136–151. [CrossRef]
3. Maggio, E.; Taj, M.; Cavallaro, A. Efficient multitarget visual tracking using random finite sets. *IEEE Trans. Circuits Syst. Video Technol.* **2008**, *18*, 1016–1027. [CrossRef]
4. Meissner, D.; Reuter, S.; Dietmayer, K. Road user tracking at intersections using a multiple-model PHD filter. In Proceedings of the 2013 IEEE Intelligent Vehicles Symposium (IV), Gold Coast, QLD, Australia, 23–26 June 2013; pp. 377–382.
5. Ristic, B.; Vo, B.N. Sensor control for multi-object state-space estimation using finite sets. *Automatica* **2010**, *46*, 1812–1818. [CrossRef]
6. Ristic, B.; Vo, B.N.; Clark, D. A note on the reward function for PHD filters with sensor control. *IEEE Trans. Aerosp. Electron. Syst.* **2011**, *47*, 1521–1529. [CrossRef]
7. Hoang, H.G.; Vo, B.T. Sensor management for multi-target tracking via multi-Bernoulli filtering. *Automatica* **2014**, *50*, 1135–1142. [CrossRef]
8. Gostar, A.K.; Hoseinnezhad, R.; Bab-Hadiashar, A. Robust multi-Bernoulli sensor selection for multi-target tracking in sensor networks. *IEEE Signal Process. Lett.* **2013**, *20*, 1167–1170. [CrossRef]
9. Hoang, H.G.; Vo, B.N.; Vo, B.T.; Mahler, R. The Cauchy–Schwarz divergence for Poisson point processes. *IEEE Trans. Inf. Theory* **2015**, *61*, 4475–4485. [CrossRef]
10. Gostar, A.K.; Hoseinnezhad, R.; Bab-Hadiashar, A. Multi-Bernoulli sensor-selection for multi-target tracking with unknown clutter and detection profiles. *Signal Process.* **2016**, *119*, 28–42. [CrossRef]
11. Gostar, A.K.; Hoseinnezhad, R.; Bab-Hadiashar, A.; Liu, W. Sensor-management for multitarget filters via minimization of posterior dispersion. *IEEE Trans. Aerosp. Electron. Syst.* **2017**, *53*, 2877–2884. [CrossRef]

12. Beard, M.; Vo, B.T.; Vo, B.N.; Arulampalam, S. Void probabilities and Cauchy–Schwarz divergence for generalized labeled multi-Bernoulli models. *IEEE Trans. Signal Process.* **2017**, *65*, 5047–5061. [CrossRef]

13. Fantacci, C.; Vo, B.N.; Vo, B.T.; Battistelli, G.; Chisci, L. Robust fusion for multisensor multiobject tracking. *IEEE Signal Process. Lett.* **2018**, *25*, 640–644. [CrossRef]

14. Li, S.; Yi, W.; Hoseinnezhad, R.; Battistelli, G.; Wang, B.; Kong, L. Robust distributed fusion with labeled random finite sets. *IEEE Trans. Signal Process.* **2017**, *66*, 278–293. [CrossRef]

15. Yi, W.; Li, S.; Wang, B. Computationally Efficient Distributed Multi-sensor Fusion with Multi-Bernoulli Filter. *IEEE Trans. Signal Process.* **2019**, *68*, 241–256.

16. Papi, F.; Vo, B.N.; Vo, B.T.; Fantacci, C.; Beard, M. Generalized labeled multi-Bernoulli approximation of multi-object densities. *IEEE Trans. Signal Process.* **2015**, *63*, 5487–5497. [CrossRef]

17. Papi, F.; Kim, D.Y. A particle multi-target tracker for superpositional measurements using labeled random finite sets. *IEEE Trans. Signal Process.* **2015**, *63*, 4348–4358. [CrossRef]

18. Beard, M.; Reuter, S.; Granström, K.; Vo, B.T.; Vo, B.N.; Scheel, A. Multiple extended target tracking with labeled random finite sets. *IEEE Trans. Signal Process.* **2015**, *64*, 1638–1653. [CrossRef]

19. Zhu, S.; Liu, W.; Cui, H. Multiple Resolvable GroupsTracking Using the GLMB Filter. *Acta Autom. Sin.* **2017**, *43*, 2178–2189.

20. Beard, M.; Vo, B.T.; Vo, B.N. Bayesian multi-target tracking with merged measurements using labelled random finite sets. *IEEE Trans. Signal Process.* **2015**, *63*, 1433–1447. [CrossRef]

21. Mahler, R. *An introduction to Multisource-Multitarget Statistics And Applications*; Lockheed Martin: Bethesda, MD, USA, 2000.

22. Mahler, R.P. Multitarget Bayes filtering via first-order multitarget moments. *IEEE Trans. Aerosp. Electron. Syst.* **2003**, *39*, 1152–1178. [CrossRef]

23. Goodman, I.R.; Mahler, R.P.; Nguyen, H.T. *Mathematics of Data Fusion*; Vol. 37, Springer Science & Business Media: Dordrecht, The Netherlands; Boston, MA, USA; London, UK, 1997.

24. Mahler, R.; Hall, D.; Llinas, J. Random Set Theory for Target Tracking and Identification. In *Data Fusion Handbook*; CRC Press: Boca Raton, FL, USA, 2001.

25. Panta K, C.D.E. Data association and track management for the gaussian mixture probability hypothesis density filter. *IEEE Trans. Aerosp. Electron. Syst.* **2009**, *45*, 1003–1016. [CrossRef]

26. Mahler, R. PHD filters of higher order in target number. *IEEE Trans. Aerosp. Electron. Syst.* **2007**, *43*, 1523–1543. [CrossRef]

27. Kumaradevan Punithakumar, T.K. A multiple-model probability hypothesis density filter for tracking maneuvering targets. *Signal Data Process. Small Targets* **2004**, *44*, 3553–3567. [CrossRef]

28. Mahler, R.P. *Statistical Multisource-Multitarget Information Fusion*; Artech House: Norwood, MA, USA, 2007.

29. Vo, B.T.; Vo, B.N.; Cantoni, A. The Cardinality Balanced Multi-Target Multi-Bernoulli Filter and Its Implementations. *IEEE Trans. Signal Process.* **2009**, *57*, 409–423. [CrossRef]

30. Vo, B.T.; Vo, B.N. Labeled random finite sets and multi-object conjugate priors. *IEEE Trans. Signal Process.* **2013**, *61*, 3460–3475. [CrossRef]

31. Vo, B.N.; Vo, B.T.; Phung, D. Labeled random finite sets and the Bayes multi-target tracking filter. *IEEE Trans. Signal Process.* **2014**, *62*, 6554–6567. [CrossRef]

32. Vo, B.N.; Vo, B.T.; Hoang, H.G. An efficient implementation of the generalized labeled multi-Bernoulli filter. *IEEE Trans. Signal Process.* **2016**, *65*, 1975–1987. [CrossRef]

33. Liu, W.; Chi, Y. Resolvable Group State Estimation with Maneuver Based on Labeled RFS and Graph Theory. *Sensors* **2019**, *19*, 1307. [CrossRef]

34. Liu, W.; Zhong, C.; Wen, L. Multi-measurement target tracking by using random sampling approach. *Acta Automatica Sinica* **2013**, *39*, 164–174. [CrossRef]

35. Zhu, S.; Liu, W.; Weng, C.; Cui, H. Multiple group targets tracking using the generalized labeled multi-Bernoulli filter. In Proceedings of the 2016 35th Chinese Control Conference (CCC), Chengdu, China, 27–29 July 2016.

36. Liu, W.; Zhu, S.; Wen, C.; Yu, Y. Structure modeling and estimation of multiple resolvable group targets via graph theory and multi-Bernoulli filter. *Automatica* **2018**, *89*, 274–289. [CrossRef]
37. Chen, L.; Liu, W. Finite mixture modeling and tracking algorithm based on GLMB filtering and Gibbs sampling. *Acta Autom. Sin.* **2019**. [CrossRef]

Article

Sea–Sky Line and Its Nearby Ships Detection Based on the Motion Attitude of Visible Light Sensors

Xiongfei Shan [1,2], Depeng Zhao [1,2], Mingyang Pan [1,2,*], Deqiang Wang [1,2] and Lining Zhao [1,2]

[1] Navigation College, Dalian Maritime University, Dalian 116026, China; shanxiongfei@dlmu.edu.cn (X.S.); dpzhao@dlmu.edu.cn (D.Z.); dqwang@dlmu.edu.cn (D.W.); zhaolining@dlmu.edu.cn (L.Z.)

[2] Vessel Navigation System National Engineering Research Center, Dalian Maritime University, Dalian 116026, China

* Correspondence: panmingyang@dlmu.edu.cn; Tel.: +18-04-1151002

Received: 10 August 2019; Accepted: 14 September 2019; Published: 16 September 2019

Abstract: In the maritime scene, visible light sensors installed on ships have difficulty accurately detecting the sea–sky line (SSL) and its nearby ships due to complex environments and six-degrees-of-freedom movement. Aimed at this problem, this paper combines the camera and inertial sensor data, and proposes a novel maritime target detection algorithm based on camera motion attitude. The algorithm mainly includes three steps, namely, SSL estimation, SSL detection, and target saliency detection. Firstly, we constructed the camera motion attitude model by analyzing the camera's six-degrees-of-freedom motion at sea, estimated the candidate region (CR) of the SSL, then applied the improved edge detection algorithm and the straight-line fitting algorithm to extract the optimal SSL in the CR. Finally, in the region of ship detection (ROSD), an improved visual saliency detection algorithm was applied to extract the target ships. In the experiment, we constructed SSL and its nearby ship detection dataset that matches the camera's motion attitude data by real ship shooting, and verified the effectiveness of each model in the algorithm through comparative experiments. Experimental results show that compared with the other maritime target detection algorithm, the proposed algorithm achieves a higher detection accuracy in the detection of the SSL and its nearby ships, and provides reliable technical support for the visual development of unmanned ships.

Keywords: SSL; six-degrees-of-freedom motion; motion attitude model; edge detection; straight-line fitting; visual saliency

1. Introduction

In recent years, with the continuous development of artificial intelligence (AI), big data, and communication technology, unmanned driving technology has made breakthrough achievements. Unmanned aerial vehicles (UAVs) have gradually entered the civil field from the military field, and unmanned ground vehicles (UGVs) are continually testing on public roads around the world. The research on unmanned ships is also developing rapidly. Major research institutions at home and abroad are investing a large amount of manpower, material resources, and financial resources to carry out theoretical research, technology research, and development of large-tonnage unmanned merchant ships. The key technologies of unmanned ships mainly include situational awareness, intelligent decision-making, motion control, maritime communication, and shore-based remote control, etc., and situational awareness is the premise of all other technologies. The advanced sensors are used to obtain the situation information around unmanned ships, provide basic data support for complex tasks such as intelligent decision-making and motion control, and ensure the autonomous operation safety of unmanned ships [1].

Currently, ships perceive the maritime environment mainly through two kinds of sensors, namely, radio detection and ranging (RADAR) and automatic identification system (AIS). They transmit the target information to the electronic chart display and information system (ECDIS), which realizes a certain degree of intelligent analysis and decision. However, the maritime navigation environment is complex and variable. RADAR and AIS cannot directly reflect the spatial information of detection targets. The situational awareness cannot be established quickly, and mariners need to confirm the situation. At the same time, RADAR detection is sensitive to meteorological conditions and the shape, size, and material of the target. AIS cannot effectively detect small targets that are not equipped with it or are not turned on. Visible light sensors are intuitive, reliable, informative, and cost-effective [2]. With the continuous development of computer vision technology, visible light cameras as important situational awareness sensors are gradually being applied to unmanned ships, providing a reliable source of information for intelligent decision-making.

The main targets for maritime detection using cameras include ships, rigs, navigation aids, and icebergs. When maritime targets appear in the field of view of the camera, they must appear in the vicinity of the sea–sky line (SSL). As the distance between the camera and the target approaches, the target gradually enters the sea area. It can be seen that extracting the SSL and performing maritime target detection in its vicinity can greatly reduce the target detection range and reduce the complexity and calculation amount of the algorithm. However, a target near the SSL has a very small area in the image, usually only a few tens or hundreds of pixels, which is easily overwhelmed by the complex sea–sky background, resulting in target missed detection or false detection [3]. Therefore, this paper proposes an algorithm based on the motion attitude model of a visible light camera for the SSL and its nearby ships.

2. Related Work

In general, maritime target detection technology mainly includes three steps, namely, SSL detection, background removal, and foreground segmentation. Based on the research status at home and abroad in recent years, this paper briefly reviews the three algorithms and proposes the main technical framework.

2.1. SSL Detection

The SSL is an important feature of maritime images, and there are many related researches, which are mainly divided into two categories. The first category is based on the combination of edge detection and straight-line fitting. The image is processed by the edge detection operator, and then the high gradient edge pixels are straight-line fitted or projected. Liu [4] proposed an SSL detection algorithm based on inertial measurement and Hough transform fusion. The inertial data of the shipboard camera are used to estimate the position of the SSL in successive frames, then Canny operator and Hough transform are used to realize SSL extraction in the detection region. Wang Bo et al. [5] proposed an SSL detection algorithm based on gradient saliency and region growth. The gradient saliency calculation effectively improves the characteristics of the SSL and suppresses the influence of complex sea conditions such as clouds and sea clutter. Kim et al. [6] proposed an algorithm for estimating the position of the SSL by camera pose and fitting it using random sampling consistency (RANSAC). Fefilatyev et al. [7] used the combination of Gaussian distribution and Hough transform to select the optimal SSL from five candidate SSLs. Santhalia et al. [8] proposed a Sobel operator edge detection algorithm based on eight directions, which effectively eliminates edge noise and has small computational complexity and strong stability. The second category is based on the method of image segmentation, which extracts the upper part of the SSL by threshold processing or background modeling. Dai et al. [9] proposed an edge detection algorithm based on local Otsu segmentation, which solves the problem of poor global threshold segmentation. Zhang et al. [10] proposed an SSL extraction algorithm based on Discrete Cosine Transform (DCT) coefficients. The image is segmented into 8×8 non-overlapping blocks, and the DCT coefficients in the block are calculated to segment the sky and sea areas. Zeng et al. [11] extracted the contour edges using the improved Canny operator of the surrounding texture suppression, and then voted the Hough transform to finely detect the horizontal or oblique SSL. Nasim et al. [12] proposed a K-means algorithm to segment the sea scene into clusters, and extract the SSL by analyzing the image segments. The above algorithms have achieved good detection results in their respective experiments, but the first category of algorithm is not able to balance SSL edge extraction and wave edge suppression according to the gradient extraction edge process, and the second category of algorithm is not able to get the SSL limited by the image segmentation accuracy.

2.2. Background Removal

In the maritime scene, we usually segment the sea–sky background by simulating the color, texture, saturation, and other features, and subtract it from the original image. Kim et al. [13] used improved mean difference filtering to improve the target signal-to-noise ratio while processing infrared images, and averaged the sea–sky background to remove sea clutter interference. However, this method only worked well for structural clutter similar to the SSL, and had a poor effect on the sea surface interference with strong light reflection. Zeng et al. [14] used the surrounding texture filter instead of mean-shift filter to improve the mean-shift image segmentation algorithm, and controlled the filter parameters to perform fast region clustering to remove the sea–sky background. However, the texture filter parameters and clustering parameters needed to be manually set, which needed a certain prior knowledge. In addition, a technique based on the visual saliency model is gradually being applied to maritime target detection. It simulates human visual features through intelligent algorithms, suppresses the sea–sky background, and extracts visual salient regions in the image—that is, regions of human interest. Fang et al. [15] applied the theory of color space and wavelet transform to extract the low frequency, high frequency, hue, saturation, and brightness characteristics of the task water image. The visual attention operator was used to fuse various features, effectively overcoming the background disturbances such as waves, wakes, and onshore buildings. Lou et al. [16] solved the small target detection problem in color images from two aspects of stability and saliency. By multiplying the stability and saliency maps by pixels, the noise interference in the background was eliminated. Agrafiotis et al. [17] designed a maritime tracking system by combining a visual saliency model with a Gaussian Mixed model (GMM) and used an adaptive online neural network tracker to further refine the tracking results. Liu et al. [18] achieved further enhancement of the visual saliency model through a two-scale detection scheme. On a larger scale, the sea surface was removed by the mean-shift filter. On a smaller scale, the target was coarsely extracted by extracting the edge of the significant region, and then the fine processing of the chroma component was used to select the output target. The above algorithm achieves background removal by reducing the background noise of the sea–sky background and enhancing the salient features of the region of interest, but when there is strong cloud, wave, or ship wake disturbance in the sea–sky background, the saliency is the same or higher than the target, which causes the visual saliency algorithm to produce large errors. In addition to the above documents, Ebadi et al. [19] proposed a modified approximated robust PCA algorithm that can handle moving cameras and takes advantage of the block sparse structure of the pixels corresponding to the moving objects.

2.3. Foreground Segmentation

After the image background is removed, we can apply morphological processing to obtain the maritime target. Westall et al. [20] applied improved morphological processing of close-minus-open (CMO) techniques to enhance target detection. Fefilatyev [21] used the Otsu algorithm to obtain global thresholds in the region above the SSL, and used the global threshold to segment the target vessel. Although features such as edges and contours are widely used in target ship detection, it is still difficult to achieve ideal results in complex sea–sky backgrounds with the above algorithms. Besides, Kumar et al. [22] and Selvi et al. [23] made full use of the target ship's color, texture, shape, and other information, and used the support vector machine to classify the target. Frost et al. [24] also applied the prior knowledge of ship shape to the level set segmentation algorithm to improve ship detection results. Loomans et al. [25] integrated a multi-scale Histogram of Oriented Gradient (HOG) detector and a hierarchical Kanade-Lucas-Tomasi (KLT) feature point tracker to track ships in the port, and achieved better detection and tracking effects. The above algorithm is not based on the background subtraction algorithm, but is based on the manually set ship characteristics for target detection. With the continuous development of deep learning technology, the feature extraction algorithm, based on convolutional neural network, is gradually dominating image classification, detection, segmentation, etc., and is gradually being applied to the field of maritime target detection. Ren et al. [26] proposed an improved Faster R-CNN system to detect small target ships in remote sensing images. The statistical algorithms were used to screen the appropriate anchors, and the detection techniques were greatly improved by using jump links and texture information. Yang et al. [27] designed a rotational density pyramid network model to extract the ship's direction while accurately detecting the target ship. Zhang et al. [28] proposed a scheme combining saliency detection and convolutional neural networks to accurately detect ships in remote sensing images of different poses, scales, and shapes. In addition to the above documents, Biondi [29] presented a complete procedure for the automatic estimation of maritime target motion parameters by evaluating the generated Kelvin waves detected in synthetic aperture radar (SAR) images. Graziano et al. [30] proposed a novel technique using X-band Synthetic Aperture Radar images provided by COSMO/SkyMed and TerraSAR-X for ship wake detection. Biondi et al. [31] proposed a new approach where the micro-motion estimation of ships, occupying thousands of pixels, was measured, processing the information given by sub-pixel tracking generated during the co-registration process of several re-synthesized time-domain and overlapped sub-apertures.

In summary, the above algorithms have achieved good application results in their respective research fields, but it is still difficult to achieve high detection accuracy for the SSL and its nearby ships in the complex sea–sky background. For the above problems, we propose the technical framework of this paper, which mainly includes two technologies, as shown in Figure 1. First, SSL detection. After the camera and the inertial sensor acquire the data synchronously, we pass the inertial data to the camera motion attitude model to obtain the image candidate region (CR) position, then cut the CR from the original image, and only perform edge detection and Hough transform in the CR to extract the optimal SSL. Finally, the CR with the optimal SSL is stitched back to the original image. Second, according to the optimal SSL position, we cut the region of ship detection (ROSD) of the image, and then only perform saliency detection and foreground segmentation on the ROSD, and finally stitch the detection result back to the original image.

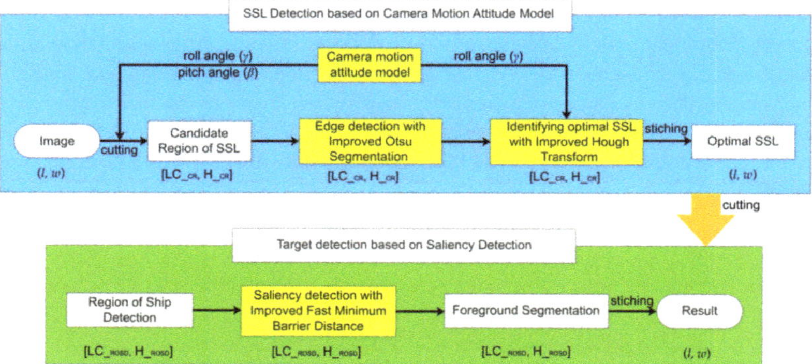

Figure 1. Technical framework of this paper. Roll angle (γ) and pitch angle (β) represent inertial data, (l, w) represents the size of the original image. [LC_CR, H_CR] represents the candidate (CR) of the image, where LC_CR represents the upper left corner of the CR, and H_CR represents the height of the CR. [LC_ROSD, H_ROSD] represents the region of ship detection (ROSD) of the image, where LC_ROSD represents the upper left corner of the ROSD, and H_ROSD represents the height of the ROSD.

In the remainder of this paper, we describe the camera motion attitude model in Section 3, the SSL detection model in Section 4, and the visual saliency detection model in Section 5. We introduce the dataset used in the experiment and compare experiments with other algorithms in Section 6. Finally, in Section 7, we summarize and draw conclusions.

3. Camera Six-Degrees-of-Freedom Motion Attitude Modeling

In navigation, the ship is sailing in a large circle at sea; the tester with an eye height of h sees that the farthest sea and the sky intersect into a circle, which is called the tester's visible horizon, that is, the SSL. In ship vision, we use cameras instead of human eyes for sea target detection and identification. Assuming that the installation position of the camera is h from the sea level, the geometric relationship can be obtained considering the curvature of the earth and the difference of atmosphere refraction, as shown in Figure 2. The circle MN represents the SSL and the blue triangle represents the camera. Before using it, we finished camera calibration and distortion correction [32]. Therefore, in this analysis, we suppose the optical axis of the camera is parallel to the horizontal plane, which is called the initial state of the camera motion. The point O is the camera center, the point K is the projection of the point O at the sea level, r represents the radius of the earth, δ represents the angle of the ball, ε represents the difference of atmosphere refraction, the difference in the navigation is $(1/13)\,\delta$, and the straight line OM represents the actual distance from camera to the SSL instead of $\hat{KM}$, which is expressed by D_e. In the triangle ΔOKM, since both $(\delta/2)$ and $(\delta/2 - \varepsilon)$ are small angles, we can approximate $\cos(\delta/2) \approx 1$ and $\sin(\delta/2 - \varepsilon) \approx \delta/2 - \varepsilon$, and D_e can be obtained by Equation (1). According to the 1 nautical mile representing 1852 meters in navigation, it can be inferred that the r is 6366707 m. The position angle θ of the SSL in the camera can be obtained by Equation (2).

$$D_e = \frac{h\cos(\delta/2)}{\sin(\delta/2 - \varepsilon)} \approx \sqrt{\frac{26rh}{11}} = 3871\sqrt{h} \tag{1}$$

$$\theta = \pi/2 - \varepsilon - \angle KOM = \frac{11D_e}{13r} = 0.0295\,\sqrt{h} \tag{2}$$

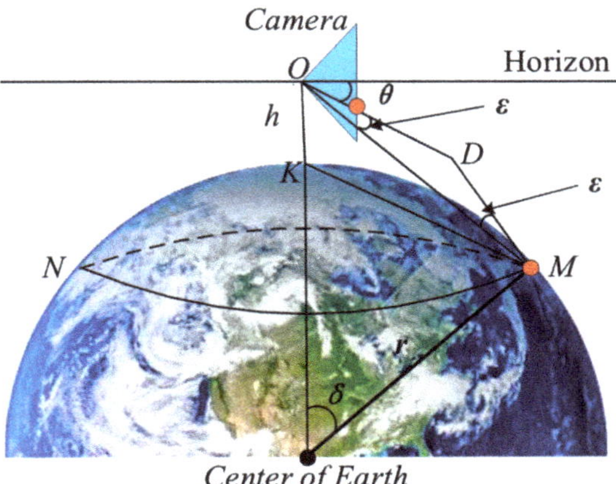

Figure 2. Geometric relationship between the big circle and the camera position.

In order to simplify the projection relationship of the camera, we assume the sea level as a plane, while ignoring the relative motion of the camera and the ship, so that the camera coordinate system coincides with the ship's motion coordinate system. Next, we model the camera's six-degrees-of-freedom motion and the SSL position according to the coordinate system projection method [33].

3.1. Influence of Camera Swaying, Surging, and Yawing Motions on the Position of the SSL

Under the condition of maintaining the initial state, the height h of the camera remains unchanged when the camera only performs the swaying, surging, and yawing motions. It can be known from Equation (1) that θ is only related to h, so the camera swaying, surging, and yawing motions have no effect on the position of the SSL on the imaging plane.

3.2. Influence of Camera Heaving and Pitching Motions on the Position of the SSL

Under the condition of maintaining the initial state, we assume the sea level as a plane according to Figure 2, and obtain the geometric relationship as shown in Figure 3a. The triangle represents the camera. In the imaging plane of the camera, the line *js* represents the sky area, the line *sg* represents the sea area, the point *s* represents the projection point of the SSL, and the point *i* represents the center point, which is also taken as the origin (0, 0) of the image coordinate system. Assuming that the pitch angle of the camera is represented by β, the camera's vertical viewing angle is represented by 2α, the longitudinal width of the imaging plane of the camera is w, and the position of the SSL in the image is represented by z_s, and z_s can be obtained by:

$$z_s = w\,(-\theta)/2\alpha \tag{3}$$

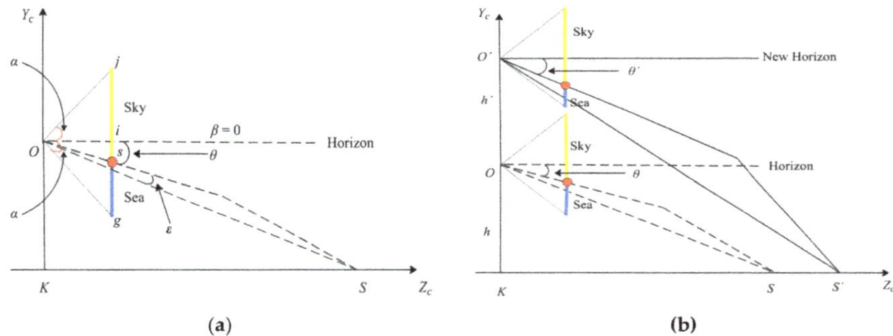

Figure 3. Geometric relationship between the sea–sky line (SSL) position and the camera position. (**a**) Geometric relationship under the initial state. (**b**) Geometric relationship under the camera heaving motion.

3.2.1. Influence of Camera Heaving Motion

Under the condition of maintaining the initial state, when the camera only performs the heaving motion, as shown in Figure 3b, it is assumed that the heaving height is h', and the point O' represents the new position of the camera center. According to Equation (1), the position angle θ' and the position $z_{h'}$ can be obtained by:

$$\theta' = 0.0295\,\sqrt{h + h'} \tag{4}$$

$$z_{h'} = w\,(-\theta')/2\alpha \tag{5}$$

3.2.2. Influence of Camera Pitching Motion

Under the condition of maintaining the initial state, when the camera only performs the pitching motion, as shown in Figure 4, it is assumed that the pitch angle β clockwise rotation is positive and the counterclockwise rotation is negative. Under β clockwise rotation, when $0 < \beta < \theta$, the SSL is located at the lower part of the imaging plane center line and gradually approaches it as β increases. When $\beta = \theta$, the SSL is located at the center line of the imaging plane. When $\theta < \beta < \theta + \alpha$, the SSL is located on the center line of the imaging plane. As the β increases, it gradually moves away from the center line and close to the top of the image. When $\beta > \theta + \alpha$, the SSL is not in the imaging plane, and only the sea area can be seen in the image. Under β counterclockwise rotation, when $\theta - \alpha < \beta < 0$, the SSL is located at the lower part of the center line of the imaging plane, and as the β increases, it gradually moves away from the center line and close to the bottom of the image. When $\beta < \theta - \alpha$, the SSL is not in the imaging plane, and only the sky area can be seen in the image. According to the above analysis, the position of the SSL after the pitching motion can be obtained by:

$$z_\beta = \begin{cases} w\,(\beta - \theta)/2\alpha & if\ \theta - \alpha < \beta < \theta + \alpha \\ invalid & else \end{cases} \tag{6}$$

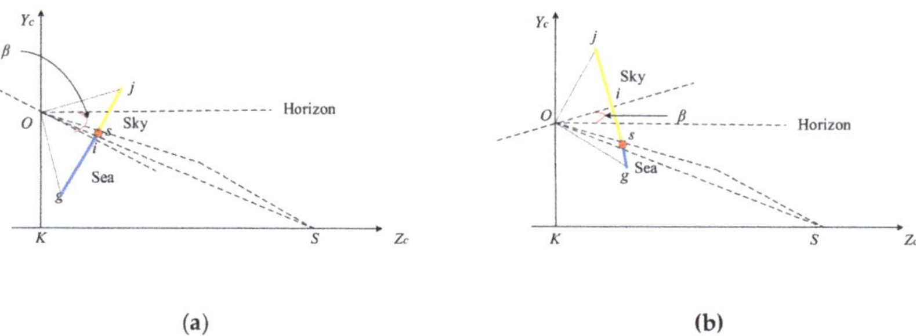

(a)   (b)

Figure 4. Geometric relationship between the SSL position and camera pitching motion. (**a**) Clockwise rotation. (**b**) Counterclockwise rotation.

3.3. Influence of Camera Rolling Motion on the Position of the SSL

Under the condition of maintaining the initial state, when the camera only performs the rolling motion, as shown in Figure 5, it is assumed that the rolling angle γ is clockwise rotated (it is the same as γ counterclockwise rotation), $x'z'$ is a new image coordinate system, and the SSL intersects the z' axis at s'. So, the SSL can be expressed by:

$$y = x\,\tan\gamma + z_s/\cos\gamma \tag{7}$$

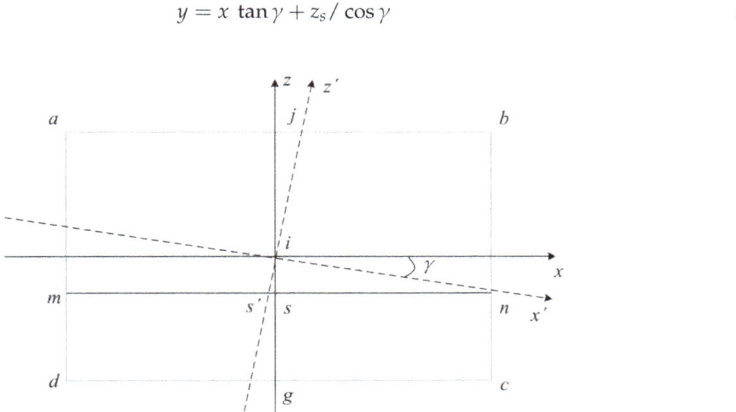

Figure 5. Geometric relationship between the SSL position and camera rolling motion.

Comprehensive analysis of the relationship between the camera six-degrees-of-freedom motion and the SSL shows that when the camera performs the swaying, surging, and yawing motions, the SSL does not change in the image coordinate system. However, when the camera performs the heaving and pitching motions, the SSL performs a translational motion up and down in the image coordinate system, and when the ship performs the rolling motion, the SSL performs a rotational motion in the image coordinate system. Equations (1)–(7) can be used to obtain the estimation equation of the SSL in the image coordinate system, as shown in Equation (8), where the range of β is $\theta - \alpha < \beta < \theta + \alpha$. It can be seen from Equation (8) that the height change h' generated by the camera's heaving motion has less influence on the position of the SSL in the image, and it is also much smaller than the installation height of the camera; so, Equation (8) can be simplified to obtain the final SSL estimation equation, as shown in Equation (9).

$$y = x\,\tan\gamma + w/2\alpha\left(-0.0295\,\sqrt{h + \Delta h}\,(1/\cos\gamma) + \left(\beta - 0.0295\,\sqrt{h}\right)\right) \tag{8}$$

$$y = x\,\tan\gamma + w/2\alpha\left(\beta - 0.0295\,\sqrt{h}(1/\cos\gamma + 1)\right) \tag{9}$$

4. Edge Detection and Hough Transform Algorithm for the Detection of the SSL

4.1. Estimating the CR of the SSL

In order to estimate the CR of the SSL, we need to change from the camera coordinate system to the image coordinate system; that is, the coordinate origin moves from the center point to the upper left corner. Then, we begin to explore the relationship between the pixel points on the image and the actual distance at sea. First, we find the position of the SSL on the image in the current coordinate system, as shown in Equation (10), where z_i represents the pixel points on the image. Then, through Equations (1) and (10), we can obtain the relationship between z_i and the actual distance at sea, as shown in Equation (11). Assuming $h = 20$ m, the camera parameters are $w = 964$ pixels and $\alpha = 3.7°$, and the relationship between D and z_i can be obtained as shown in Figure 6, where horizontal and vertical coordinates represent D and z_i, respectively. Since we only want to show the relationship between the SSL and the sea area, the value of the ordinate is from the center of the image to the bottom, so the range is [482, 964]. From Figure 6, it can be seen that the closer to the SSL, the larger the actual distance represented by each pixel. A distance of 2.55 nm or beyond from the camera can be represented by 30 pixels on the image.

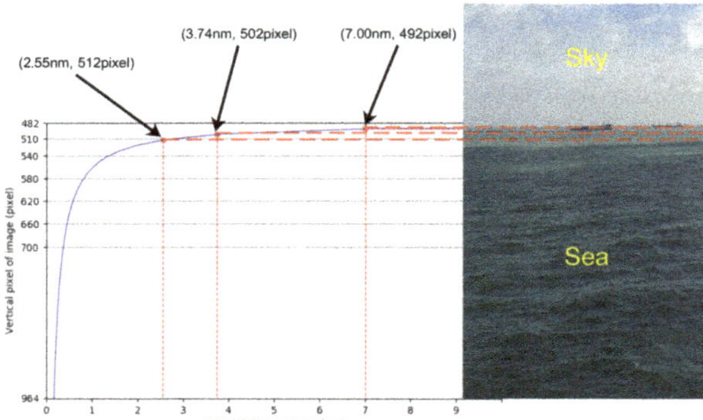

Figure 6. Relationship between the pixel points on the image and the actual distance at sea.

Considering that the SSL is usually a straight line that runs through the entire image and generally has a certain angle of inclination, we use a rectangle to describe the SSL. The upper left corner and the height of the rectangle are represented by LC and H, respectively. In order to reduce the estimation error, we add a yellow area with a height of 30 pixels to the upper and lower sides of the rectangular area as the CR of the SSL. The parameter value can be obtained by Equation (12), as shown in Figure 7a.

$$z_i = w\,(\alpha + \theta)/2\alpha \tag{10}$$

$$D = \frac{h\cos 2.36(2\alpha z_i/w - \alpha)}{\sin(2\alpha z_i/w - \alpha)} \tag{11}$$

$$CR = [LC - (0,\ 30), H + 60] \tag{12}$$

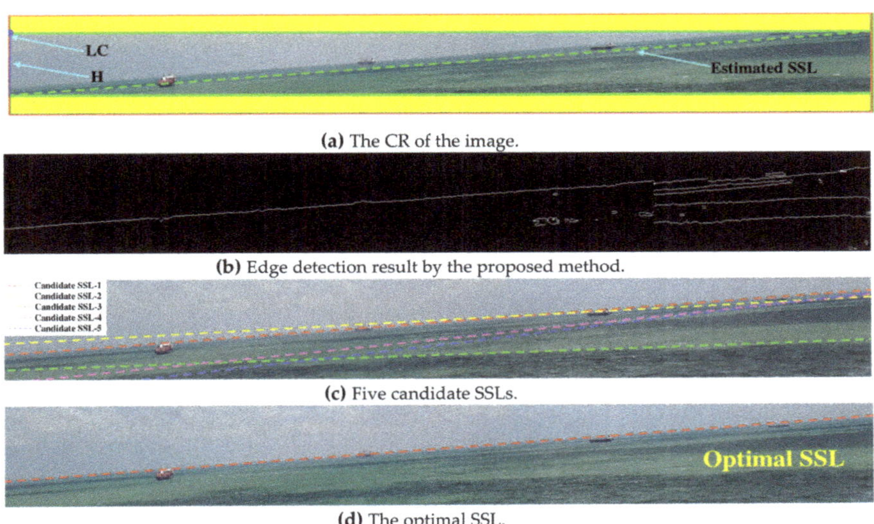

(a) The CR of the image.

(b) Edge detection result by the proposed method.

(c) Five candidate SSLs.

(d) The optimal SSL.

Figure 7. Figures of each stage in the SSL extraction algorithm. In (a), the camera parameters are $h = 20$ m, $\alpha = 3.7°$, $l = 1288$ pixels, $w = 964$ pixels; the inertial sensor parameters are $\beta = 0°$, $\gamma = 5.0°$; and the green dotted line represents the estimated SSL obtained by the camera motion attitude model. In (b), $N = 16$. In (c), the clustering parameter is 5. In (d), $\omega = 0.4$.

4.2. Edge Detection in the CR

After acquiring the CR, it is only necessary to process the image edges in the region, which can effectively reduce the calculation amount of the image processing work. In this paper, a novel edge detection algorithm based on the local Otsu segmentation is designed in the CR. The specific algorithm is shown as follows:

1. Preprocessing. The CR is grayscaled and smoothed with median filtering to filter out noise. Median filtering is often used to remove salt and pepper noise, which has a good smoothing effect on the sea surface reflected by strong light, and can maximize edge information.

2. Obtaining a binary map by the local Otsu algorithm. According to the gray information of the CR, the Otsu algorithm automatically selects the threshold that maximizes the variance between the two types of pixel as the optimal threshold. For the whole sea–sky image, besides the sea and the sky, there are many other types of pixels, such as clouds, waves, and strong light reflections. It is difficult to obtain an accurate global threshold for the whole image by Otsu algorithm, and the image segmentation accuracy is poor. In this paper, the CR after median filtering is processed into N adjacent image blocks. The pixel distribution of the sky and the sea region in most image blocks has significant differences. Then, the Otsu algorithm is applied to each image block to obtain a local binary image, and finally the binarized image blocks are spliced back to the CR.

3. Edge extraction. For the binary image of the CR, we check the position of the pixel mutation in the vertical direction line-by-line, and the position of the pixel mutation is the edge of the binary image, as shown in Equation (13), where I_{edge} $(\cdot)$ represents the edge image, I_{bm} $(\cdot)$ represents the binary image, w_i represents the position along any line of pixels, and $\bigotimes$ represents the morphological XOR operation. The edge extraction effect is shown in Figure 7b.

$$I_{edge}(w_i, l) = I_{bm}(w_i, l) \bigotimes I_{bm}(w_{i+1}, l) \tag{13}$$

4.3. Identifying the Optimal SSL with Improved Hough Transform

The Hough transform is used to display the edge detection result in the accumulator. As shown in Figure 8, the horizontal coordinate represents the polar angle (θ) and the vertical coordinate represents the polar diameter (ρ), and each curve represents a point in the edge image. The brighter the point, the more the number of curves (represented by τ) that pass through this point, indicating that the more points are collinear in the edge image. In this paper, we optimize the SSL extraction algorithm by combining the prediction results calculated by Hough transform according to the SSL length, with the measurement results provided by the inertial sensor according to the SSL angle. The specific algorithm is as follows:

1. Prediction of SSL using Hough transform. In order to avoid the sample dispersion problem caused by the excessive voting range of the accumulator, we use the Kernel-based Hough transform [34] algorithm to smooth the accumulator. First, we calculate five clusters of pixel points with collinear features, and then find the best fitting line and model the uncertainty for each cluster, and last, vote for the main lines using elliptical-Gaussian kernels computed from the lines associated uncertainties. We obtain the following parameters using Figure 7b processed by Hough transform, as shown in Table 1. In order to present five SSLs visually, we use five colors to mark them in the original image, as show in Figure 7c.
2. Measurement results with the inertial sensor. In an ideal state, the roll angle γ obtained by the inertial sensor is the tilt angle of the SSL, and there is a mutual residual between γ and θ. Therefore, we can take θ into the Hough space to find the SSL. However, there is a certain measurement error in the inertial sensor data, so it is necessary to comprehensively consider the prediction result of Hough transform and the measurement result of inertial sensor.
3. Defining cost function. Firstly, we use τ and φ to represent the prediction of length and angle, $(l/\cos\gamma)$ and γ represent the measurement of length and angle, ω and $(1-\omega)$ represent the influence factors of length and angle, respectively. The cost function can be obtained from Equation (14). Then, we need to eliminate the effect of the angle's direction on the cost function. If $\gamma \geq 0$, we need to remove the SSL with a negative angle in the predicted value; the processing method is the same if $\gamma < 0$. Finally, in order to eliminate the dimensional influence between the evaluation metrics, the min–max normalization processing method is used to map the result values between [0, 1], and the conversion function is shown in Equation (15).

$$J_{min} = \omega(\tau - (l/\cos\gamma))^2 + (1-\omega)(\varphi - \gamma)^2 \tag{14}$$

$$x^* = \frac{x - min}{max - min} \tag{15}$$

Assuming $l = 1288$ pixels, $\gamma = -5.0°$, and $\omega = 0.4$, the relevant parameters of the cost function are shown in Table 2. We know SSL-1 is the optimal SSL and display it in the original image as shown in Figure 7d.

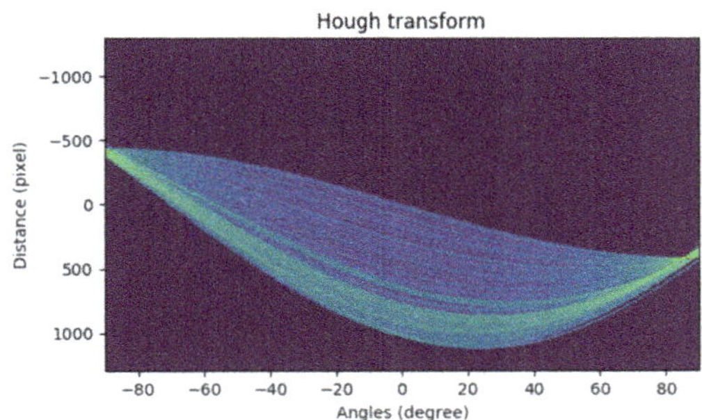

Figure 8. Hough space of the edge detection image.

Table 1. Hough spatial parameters of candidate SSLs ranked top five.

Rank	(θ, ρ)	τ
SSL-1	$(85.90°, 528.66)$	606
SSL-2	$(86.92°, 512.66)$	271
SSL-3	$(87.95°, 557.67)$	183
SSL-4	$(83.85°, 572.68)$	165
SSL-5	$(82.82°, 592.68)$	162

Table 2. The relevant parameters of the cost function.

Rank	τ	$(l/cos\gamma - \tau)^*$	$\varphi\ (°)$	$(\gamma - \varphi)^*$	J_{min}
SSL-1	606	0	4.10	0	0
SSL-2	271	0.75	3.08	0.50	0.38
SSL-3	183	0.95	2.05	1	0.96
SSL-4	165	0.99	6.15	0.12	0.40
SSL-5	162	1	7.18	0.62	0.63

5. Visual Saliency Detection in the ROSD of the SSL

After obtaining the optimal SSL, we add 30 pixels to the rectangle where the optimal SSL is located, cut it, and define it as the ROSD. In the ROSD, the influence of clouds and sea clutter is small. The long-distance ship is mainly near the SSL, and the sea–sky background is relatively uniform and connected with the boundary part of the area. According to the characteristics of the ROSD, we use the fast minimum barrier distance (FMBD) [35] to measure the connectivity of the pixel and the region boundary. The algorithm operates directly on the original pixel, and does not have to acquire the superpixel of the image through the region abstraction [36–39], which improves the detection performance of the saliency map.

The FMBD algorithm mainly consists of three steps, namely, obtaining the minimum barrier distance (MBD) distance map, backgroundness, and post-processing. We used the same approach as FMBD in the first two steps, but we made appropriate improvements in the post-processing step. The specific algorithm is as follows:

Firstly, we convert the color space of the ROSD from RGB to Lab to better simulate the human visual perception. In each channel, we select a pixel-wide row and column as the seed set S in the upper, lower, left, and right boundaries of the ROSD region. Then, the FMBD algorithm is used to calculate the path cost function of each pixel in the ROSD region to the set S, as shown in Equation (16), where i represents any pixel other than the boundary in the image, and $\pi(i)$ represents the path of the pixel to the set S. In this paper, we consider four paths adjacent to each pixel point; $\mathcal{I}(\cdot)$ represents the pixel value of a point, and the cost function $\beta_I(\pi)$ represents the distance between the highest pixel value and the lowest pixel value on a path.

$$\beta_I(\pi) = \max_{i=[0,k]} I(\pi(i)) - \min_{i=[0,k]} I(\pi(i)) \tag{16}$$

We scan the ROSD area three times, which are raster scan, inverse raster scan, and raster scan. In each scan, half of the four neighborhoods of each pixel are used; that is, the upper neighborhood and the left neighborhood pixel. The path minimization operation is shown in Equation (17), where $\mathcal{P}(y)$ represents the path currently assigned to pixel y, $\langle y, x \rangle$ represents the edge from pixel y to pixel x, $\mathcal{P}(y) \cdot \langle y, x \rangle$ represents the path of x, and the direction is from y to x. Assuming $\mathcal{P}_y(x) = \mathcal{P}(y) \cdot \langle y, x \rangle$, you can get Equation (18), where $\mathcal{U}(y)$ and $\mathcal{L}(y)$ are the maximum and minimum values on the path, respectively.

$$\mathcal{D}(x) = \min \left\{ \begin{array}{l} D(x) \\ \beta_I(\mathcal{P}(y) \cdot \langle y, x \rangle) \end{array} \right. \tag{17}$$

$$\beta_I\big(\mathcal{P}_y(x)\big) = \max\{\mathcal{U}(y), I(x)\} - \min\mathcal{L}(y), I(x) \tag{18}$$

In summary, when a pixel appears in the region of the salient target, its pixel value should be close to the maximum pixel value on each path, and the cost function here is relatively large. When a pixel appears in the background area, its pixel value should be close to the minimum pixel value on each path, and the cost function here is relatively small. Thereby, the highlighting area can be realized, the background area can be darkened, and the target saliency detection can be completed.

Secondly, after obtaining the FMBD distance maps accumulated in the three-color spaces, we apply the backgroundness cue of the ROSD region to enhance the brightness of the saliency map. In the ROSD, the boundary of the image is the sea–sky background. According to this feature, first, we select 10% of the area in the upper, lower, left, and right directions of the ROSD as the boundary part, and then calculate the Mahalanobis Distance of the color mean between all the pixels and the four boundary areas. Finally, the maximum value of the boundary information is subtracted from the sum of the boundary information obtained from the four regions to obtain a boundary comparison map. Therefore, we can exclude the case where a region may contain a foreground region, as shown in Equation (19), where $\bar{x}$ and Q^{-1} represent the color mean and covariance of each boundary part, respectively.

$$\left\{ \begin{array}{l} u_k^{ij} = \sqrt{\left(x_k^{ij} - \bar{x}\right)Q^{-1}\left(x_k^{ij} - \bar{x}\right)^T} \\ u^{ij} = \sum\limits_{k=1}^{4} u_k^{ij} - \max\limits_k u_k^{ij} \end{array} \right. \tag{19}$$

Finally, in the post-processing section, the three processing techniques of the original article do not adapt to ship detection near the SSL, so we make appropriate improvements. For the first processing, we replace the previous morphological filtering with morphological reconstruction with opening operation. The specific operation is that we use the structural element b to erode the saliency map (the saliency map is represented by F) n times to obtain the erosion map F', then use b to dilate F'. Next, we take the minimum value of the dilation map and the original map F, and iterate the process until F' no longer changes. The results of our processing can be obtained by Equation (20), where $\oplus$ and $\ominus$ represent the dilation and erosion operations in morphology, respectively. For the second processing, the original processing utilizes the image enhancement technique in the middle of the image, but it is easy to ignore the small targets around, so this paper directly removes this technology. The third processing is consistent with the original article; the sigmoid function is used to increase the contrast between the target and the background region, as shown in Equation (21), where parameter a is used to control the contrast level of the target and the background.

$$\begin{cases} F' = (F \ominus nb) \\ O_R^{(n)}(F) = min(F' \oplus b), F \end{cases} \tag{20}$$

$$f(x) = \frac{1}{1 + e^{-a(x-0.5)}} \tag{21}$$

The saliency feature map obtained by the proposed algorithm has the following characteristics: The target part is highlighted, the background part is darkened, and the contrast is obvious. We select the appropriate threshold to test the saliency map, and use the area threshold to extract the final l target ship, eliminating trivial small area interference. The processing of target detection is shown in Figure 9.

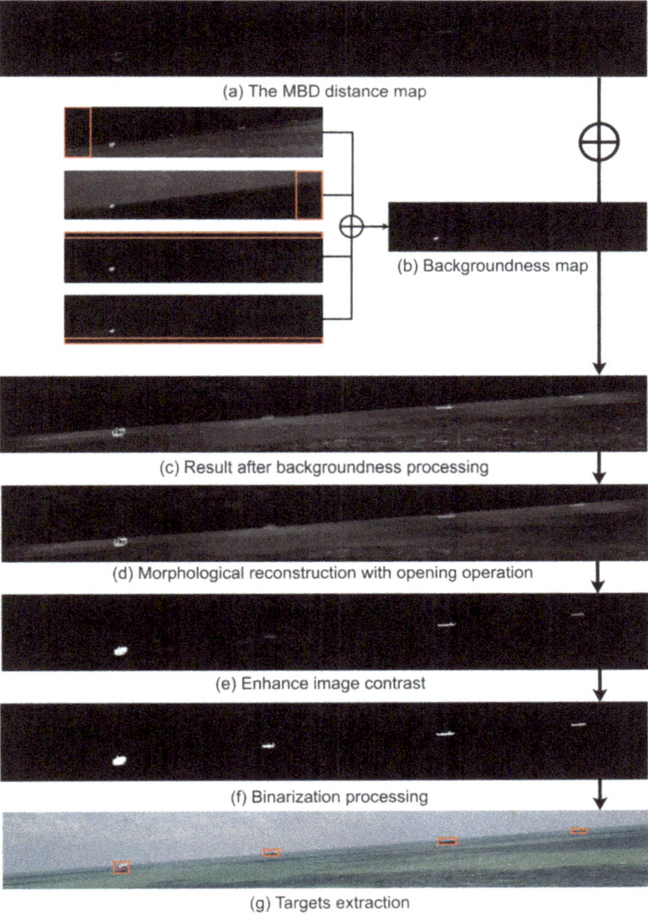

Figure 9. A processing case for targets detection. (**a**) is obtained by fusing the average values of the MBD maps of the three channels L, a, and b. In (**b**), each red box represents 10% of the image area in the four directions of up, down, left, and right, and $\oplus$ represents the average of the four images after adding. (**d,e**) represent post-processing, where $n = 8$ and $a = 10$. (**f,g**) represent foreground segmentation, where the binarization threshold is 5 times the average intensity of (**e**), and the area threshold is 100 pixels.

6. Experimental Results and Discussion

This paper conducts a real ship experiment on the "YUKUN" of Dalian Maritime University's special teaching practice ship, and uses an inertial sensor and a visible light camera for data acquisition. The inertial sensor adopts the MTi-G-700 MEMS inertial measurement system produced by Xsens Company of the Netherlands. The measurement range of roll angle and pitch angle is [–180°, 180°], and the accuracy of measurement is less than 0.1°. The camera uses the Blackfly U3-13S2C/M-CS camera from PointGrey, Canada. The chip size is 4.8 × 3.6 mm, and the number of pixels on the target surface is 1288 × 964. The camera focal length during the experiment was 27.82 mm. All the experiments in this paper were tested on an Intel i5 processor, 8G memory MacBook Pro, and programmed in Python.

6.1. Dataset and Evaluation Indicators

6.1.1. Dataset

For maritime target detection, there is currently no authoritative dataset to verify the validity of the algorithm. A few datasets that have been opened do not include camera-related attitude data. Therefore, the images in our dataset were obtained by the Blackfly U3-13S2C/M-CS camera installed on the "YUKUN" ship. The image size is 1288 × 964 pixels. We used the inertial sensor and visible camera synchronization processing algorithm to obtain the camera motion attitude data. The detailed information of the experiment images is shown in Table 3.

Table 3. Detailed information of the experiment images.

Type of Images	Conditions	Train Set	Test Set	Total
Images with SSL	Sunny	350	150	500
	Glare	350	150	500
	Hazy	350	150	500
	Occlusion	350	150	500
Images with ships	One ship	420	180	600
	Two ships	210	90	300
	Multiple ships	280	120	400
	Multiple ships + Occlusion	140	60	200

6.1.2. Evaluation Metrics

As can be seen from Section 4.1, each SSL can be represented by a rectangle, so we can describe the SSL by Equation (22). The true value can be obtained by manually marking SSLs in the images. In the experiment, to verify the camera's motion attitude model, we can calculate the difference between the estimate values and the actual values of LC and H to obtain the model accuracy. In evaluating the detection performance of the SSL, if the difference of LC is less than 5 pixels and the difference of H is less than 10 pixels, we believe that the SSL is correctly detected.

In evaluating the performance of detection, we used the confusion matrix of classification result to represent the detection results, namely, the true positive (TP), false positive (FP), true negative (TN), and false negative (FN). Precision and recall were obtained by Equation (23). Intersection over Union (IoU) was also used as the evaluation metrics; that is, the intersection of the detection result and the true value was compared to their union. When the IoU was greater than or equal to 0.5, the test result was marked as TP. When IoU was less than 0.5, the test result was marked as FP.

$$L = [LC, H] \tag{22}$$

$$P = \frac{TP}{TP + FP}, \ R = \frac{TP}{TP + FN} \tag{23}$$

6.2. Experimental Results and Discussion on SSL Detection

The SSL extraction algorithm in this paper mainly includes three models, namely, camera motion attitude model, improved edge detection model, and improved Hough transform model. In order to verify the performance of each model separately, the following three experiments were designed. Some parameter settings in the experiment are shown in Table 4.

1   Experiment 1—Verification of the camera motion attitude model

Table 4. Some parameter settings in the test.

Parameter	Description	Value
h(m)	The height of the camera position	20
l(pixel)	The width of the image	1288
w(pixel)	The height of the image	964
N	Number of adjacent image blocks in local Otsu algorithm	16

The camera motion attitude model uses the pitch angle β and roll angle γ provided by the inertial sensor to estimate the CR of the SSL in the image, which can effectively narrow the detection range and is of great significance for subsequent algorithms. In this experiment, we used the difference between LC and H of the SSL candidate region and the real region to describe the estimation accuracy.

Eight experiment results of the model accuracy are shown in Table 5. The total number of detected images is 2000. The analysis results show that the LC estimation accuracy of the camera motion attitude model is 6–13 pixels, and the H estimation accuracy is 7–19 pixels. It can be seen from the experimental results that it is reasonable to estimate the rectangular area of the SSL by using the camera motion attitude model, and then increase the height of 30 pixels above and below the estimate rectangular as the CR of the SSL, which can effectively ensure that the real SSL is in the CR.

Table 5. Location estimated accuracy of camera motion attitude model.

Dataset	Number of Images	Mean Difference of LC Coordinate (Pixels)	Mean Difference of H (Pixels)
Sunny (Train)	350	7.09	8.82
Glare (Train)	350	8.53	13.46
Hazy (Train)	350	6.09	7.14
Occlusion (Train)	350	11.45	18.26
Sunny (Test)	150	6.81	10.55
Glare (Test)	150	9.41	17.32
Hazy (Test)	150	12.75	18.89
Occlusion (Test)	150	10.46	14.21

2   Experiment 2—Verification of the improved edge detection model

This experiment was carried out in sunny, glare, hazy, and occlusion conditions from the train set, and compared the performance with the Canny operator and the deep learning-based holistically-nested edge detection (HED) algorithm [40]. In order to better illustrate the detection performance of various algorithms, one image was selected for description under four conditions, as shown in Figure 10.

First of all, we used the data provided by the inertial sensor to obtain the CR of the SSL through the camera motion attitude model, as shown in Figure 10a–d, then used three algorithms to process the image separately. We can see that the Canny operator had the worst detection performance of the SSL, since the threshold was not adaptive. In sunny conditions, only part of SSL could be detected, as shown in Figure 10(a1). When the glare conditions or the sea–sky background was hazy, the Canny operator failed to detect the edge of SSL, as shown in Figure 10(b1,c1). When there were obstacles such as ships or islands blocking the SSL, the Canny operator detected the edge of the obstacle and this affected the performance, as shown in Figure 10(d1). The HED algorithm achieved better performance under any condition, but it was easy to cause over-detection. In addition to identifying the SSL, sea clutter was detected in Figure 10(c2), and the edge of the obstruction was added to the SSL in Figure 10(d2). The proposed algorithm achieved the best performance, since the binary division was performed in the adjacent small blocks, over-detection was effectively prevented while ensuring the threshold adaptive. Although part of the spot was detected in Figure 10(a3), it did not affect the extraction of the SSL.

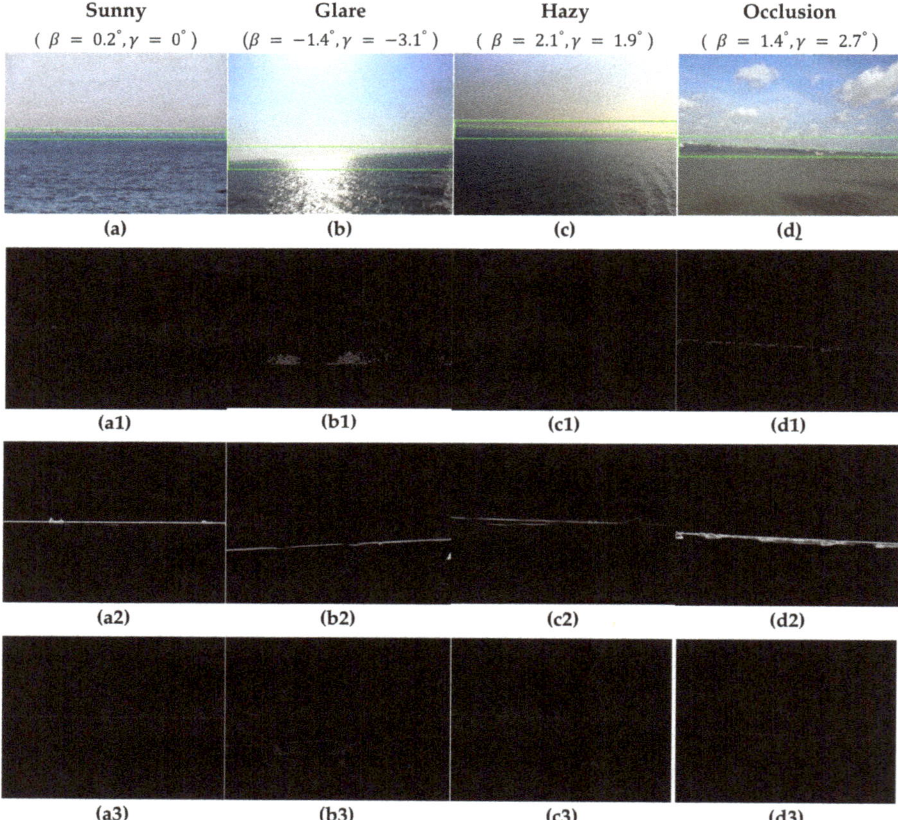

Figure 10. Edge detection results by the four method. (**a–d**) Original images with CR. (**a1–d1**) Edge detection results by Canny (30, 150). (**a2–d2**) Edge detection results by HED. (**a3–d3**) Edge detection results by the proposed method.

3   Experiment 3—Verification of the improved Hough transform model

This experiment was mainly to verify the effect of length and angle on the cost function at the stage of SSL extraction. In the train set, we set the value of ω to [0, 0.1, 0.2, 0.3, 0.4, 0.5, 0.6, 0.7, 0.8, 0.9, 1.0], found the minimum cost function for each value, and drew the corresponding SSL. According to the evaluation metrics of the SSL, the average precision (AP) of the SSL under each ω is shown in Figure 11. It can be seen that when $\omega = 0.4$, the extracted SSL had the highest AP, reaching 99.5%, but when only the length factor of the SSL was considered, the AP was the lowest with just 81.23%. Therefore, it can be concluded that the influence of the angle is greater than the length in the cost function extracted by the SSL.

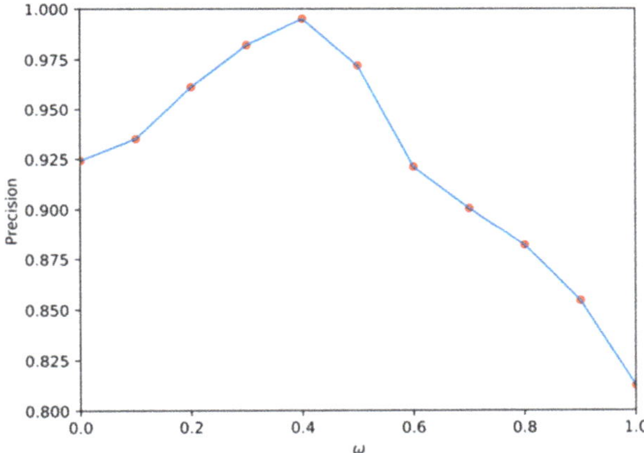

Figure 11. Relationship between the average precision (AP) of SSL detection and ω.

The above three experiments were fully verified for each model of the SSL detection. According to the proposed algorithm, we processed 600 images in the test set, and compared with Fefilatyev's method [7] and Zhang's method [10]. The precision and recall rates are as shown in Table 6. We can see that all of the three methods achieved good performance in SSL detection, but the proposed method was still better than the other two methods, whose average precision (AP) and average recall (AR) in the test set reached 99.67% and 100%, respectively.

Table 6. Precision and recall scores for the three methods.

Dataset (Test)	Number of Images	Fefilatyev's Method (%)		Zhang's Method (%)		Proposed Method (%)	
		P	R	P	R	P	R
Sunny	150	100	100	100	100	100	100
Glare	150	91.10	97.08	97.67	85.71	100	100
Hazy	150	94.37	94.37	99.28	91.95	99.33	100
Occlusion	150	96.62	98.62	98.58	93.92	99.33	100
Total	600	95.56	97.56	98.75	92.91	99.67	100

6.3. Experimental Results of Ship Detection in the Train Set

In this experiment, there were a total of 1050 images in the train set. In order to verify the performance of the proposed method, the other three saliency detection algorithms, SR [41], RBD, and traditional FMBD, were used as the comparison experiments. The parameters of the proposed method are in Table 7.

Table 7. Experiment parameters of saliency detection.

Parameter	Description	Value
n	Number of erode operations in post-processing	8
a	The parameter of function sigmoid in post-processing	10

When evaluating the performance of ship detection, the binarization threshold T and the area threshold S determine the satisfaction from two aspects of the pixel intensity and the number of pixels connected. If the T and S are too large, the target will be submerged in the background. If the T and S are too small, false target interference will occur. When determining the range of T, the target pixel intensity significantly exceeds the average intensity $\overline{T}$, so the value of T represents by times of the average intensity $(\overline{T})$ of the salient map, and a combination of T and S is shown in Table 8.

Table 8. Combination of T and S.

Threshold	Value									
$T\,(\overline{T})$	1	2	3	4	5	6	7	8	9	10
$S\,(\text{pixel}^2)$	20	40	60	80	100	120	140	160	180	200

According to the above threshold combination, we can draw the precision-recall graphs of the four detection methods, as shown in Figure 12. It can be seen that the proposed method is superior to the other three saliency detection methods.

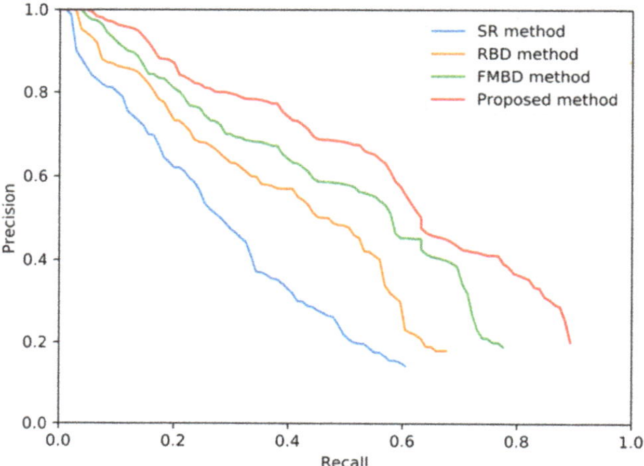

Figure 12. Precision-recall curves of the four object detection methods.

Figure 13 shows the detection performance of several methods on different datasets of the train set. It can be seen that although the residual spectrum (Figure 13(a1–d1)) obtained by the SR method can detect the ship, it does not accurately indicate the position and shape of the target ship, and is liable to cause false detection. The RBD method achieves better detection results when the number of targets in the image is small, as shown in Figure 13(a2,b2), but it is easy to cause missed detection when there are many targets, such as Figure 13(c2,d2). The traditional FMBD method (Figure 13(a3–d3)) can detect the salient targets well, but the contrast between the target and the background is not obvious enough, which is not conducive to subsequent target extraction. The proposed method (Figure 13(a4–d4)) in this paper can clearly distinguish the target and background, accurately detect the shape and position of the target, and the detection performance is the best.

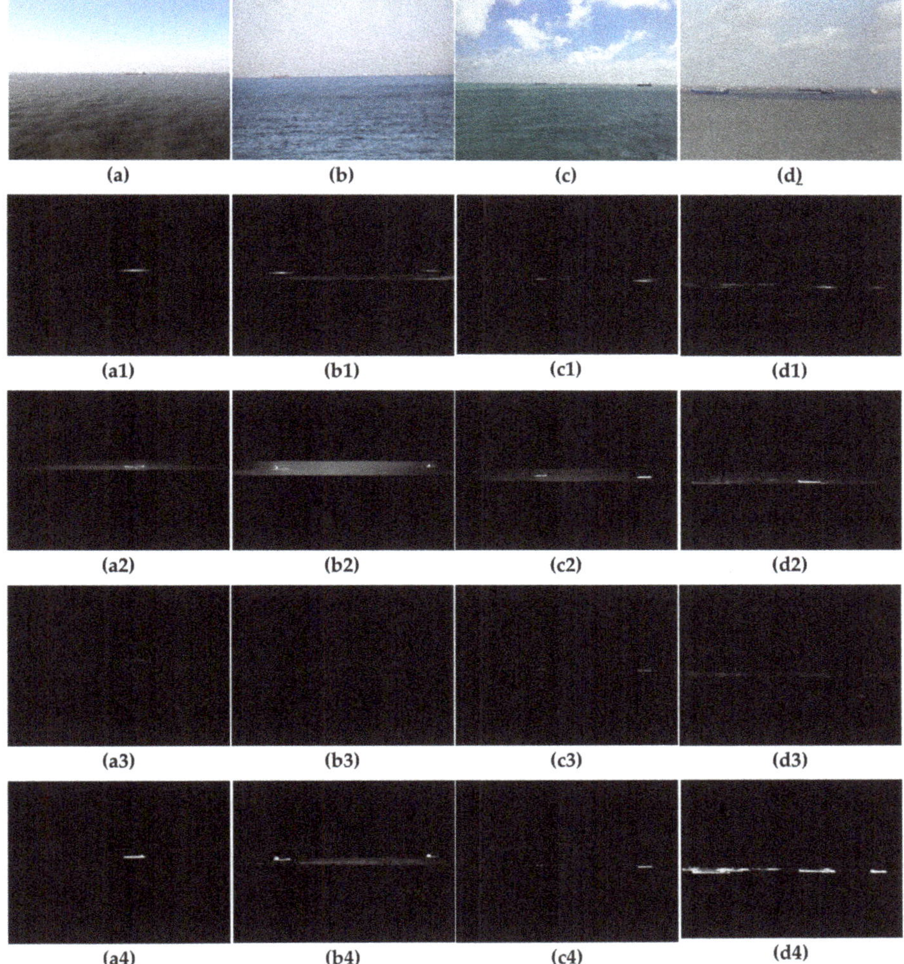

Figure 13. Saliency detection results by the four methods. (**a–d**) Original images. (**a1–d1**) Saliency detection results by SR. (**a2–d2**) Saliency detection results by RBD. (**a3–d3**) Saliency detection results by FMBD. (**a4–d4**) Saliency detection results by the proposed method.

Figure 14 shows segmentation results of the target ships from the salient feature map where the T is 5 times of $\overline{T}$ and S is set to 100 pixels. It can be seen from Figure 14(a1–d1) that the SR method has the worst segmentation result, the detection result has many missed detections and false detections, and the target ship positioning accuracy is also poor. The RBD method is more powerful than the SR method, but with a certain degree of target missed detection, such as Figure 14(c2,d2). The traditional FMBD method has a good segmentation result, but it still has some shortcomings in target positioning accuracy and missed detection, such as Figure 14(a3–c3). The proposed method accurately detects the target in Figure 14(a4–c4), but the method fails to perform accurate target segmentation when the target ship appears to be covered, as shown in Figure 14(d4). This is also a direction we will focus on in the future.

Figure 14. Object segmentation results by the four methods. (**a1–d1**) Object segmentation results by the SR method. (**a2–d2**) Object segmentation results by the RBD method. (**a3–d3**) Object detection results by the FMBD method. (**a4–d4**) Object segmentation results by the proposed method.

6.4. Experimental Results of Ship Detection in the Test Set

In this experiment, we verified the proposed target detection method in the test set and compared it with Fefilatyev's and Zhang's methods. The precision and recall rates are shown in Table 9. Since Fefilatyev's method only detects the ship above the SSL, both AP and AR are relatively low. Zhang's method is relatively good, as the AP and AR reached 59.21% and 73.25%. However, the proposed method achieved the best scores, with an AP and AR of 68.50% and 88.32%, respectively.

Table 9. Precision and recall scores of the three ship detection methods.

Dataset (Test)	Number of Images	Number of Ships	Fefilatyev's Method (%)		Zhang's Method (%)		Proposed Method (%)	
			AP	AR	AP	AR	AP	AR
Total	450	1023	19.15	35.26	59.21	73.25	68.50	88.32

7. Conclusions

This paper proposes a novel maritime target detection algorithm based on the motion attitude of visible light camera. The camera was fixed on the "YUKUN" ship, and the camera's motion attitude data was acquired synchronously by the inertial sensor, so that the CR of the SSL on the image could be estimated. Then, the improved local Otsu algorithm was applied to the edge detection in the CR, and the Hough transform was improved to extract the optimal SSL. Finally, the improved FMBD algorithm was used to detect the target ships in the vicinity of the SSL. The experimental results show that the proposed algorithm has obvious advantages compared with the other maritime target detection algorithms. In the test set, the detection precision of the SSL reached 99.67%, effectively overcoming the complex maritime environment. The ship detection precision and recall rates were 68.50% and 88.32%, respectively, which improved the detection precision while avoiding the ship's missed detection.

The main contribution of this paper is the construction of a camera motion attitude model by analyzing the six-degrees-of-freedom motion of the camera at sea, combined with the maritime target detection algorithm, which narrowed the detection range and improved the detection accuracy. The edge detection algorithm was improved. The local Otsu algorithm was used for edge processing in the CR, which effectively overcame the complex maritime environment. The Hough transform algorithm was improved. The length and angle of the SSL were simultaneously considered as evaluation metrics of the cost function, which effectively improved the accuracy of SSL extraction. The ROSD was detected by the improved the FMBD algorithm. In the post-processing part of the algorithm, the morphological reconstruction with opening operation, was used to replace the previous processing method to smooth the sea–sky background, which effectively improved the target ship's saliency detection effect.

Author Contributions: Conceptualization, X.S. and M.P.; Methodology, X.S.; Software, X.S.; Validation, X.S.; Resources, M.P.; Data curation, L.Z.; Writing—original draft preparation, X.S.; Writing—review and editing, X.S., D.W. and L.Z.; Supervision, D.Z.; Funding acquisition, D.W.

Funding: This research was financially supported by National Natural Science Foundation of China (61772102) and the Fundamental Research Funds for the Central Universities (3132019400).

Conflicts of Interest: The authors declare no conflict of interest.

References

1. Porathe, T.; Prison, J.; Yemao, M. Situation awareness in remote control centers for unmanned ships. In Proceedings of the Human Factors in Ship Design & Operation, London, UK, 26–27 February 2014; pp. 93–101.
2. Prasad, D.K.; Rajan, D.; Rachmawati, L.; Rajabally, E.; Quek, C. Video Processing from Electro-Optical Sensors for Object Detection and Tracking in a Maritime Environment: A Survey. *IEEE Trans. Intell. Transp. Syst.* **2017**, *18*, 1993–2016. [CrossRef]
3. Yuxing, D.; Weining, L. Detection of sea-sky line in complicated background on grey characteristics. *Chin. J. Opt. Appl. Opt.* **2010**, *3*, 253–256.
4. Liu, W. *Research on the Method of Electronic Image Stabilization for Shipborne Mobile Video*; Dalian Maritime University: Dalian, China, 2017; pp. 96–101.
5. Wang, B.; Su, Y.; Wan, L. A Sea-Sky Line Detection Method for Unmanned Surface Vehicles Based on Gradient Saliency. *Sensors* **2016**, *16*, 543. [CrossRef]
6. Kim, S.; Lee, J. Small infrared target detection by region-adaptive clutter rejection for sea-based infrared search and track. *Sensors* **2014**, *14*, 13210–13242. [CrossRef] [PubMed]
7. Fefilatyev, S.; Goldgof, D.; Shreve, M. Detection and tracking of ships in open sea with rapidly moving buoy-mounted camera system. *Ocean Eng.* **2012**, *54*, 1–12. [CrossRef]
8. Santhalia, G.K.; Sharma, N.; Singh, S. A Method to Extract Future Warships in Complex Sea-sky background which May Be Virtually Invisible. In Proceedings of the Third Asia International Conference on Modelling & Simulation, Bali, Indonesia, 25–29 May 2009; pp. 533–539.
9. Yongshou, D.; Bowen, L.; Ligang, L.; Jiucai, J.; Weifeng, S.; Feng, S. Sea-sky-line detection based on local Otsu segmentation and Hough transform. *Opto Electron. Eng.* **2018**, *45*, 180039.
10. Zhang, Y.; Li, Q.Z.; Zang, F.N. Ship detection for visual maritime surveillance from non-stationary platforms. *Ocean Eng.* **2017**, *141*, 53–63. [CrossRef]
11. Zeng, W.J.; Wan, L.; Zhang, T.D.; Xu, Y.R. Fast Detection of Sea Line Based on the Visible Characteristics of MaritneImages. *Acta Optica Sinica.* **2012**, *32*, 01110011–01110018.

12. Boroujeni, N.S.; Etemad, S.; Ali, W.A. Robust horizon detection using segmentation for UAV applications. In Proceedings of the 2012 9th Conference on Computer and Robot Vision, Toronto, ON, Canada, 28–30 May 2012; pp. 28–30.
13. Kim, S. High-Speed Incoming Infrared Target Detection by Fusion of Spatial and Temporal Detectors. *Sensors* **2015**, *15*, 7267–7293. [CrossRef]
14. Zeng, W.J.; Wan, L.; Zhang, T.D.; Xu, Y.R. Fast detection of weak targets in complex sea-sky background. *Opt. Precis. Eng.* **2012**, *20*, 403–412. [CrossRef]
15. Fang, J.; Feng, S.; Feng, Y. Image algorithm of ship detection for surface vehicle. *Trans. Beijing Inst. Technol.* **2017**, *37*, 1235–1240.
16. Lou, J.; Zhu, W.; Wang, H.; Ren, M. Small Target Detection Combining Regional Stability and Saliency in a Color Image. *Multimed. Tools Appl.* **2017**, *76*, 14781–14798. [CrossRef]
17. Agrafiotis, P.; Doulamis, A.; Doulamis, N.; Georgopoulos, A. Multi-sensor target detection and tracking system for sea ground borders surveillance. In Proceedings of the 7th International Conference on PErvasive Technologies Related to Assistive Environments, Rhodes, Greece, 27–30 May 2014.
18. Liu, Z.; Zhou, F.; Bai, X.; Yu, X. Automatic detection of ship target and motion direction in visual images. *Int. J. Electron.* **2013**, *100*, 94–111. [CrossRef]
19. Ebadi, S.E.; Ones, V.G.; Izquierdo, E. Efficient background subtraction with low-rank and sparse matrix decomposition. In Proceedings of the 2015 IEEE International Conference on Image Processing, Quebec City, QC, Canada, 27–30 September 2015.
20. Westall, P.; Ford, J.J.; O'Shea, P.; Hrabar, S. Evaluation of maritime vision techniques for aerial search of humans in maritime environments. In Proceedings of the Digital Image Computing: Techniques and Applications, Canberra, Australia, 1–3 Decemcer 2008; pp. 176–183.
21. Fefilatyev, S. Algorithms for Visual Maritime Surveillance with Rapidly Moving Camera. Ph.D. Thesis, University of South Florida, Tampa, FL, USA, 2012.
22. Kumar, S.S.; Selvi, M.U. Sea objects detection using color and texture classification. *Int. J. Comput. Appl. Eng. Sci.* **2011**, *1*, 59–63.
23. Selvi, M.U.; Kumar, S.S. Sea object detection using shape and hybrid color texture classification. In *Trends in Computer Science, Engineering and Information Technology*; Springer: Berlin/Heidelberg, Germany, 2011; pp. 19–31.
24. Frost, D.; Tapamo, J.R. Detection and tracking of moving objects in a maritime environment using level set with shape priors. *J. Image Video Process.* **2013**, *2013*, 42. [CrossRef]
25. Loomans, M.; de With, P.; Wijnhoven, R. Robust automatic ship tracking in harbors using active cameras. In Proceedings of the 20th IEEE International Conference on Image Processing, Melbourne, Australia, 15–18 September 2013; pp. 4117–4121.
26. Ren, Y.; Zhu, C.; Xiao, S. Small Object Detection in Optical Remote Sensing Images via Modified Faster R-CNN. *Appl. Sci.* **2018**, *8*, 813. [CrossRef]
27. Yang, X.; Sun, H.; Fu, K.; Yang, J.; Sun, X.; Yan, M.; Guo, Z. Automatic Ship Detection in Remote Sensing Images from Google Earth of Complex Scenes Based on Multiscale Rotation Dense Feature Pyramid Networks. *Remote Sens.* **2018**, *10*, 132. [CrossRef]
28. Zhang, R.; Yao, J.; Zhang, K.; Feng, C.; Zhang, J. S-CNN-based ship detection from high-resolution remote sensing images. In Proceedings of the International Archives of the Photogrammetry, Remote Sensing and Spatial Information Sciences, Prague, Czech Republic, 18 July 2016; pp. 423–430.
29. Biondi, F. Low-Rank Plus Sparse Decomposition and Localized Radon Transform for Ship-Wake Detection in Synthetic Aperture Radar Images. *IEEE Geosci. Remote Sens. Lett.* **2018**, *15*, 117–121. [CrossRef]
30. Graziano, M.D.; Grasso, M.; D'Errico, M. Performance Analysis of Ship Wake Detection on Sentinel-1 SAR Images. *Remote Sens.* **2017**, *9*, 1107. [CrossRef]
31. Biondi, F.; Addabbo, P.; Orlando, D.; Clemente, C. Micro-Motion Estimation of Maritime Targets Using Pixel Tracking in Cosmo-Skymed Synthetic Aperture Radar Data—An Operative Assessment. *Remote Sens.* **2019**, *11*, 1637. [CrossRef]
32. Zhang, Z. A flexible new technique for camera calibration. *IEEE Trans. Pattern Anal. Mach. Intell.* **2000**, *22*, 1330–1334. [CrossRef]
33. Guo, L.; Xu, Y.C.; Li, K.Q.; Lian, X.M. Study on real-time distance detection based on monocular vision technique. *J. Image Graph.* **2016**, *211*, 74–81.

34. Fernandes, L.A.F.; Oliveira, M.M. Real-time line detection through an improved Hough transform voting scheme. *Pattern Recognit.* **2008**, *41*, 299–314.
35. Zhang, J.; Sclaroff, S.; Lin, Z.; Shen, X.; Price, B.; Mech, R. Minimum Barrier Salient Object Detection at 80 FPS. In Proceedings of the 2015 IEEE International Conference on Computer Vision (ICCV), Santiago, Chile, 7–13 December 2015; pp. 1404–1412.
36. Wei, Y.; Wen, F.; Zhu, W.; Sun, J. Geodesic saliency using background priors. In *ECCV*; Springer: Berlin/Heidelberg, Germany, 2012; pp. 29–42.
37. Zhu, W.; Liang, S.; Wei, Y.; Sun, J. Saliency optimization from robust background detection. In Proceedings of the IEEE Conference on Computer Vision and Pattern Recognition, Columbus, OH, USA, 23–28 June 2014; pp. 2814–2821.
38. Jiang, B.; Zhang, L.; Lu, H.; Yang, C.; Yang, M.-H. Saliency detection via absorbing markov chain. In Proceedings of the IEEE International Conference on Computer Vision, Sydney, Australia, 1–8 December 2013.
39. Yang, C.; Zhang, L.; Lu, H.; Ruan, X.; Yang, M.-H. Saliency detection via graph-based manifold ranking. In Proceedings of the IEEE Conference on Computer Vision and Pattern Recognition, Portland, OR, USA, 23–28 June 2013; pp. 3166–3173.
40. Xie, S.; Tu, Z. Holistically-Nested Edge Detection. In Proceedings of the 2015 IEEE International Conference on Computer Vision (ICCV), Santiago, Chile, 7–13 December 2015; pp. 3–18.
41. Hou, X.; Zhang, L. Saliency Detection: A Spectral Residual Approach. In Proceedings of the 2007 IEEE Conference on Computer Vision and Pattern Recognition, Minneapolis, MN, USA, 17 June 2007; pp. 1–8.

Article

AMARO—An On-Board Ship Detection and Real-Time Information System

Katharina Willburger [1,*]**, Kurt Schwenk** [1] **and Jörg Brauchle** [2]

[1] German Space Operations Center (GSOC), German Aerospace Center (DLR), Münchener Straße 20, 82234 Weßling, Germany; kurt.schwenk@dlr.de

[2] Institute of Optical Sensor Systems, German Aerospace Center (DLR), Rutherfordstr. 2, 12489 Berlin, Germany; joerg.brauchle@dlr.de

* Correspondence: katharina.willburger@dlr.de; Tel.: +49-8153-28-2273

Received: 30 January 2020; Accepted: 24 February 2020; Published: 29 February 2020

Abstract: The monitoring of worldwide ship traffic is a field of high topicality. Activities like piracy, ocean dumping, and refugee transportation are in the news every day. The detection of ships in remotely sensed data from airplanes, drones, or spacecraft contributes to maritime situational awareness. However, the crucial factor is the up-to-dateness of the extracted information. With ground-based processing, the time between image acquisition and delivery of the extracted product data is in the range of several hours, mainly due to the time consumed by storing and transmission of the large image data. By processing and analyzing them on-board and transmitting the product data directly as ship position, heading, and velocity, the delay can be shortened to some minutes. Real-time connections via satellite telecommunication services allow small packets of information to be sent directly to the user without significant delay. The AMARO (Autonomous Real-Time Detection of Moving Maritime Objects) project at DLR is a feasibility study of an on-board ship detection system involving on-board processing and real-time communication. The operation of a prototype system was successfully demonstrated on an airborne platform in spring 2018. The on-ground user could be informed about detected vessels within minutes after sighting without a direct communication link. In this article, the scope, aim, and design of the AMARO system are described, and the results of the flight experiment are presented in detail.

Keywords: real-time communication; maritime situational awareness; ship detection; Iridium; on-board; image processing; flight campaign

---

1. Introduction

Nowadays, about 90% of the world's volume of cargo is seaborne [1]. An enormous amount of money depends on reliable transportation routes. However, safeguarding the seaways is not only essential for the carriage of goods, but especially for the integrity of humans' lives. Piracy, illegal fishery, ocean dumping, and refugee transportation are daily occurrences.

Due to these reasons, maritime surveillance is an important factor for government and private organizations. The European Maritime Safety Agency (EMSA), for example, has set up a vessel traffic monitoring and information system to be able to receive information on ships, ship movements, and hazardous cargoes [2]. General information around maritime domain awareness and how it is handled today can be found in [3,4].

One major issue of maritime surveillance is the vast expanse of the sea on the Earth's surface, which makes observation of ship traffic difficult [5]. The only method to globally get general reliable information about a ship's current position in near-real-time is by using satellite-based AIS (Automatic Identification System) information services [6]. AIS is a cooperative system, primarily intended for collision avoidance. Ships send out their identification, position, course, speed, and several other traffic-related data. This data is then received by other ships and ground stations in close range. Nowadays, to be able to track ships globally in real-time, satellites are also used to receive AIS data [7]. However, based on AIS data only, the detection of illegal activities like water pollution, illegal fishing, or smuggling is limited.

To improve maritime domain awareness, Earth observation (EO) satellite data is a valuable source of information. Great efforts are made in researching the potential of vessel detection in optical and radar satellite images [8–10]. However, in most cases, these images are analyzed long after the data have been acquired [11]. To tackle this bottleneck, there is also promising progress in establishing near-real-time services on the ground, which today can provide information in, at best, the range of 15 min, measured from on-ground data reception [12,13]. However, the most significant time delay occurs between data acquisition on-board and data reception on-ground, since image data are comparatively huge and their downlink requires a direct contact to a ground station. This delay can amount to hours or even days [14].

A second drawback of EO satellites for time-critical applications is their inability to continuously monitor a defined region of interest. Satellites with a reasonable spatial resolution for ship detection orbit in LEO (low Earth orbit) with speeds of approximately 7 km/s over ground, and typically have a revisit cycle of several days [15].

Promising upcoming observation platforms are unmanned autonomous vehicles [16]. For instance, with their Remotely Piloted Aircraft Systems (RPAS), the European Maritime Safety Agency operates a number of services supporting maritime surveillance [17]. These vehicles are small, lightweight, and ready to take off within minutes [16]. However, their operational flight duration, and thus the range of their geographical applicability, is limited. High-altitude pseudo-satellites (HAPS) are the perfect fit for long-endurance wide-area monitoring tasks. Although there is still a significant portion of development left, major progress has been achieved during recent years. One of the most famous HAPS, the Airbus Zephyr S, can carry a payload of up to 20 kg; with nearly 26 days, it holds the world record for the longest uninterrupted flight [18]. However, if HAPS shall be flexibly and rapidly deployable, even in remote areas, they have to overcome a similar problem to that of satellites: Downlinking time-critical information as fast as possible and informing the user immediately without a direct link to a ground station.

To reduce the time between acquiring data with an Earth observation (EO) platform and delivering meaningful information to the user, the capability of real-time communication from the satellite to the ground is needed. One option is using satellite communication services like Iridium or Orbcomm [19]. These services are able to transfer data 24/7 nearly globally within a few minutes, but offer only restricted bandwidth, which is insufficient to send the raw sensor data continuously down to a ground station for on-ground processing. However, the product data that should reach the user within the shortest possible time typically comprises small information like position, heading, velocity, type, and status of the ship. With on-board processing, this information can be extracted directly after acquisition. Since its size amounts to only a fraction compared to the raw sensor data, it can be sent to the user via the mentioned satellite communication services.

A challenge for on-board data processing is that of the limited computer resources that are available on satellites or other autonomous platforms. Furthermore, the special hardware that is used for on-board systems often differs from the mature technology in on-ground data centers, which makes it unfeasible to simply let the on-ground algorithm run on-board. This problem is discussed widely in the literature. In [20], Yuan Yao et al., present a computing system for on-board vessel detection targeting micro- and nano-satellites. This ship detection system extracts image patches and position information from acquisitions using deep learning methods, with the goal of decreasing data size. The authors were able to reduce an image with a size of 90 MB to product data below 1 MB within 1.25 s with a Commercial Off-the-Shelf (COTS) NVIDIA Jetson TX2. In [21], Yu Ji-yang et al., proposed a real-time on-board ship detection method based on FPGA hardware. They used statistical analysis and shape information for extracting images by marking their pixels. On an 8 bit image with 1024 $\times$ 1024 pixels, they were able to extract ships within 10 s with a precision and recall of over 90%.

Another question which seems to be disregarded so far is that of how a modern on-board computing information system should operate as a whole. With the on-board processing systems mentioned above, data are only analyzed and product data are sent to the ground. This approach is a static concept, not allowing user interaction. What we are targeting is an overall and more flexible system, where the user is able to order data, as is done in a web query. They should also be free to choose when and about what to be informed, and to be able to set automated alarms, which are pushed to them in the case of the occurrence of predefined events.

Within this paper, we present the results of a feasibility study of a comprehensive concept for a real-time on-board ship detection system for satellites and other kinds of unmanned flying vehicles. The study involves the development of a prototype system, called AMARO (Autonomous Real-Time Detection of Moving Maritime Objects), and its testing within an aircraft flight experiment campaign. The focus of the study was on how to design a flexible real-time ship detection system for on-board operation, how to realize it, and what performance, especially regarding real-time information capability, can be expected. The prototype system was designed and built using COTS hardware adequate for the aircraft test campaign. The prototype processes image data on-board and communicates the extracted information to the user immediately and without geographical bounds. The system provides product data like the position, heading, velocity, and shape of ships within minutes after sighting. Furthermore, this product data can be individually requested by the user via email on any smart device on the ground, independently of its locality. AMARO was tested in a flight experiment which took place in April 2018.

2. Materials and Methods

2.1. Conceptualization

The initial situation we assume involves an EO platform and a user on-ground who demands to be informed about ship-related events in the shortest possible time. Evaluating various usecase scenarios, the following requirements were identified:

1. The user shall be able to post user-defined requests.
2. The user shall be able to define events about which he/she is informed automatically.
3. The user shall be able to get information with a topicality of at least five minutes.
4. The communication to the user shall be location-independent (e.g., open sea).
5. The information available shall include the object's position, classification, shape attributes (e.g., size, perimeter), trajectory, and estimated heading and velocity.
6. The information available shall include a small preview image of the object.

Requirements 1 and 2 demand a bidirectional communication link, where users can interactively exchange custom-tailored information with the on-board system. Requirements 3 and 4 imply that the images are processed on-board and messages are linked via a satellite-based communication system, since direct links are ineligible due to their limited range. Requirement 5 suggests the usage of a database for storing and managing information.

Based on these deliberations, the concept of the AMARO on-board ship detection system was developed. It consists of one or more Earth observing platforms carrying a camera, a GNSS receiver, an on-board computer, and a modem for real-time communication. An AIS receiver can be mounted on board, and its signals can be synchronized with the image data. Ships in the observation area, which send no signal—possibly on purpose—can thus be identified.

On board, ships are detected from the image data by means of remote sensing algorithms. Product data like position, heading, velocity, type, and status of the ship are extracted. These data—some kilobytes in size—can be sent from the EO platform to the network of communication satellites, which forwards the message until it can be delivered. A small quicklook of the detected object can be included for visual inspection. At most, this procedure will take a few minutes.

On operation, sensor data are acquired continuously by the camera system. These data are immediately evaluated on-board the flying platform, and the product data are stored in a database on the satellite. The user shall be able to query this database by using real-time communication. Furthermore, the user shall be able to define events about which he/she is automatically informed.

Figure 1 shows an exemplary sequence of events involving automatically transmitted and manually requested information: A user is interested in ships that send no AIS signals. He/she therefore requests to be informed automatically if a corresponding event occurs. He/she will not be spammed with information about other detected objects that do not fulfill his/her requirements. As soon as a ship without AIS is detected, a message is sent automatically from the Earth observing platform to the user via the satellite communication network. Since the user is further interested in this ship, he/she requests details via a one-time order. Among others, these details can include a small image of the detected object in order to verify it by visual inspection.

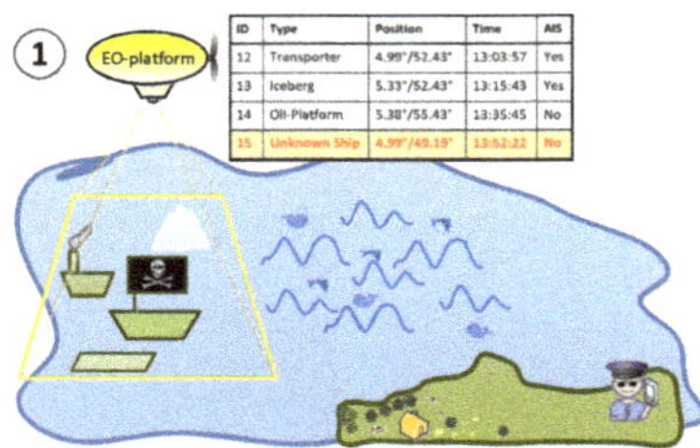

(**a**) Images are taken by the Earth observing platform and processed autonomously on board.

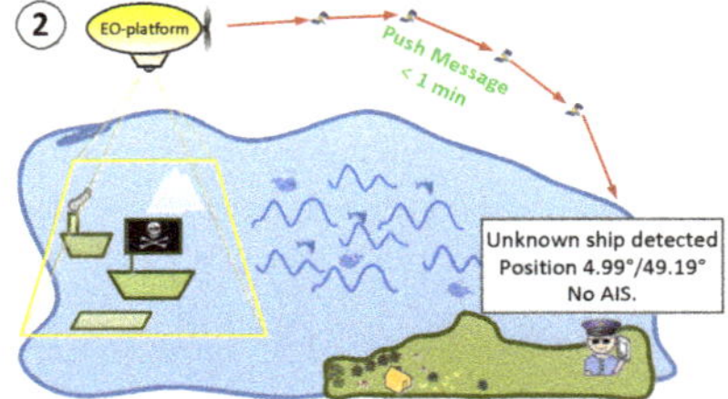

(**b**) The user has requested automatic notifications of whenever a ship is detected that does not send AIS (Automatic Identification System) information.

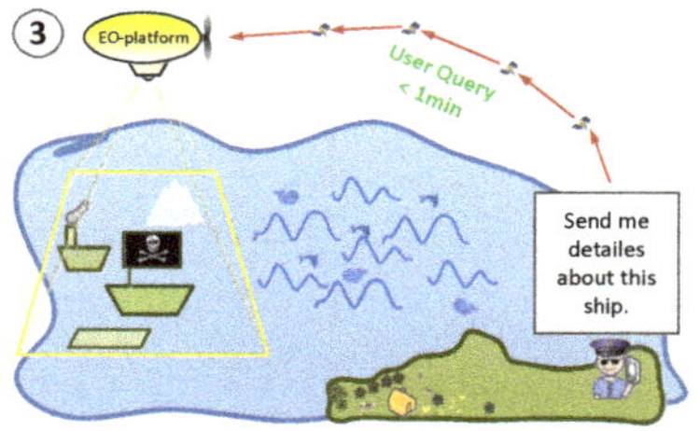

(**c**) User requests detailed information.

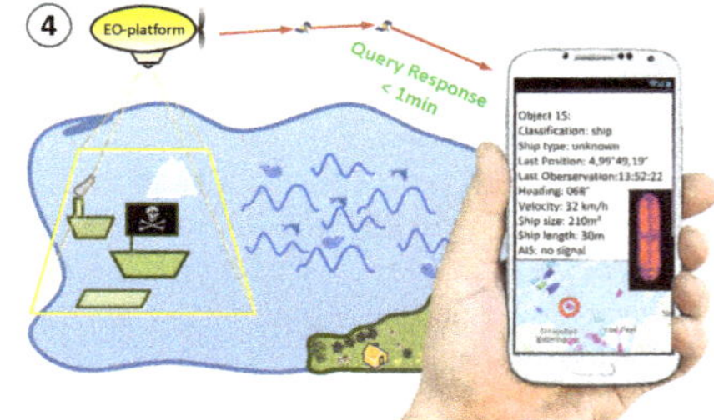

(**d**) User receives detailed information about the unknown ship.

Figure 1. An exemplary user story.

2.2. Hardware Architecture

The AMARO-Box and its contents were specially built for the airborne test campaign. The different hardware devices are therefore not necessarily suitable for an operation on another EO platform. An image of the box during assembly is shown in Figure 2. In the following, the components of the AMARO-Box are explained in detail. Since the camera, which was used in the experiment, and the corresponding image data are an essential part of the flight campaign, but not of the AMARO-Box, they are explained later on, in Section 2.4.

Figure 2. Autonomous Real-Time Detection of Moving Maritime Objects (AMARO)-Box with hardware components during assembly.

2.2.1. Communication System

As mentioned before, the main goal of AMARO is generating and delivering the product information to the user as fast as possible. The user may be a crisis intervention center with a high-rate internet connection or a single person off any ground-based connection for communication. Similarly, the carrier platform of the AMARO-Box may be off any connection to ground-based communication facilities. Therefore, in order to facilitate permanent and locally independent communication, the use of satellite connections was considered mandatory. The following criteria for the real-time transmission of the product data have been collected: Low latency, global coverage, easy to obtain, easy to maintain, easy to integrate, and easy to operate.

After a careful deliberation over various options, we decided to use the Iridium Short Burst Data (SBD) service. Iridium is a satellite communication network consisting of 66 active satellites, which provides almost 100% global coverage continuously at almost 24/7. With SBD, Iridium offers a simple and efficient service for transmitting small data packages between equipment and centralized host computer systems [22], commonly used for asset tracking. Messages with a size of around 300 B can be exchanged between the on-board device and the on-ground user. For sending and receiving messages from the device, standard email is used. The email is sent to Iridium with the device's serial number as the subject. The message itself is attached to the email as a normal text file with extension *.sbd and can be of individual content. In the case of AMARO, this *.sbd file contained a database query in the sql-language.

The latency for data exchange is specified as less than one minute worldwide [23]. The size of the transceiver device is (31.5 mm $\times$ 29.6 mm $\times$ 8.1 mm w/h/d). The average power consumption is below 0.8 W.

One big advantage of the Iridium system is that the antenna needs no exact pointing alignment to a determined direction. On an aircraft, it is sufficient that the antenna points approximately to the sky. This may not apply to other platforms. The satellites of the Iridium network fly in orbits of approximately 780 km height, and their signals are broadcast such that the regions of reception overlap on the Earth's surface. However, a loss of coverage is supposable for platforms in higher altitudes.

The Iridium SBD modem and antenna are available custom off the shelf; hence, the purchase is fast and uncomplicated. We decided to buy a MiChroBurst-Q modem from Wireless Innovation [24]. It houses an Iridium 9602 modem and comes development-ready with connection ports for power supply and data transfer via RS-232. The whole box is sized 110 mm × 35 mm × 85 mm w/h/d, which is comparable to a packet of cigarettes.

As an Iridium antenna, we bought an AeroAntenna AT2775-110 [25]. Since the antenna had to be specially mounted on the plane's roof, the owner required an aircraft-certified device and its installation by a specialist. The antenna is flat and streamlined to fulfill the aerodynamic requirements, as can be seen in Figure 3. It operates in the frequency band of (1595 ± 30) MHz and consumes approximately 10 W.

While implementing and testing the AIS subsystem, we had access to an AIS simulator. With its ability to generate fake AIS messages that can be received by our system, validation was greatly eased.

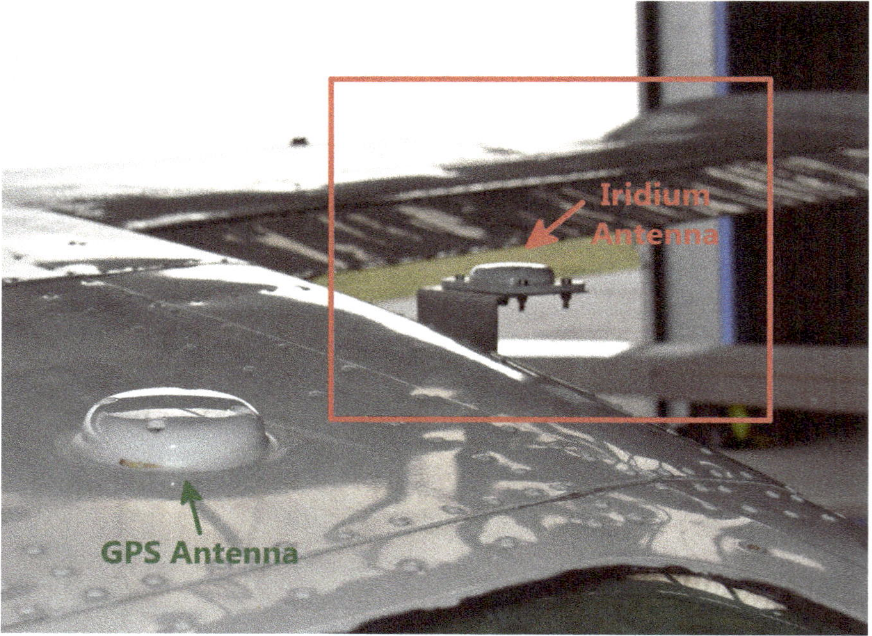

Figure 3. AeroAntenna AT2775-110 Iridium Antenna mounted on the airplane's roof.

The operation cost of the SBD service was a minor factor. In the time of operation, we paid around $20 for a monthly data volume of around 12 kB. This is depending on the size of the messages, equivalent to 40 to 120 messages.

2.2.2. AIS Receiver

To able to receive AIS navigation data from accordingly equipped vessels, an AIS receiver and antenna were installed on the airplane. The receiver we used, AMTEC CYPHO-150, is a custom version of the shelf standard device, whose primary usecase is to be installed on recreational boats, which do not need to send out AIS information. The AMTEC CYPHO is not especially qualified for deployment on airplanes and is hence available at a fraction of the price of a dedicated device. However, it worked perfectly without any trouble in installation or loss in performance.

This AIS receiver is capable of receiving AIS messages of classes A and B [26], sent out by commercial and private vessels, respectively. Receiving several other AIS formats is also possible, but was not in our interests.

It is lightweight, small in form factor (128 mm × 36 mm × 88 mm w/h/d), and has a power consumption below 1.50 W; hence, it was perfectly suited to be installed in the AMARO computing box [27].

The AIS receiver can be connected to the on-board computer via a serial or USB interface. We chose the latter, because it can also be used to power the device. To encode AIS messages, the AMTEC CYPHO-150 uses a serial text-based transmitting protocol specified by the NMEA 0183 interface standard. Typically, AIS messages contain the Maritime Mobile Service Identity (MMSI) number, the call sing and name, the type, the length and beam, the cargo information, the position of the vessel, the Course Over Ground (COG), the Speed Over Ground (SOG), the heading, the speed of the ship, and the status of the ship. The AIS messages were parsed by our on-board software and then directly inserted in the database. Our parser was based on the libais library (see [28]) and modified to meet our needs.

As the AIS antenna, a standard PROCOM HX2 was used [29]. It is a flexible 1/4 λ helix antenna for the two AIS channels in the frequencies 161.975 and 162.025 MHz. It is around 150 mm long, and was installed together with the camera in the downward-looking hole of the aircraft's fuselage. While implementing and testing the AIS subsystem, we had access to an AIS simulator. With its ability to generate fake AIS messages that could be received by our the system, validation was greatly eased.

2.2.3. On-Board Computer

The on-board computer is the core component of AMARO. It obtains the camera data, controls the Iridium transceiver and the AIS receiver, performs data analysis, and manages inner and outer communication. The requirements for the on-board computer were the following: It had to be small to fit in a 19 inch rack box together with the other components. It had to provide sufficient computing power for data processing. Moreover, it had to be physically and thermally robust for reliable operation on the aircraft (passenger cabin).

After studying the market, we decided to buy a 1.3 l slim standard personal computer (Shuttle DQ170), which is equipped with standard up-to-date desktop PC components. The computer is robust enough to handle 24/7 operation and up to 50 °C ambient air temperature. All interfaces needed for attaching the other devices are present. Equipped with modern desktop PC components, that is, Intel Core i7-6700, 16 GB RAM, and a 512 GB SSD, the system may be luxurious compared to today's or even future computer solutions, deployable on-board HAPS or satellites. Power usage, thermal output, space limitations, and radiation impact were insignificant for the demonstration of our prototype. Therefore, for the first proof of concept, we determined a restriction in this regard to be unnecessary. Nevertheless, since we are also involved in building a next-generation space-computing platform [30], we assume that it is possible to integrate the software on a future computer mounted on an autonomous carrier platform.

2.3. Software Architecture

The software is the most important and labor-intensive component of the AMARO system. Whereas most hardware components could be bought off the shelf, the software system has been developed from scratch. It is designed to be modular and flexible, such that it can modified for a variety of scenarios and deployed on an arbitrary carrier platform.

2.3.1. Software Requirements

The AMARO software system has to handle two main tasks: Data analysis and communication. The data analysis process shall extract useful information from the image data or other sources. The system shall be capable of processing as much data as possible to reach a high situational awareness. Due to the complexity of the ship detection algorithm and the high amount of image data, processing may be computationally intensive. The AMARO communication system has to be readily responsive, and the available bandwidth has to be used efficiently. Since there are only limited maintenance options at runtime and the system is deployed on-board a flying platform, it has to be absolutely reliable. Interruptions of operations are undesired, and in the case of an error, the system has to mitigate it and get back to operation with the least possible loss of information.

2.3.2. Software Infrastructure

To enable fast and efficient software development, a standard x86-64 Linux desktop distribution was selected as the operating system. As the main programming language, C++14 was chosen. Based on these conventions, a lot of up-to-date software development tools and libraries are available to help minimize development costs. The effort to deploy the software is further minimized, since the development and runtime operating systems are identical.

We want to stress that the goal of the development was to build a software system that proves the concept of a real-time on-board ship detection system within an experimental flight. Nevertheless, as C++ and Linux are also used for future on-board systems, we trust that our software is, in principle, implementable on an on-board platform without fundamental changes. In fact, we have already ported essential parts of our software to an on-board computer within the project ScOSA (Scalable On-Board Computing for Space Avionics), which has the goal of developing a high-performance on-board platform for the deployment on satellites [31].

2.3.3. Software Design

As mentioned in Section 2.3.1, the system has to be high-performance, responsive, and reliable. To meet all of these requirements, a service-based architecture was chosen. A top-level view of the service architecture is presented in Figure 4. Every task is carried out by a unique service that can operate independently of other services. As every service is its own Linux process, high responsiveness for the communication services and, at the same time, a high amount of computation time for the ship detection application can be provided. In case of an error, aborting a service has no direct effect on other processes, and the service can be restarted individually.

For inter-service communication and for data storage, we chose the file-based database SQLite [32]. SQLite can be easily implemented without the need for a dedicated database server. The (asynchronous) communication of the services is handled by the database engine itself. Furthermore, the validity of the database is warranted by SQLite in case of a writing failure. The most functional advantage of using an SQL database is the availability of the language SQL (Structured Query Language). The SQL programming language is the key enabling element used for the implementation of the user interaction with the system. In general, SQLite is not recommend for distributed systems (e.g., network file systems) and is not well suited for heavy simultaneous writing on one database file. However, for the experimental demonstration of our prototype system, all services were located on the same computer, and data were written simultaneously to one database file only sparsely. For a future operational system, the usage of a server-based database is strongly recommended.

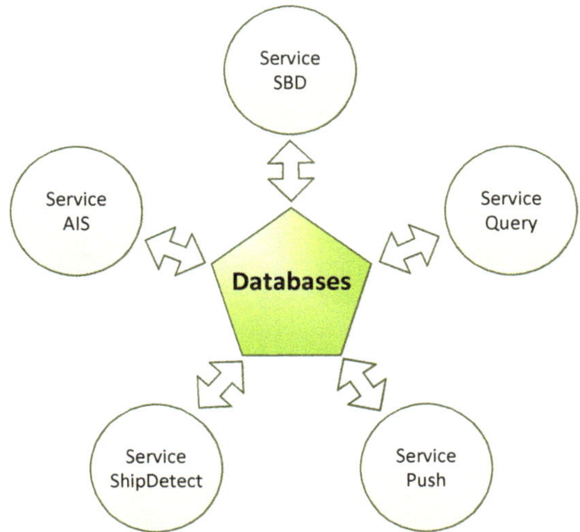

Figure 4. Top level overview of the AMARO software architecture.

The following subsections describe the different independent services of the AMARO software.

2.3.4. Service SBD Message

Our serviceSBD is a messaging service that allows other services to send and receive messages over the Iridum SBD. A service that wants to send a message adds it to the so-called *toSendMsg* table of the database. When a sending slot is available, serviceSBD checks this table and tries to send the most prioritized message. Received messages are inserted into the *msgReceived* table. As sending and receiving of messages is encapsulated in its own (Linux) process, it can be accomplished independently of other services. This guarantees the best usage of bandwidth and very good responsiveness.

2.3.5. Service Query

The serviceQuery is a query response service. A user on ground can send a one-time query over the Iridium SBD to the database. The serviceQuery tries to answer it and generates a response message.

In detail, a user can send a one-time query request via email to the on-board device using the following format:

```
<id-service-query>:<priority>:<database> <SQL-statement>
```

Two example query requests are given below:

```
5:4:system.db SELECT * FROM Log
5:2:asd.db SELECT shipID FROM shape WHERE shipArea >= 50
```

The query request is received by the serviceSBD and saved in the *msgReceived* table. The serviceQuery checks the *msgReceived* table periodically. If query requests have arrived, the most prioritized is executed, and the query request is moved from the *msgReceived* table to the *msgReceivedArchive* table. The result of the one time is then put into an SBD message and inserted into the *msgToSend* table.

With serviceQuery, the user can access all on-board databases. As a typical example, the user can request a list of objects which have a defined size and have been detected within a defined time interval.

2.3.6. Service Push

The servicePush is a messaging service that sends automatic notifications if a predefined event occurs. Events can be added and deleted during operation. Examples for such events could be the detection of oil near a ship (ocean dumping), ships entering a restricted area, ships sending no AIS signals, etc.

In detail, an event is defined as an SQL query with timing information. The timing information contains a time window and a period specifying the time points of execution of the SQL query. All activated events are saved in the *push* table. Events can be added or deleted by modifying the *push* table.

The following example shows how an event can be added to the *push* table via a query request:

```
5:3:system.db INSERT INTO PushTable
(Start ,Stop ,Periode_s ,Priority ,Category ,Db,Query) VALUES
(''2018−04−12 08:36:00'',''2018−04−12 20:45:00'',''300'',''5'',''107'',
''asd_DB.db'',
''SELECT shipID ,course ,speed FROM ships ORDER BY shipID DESC'')
```

In normal words, within the time window, every 300 s, AMARO shall try to send information about the IDs, courses, and speeds of the latest detected ships. If the query is successful, servicePush generates a result message and inserts it into the *msgToSend* table.

2.3.7. Service Ship Detection

The serviceShipDetect is responsible for data analysis. It receives the image data from the camera, analyzes them, and enters the results into a database table. Within the flight experiment, the image data are acquired with a frequency of 1 Hz (one acquisition per second) and are sent from the camera control computer to the AMARO system over ethernet. Since subsequent acquisitions will have overlapping content of around 90%, more than one observation will be made for one and the same object. For another mission with other conditions of image acquisition, these values may differ. The detected objects are examined and filtered out if they are too small or too big, or if one of the shape attributes does not match the defined constraints for being a ship. The considered shape attributes are: Size, perimeter, long axis, short axis, axes ratio, circularity, rectangularity, convexity, and solidity. More information about definitions and methods of calculation of these attributes can be found in [33].

If a ship-like object is detected in one acquisition, the following characteristics are extracted and stored in the database:

1. Time stamp of each observation;
2. location of each observation in geographic coordinates;
3. shape attributes, as mentioned above.

Two ship-like objects are considered "similar" when they have both appeared within a limited geographical range and a limited time range, and when both have similar shape attributes, as defined above. If, in two or more subsequent acquisitions, "similar" ship-like objects are detected, they are grouped together and treated as a possible ship. The single objects are marked as assigned in order to not check them again. If, in at least four subsequent acquisitions, "similar" ship-like objects are detected, they are treated confidently as ships, and the following characteristics are extracted additionally:

1. Number of observations,
2. heading, and
3. velocity.

The object data can be directly accessed by the end-user via a query message (serviceQuery) or by defining an event (servicePush). In the current version, only the thermal channel was used.

The computational steps involved are correction and normalization of the image data, water–land classification, connected component labelling [34], object analysis, and data comparison on the object's metadata. For further reading, see [35]. As the data analysis is relatively complex and a high amount of data has to be processed, the serviceShipDetect can be run up to eight times in parallel.

2.4. MACS and Image Data

Images were acquired using the instrument MACS (Modular Aerial Camera System), cf. [36,37]. A picture of the MACS camera system can be found on figure 5. Using the MACS camera, the photos were calibrated for radiometric correction and georeferenced, providing geographic coordinates, position accuracies, and absolute time for every image pixel. For the AMARO experiment, the system was equipped with a passive optical multi-sensor configuration to cover human-visible (RGB), near-infrared (NIR), and thermal infrared (TIR) spectra, as summarized in Table 1, but eventually, only the TIR channel was transmitted to the AMARO-Box. The image rate can be up to four full frames per second simultaneously for all sensors, and was set to 1 Hz during the flight experiment.

Through a hole in the aircraft fuselage, the lenses have an unobstructed view downwards. An embedded desktop class computer enables raw data recording, preprocessing, and immediate data forwarding. The MACS main computer is connected to the AMARO on-board computer through a Gigabit Ethernet link. Data of the selected image sensor are continuously fed as a byte stream. On this real-time stream, the object classification is executed in-memory, hence, without any image storage. Additionally, a function runs on the AMARO computer to re-establish geographic coordinates: Depending on the aircraft position and altitude the images are projected on sea level. The elevation of this plane is derived from the SRTM database. Because the scenery is completely flat over the sea, an image-edge four-point projection is sufficient. For a given image pixel, i.e., corresponding to a matched object, the function interpolates the edge coordinates and provides the geographic coordinates for the particular pixel.

Table 1. Modular Aerial Camera System (MACS) sensor setup.

	RGB (Bayer Color Pattern)	Near Infrared	Thermal Infrared
Spectral Bands (nm)	400–520 (blue) 500–590 (green) 590–680 (red)	700–950	7500–14,000
Resolution (pixels)	4864 × 3232	3296 × 2472	1024 × 768
Focal length (mm)	50.0	29.3	30.0
Pixel pitch (µm)	7.4	5.5	17.0
GSD @ 820 m above sea level (cm)	12.1	15.4	46.5
GSD @ 2500 m above sea level (cm)	37.0	49	141.7
Field of view across track (deg)	39.6	34.6	32.4

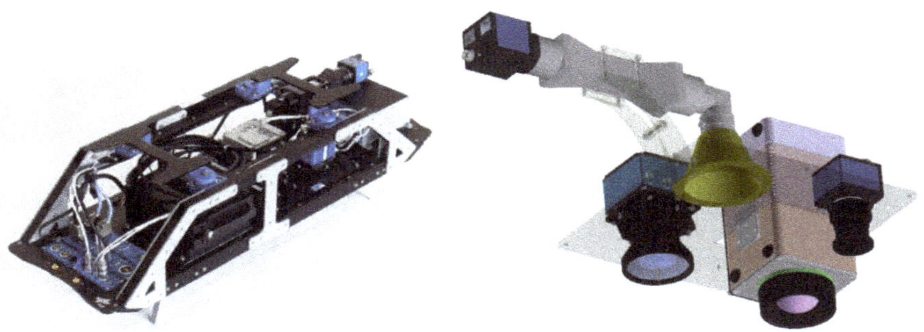

(a) Above view on the MACS system   (b) Illustration of the sensor head

Figure 5. Modular Aerial Camera System (MACS).

3. Results

3.1. Experimental Flight

The experimental flight was conducted on the 12th of April in 2018. The AMARO-Box, the antennas for Iridium and the AIS, and the MACS camera were installed into a small science aircraft, a Cesna 207T, provided by the Freie Universität Berlin. The flight started from the airfield Schönhagen, located 50 km south of Berlin, Germany, at 09:15 a.m. UTC, and ended ibidem at 03:21 p.m. UTC. From there, the route lead over northern Germany to the mouth of the Elbe in Hamburg, where the actual experiment was conducted. The flight path is depicted in Figure 6.

In the time between 11:10 a.m. and 11:54 a.m. UTC, the main naval traffic route to enter the port of Hamburg was flown forward and backward (see Figure 7). This is called the experimental core time. Afterwards, the flight was interrupted to refuel the aircraft from 11:59 a.m. to 01:10 p.m. UTC. An overview of the different phases of the experimental flight is given in Table 2.

Table 2. Overview of the timing of the different phases of the experimental flight.

Time Intervals	UTC	Local Time (MESZ = UTC + 2 h)
Flight time	09:15 a.m.–03:21 p.m.	11:15 a.m.–05:21 p.m.
Refuel stop	11:59 a.m.–01:10 p.m.	01:59 p.m.–03:10 p.m.
Operating time	09:15 a.m.–11:59 a.m.	11:15 a.m.–01:59 p.m.
	01:10 p.m.–03:21 p.m.	03:10 p.m.–05:21 p.m.
Experimental core time	11:10 a.m.–11:54 a.m.	01:10 p.m.–01:54 p.m.

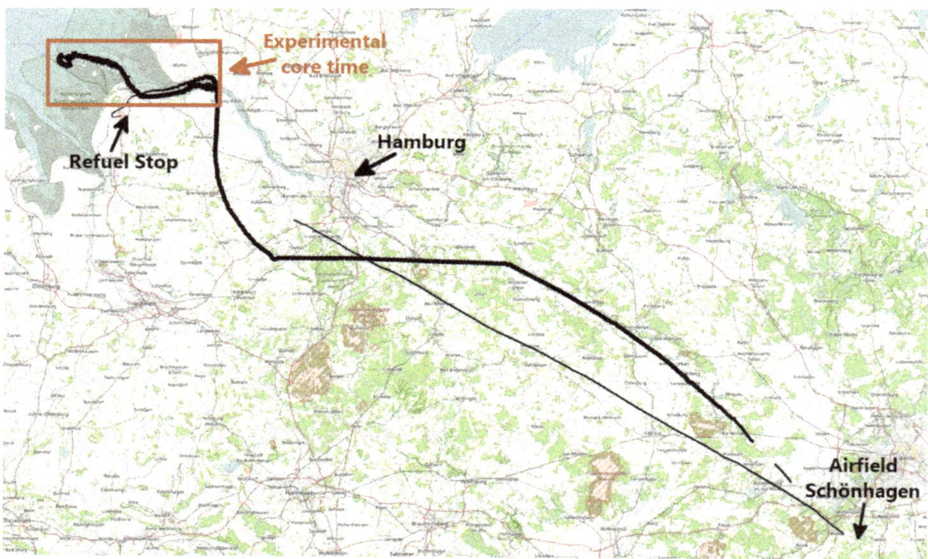

Figure 6. Flight path from the airfield Schönhagen to the North Sea and back.

On-board were the pilot and two scientists, one to supervise the AMARO-Box, the other to control the MACS camera and to support the pilot. The supervision of the AMARO-Box was actually not necessary, since it was designed to operate autonomously. However, to be on the safe side for the first in-flight test, we considered supervision to be beneficial in case of unforeseen misbehavior. For controlling purposes, we connected the AMARO-Box with an external terminal PC. On-ground, two more people were assisting to install the camera and the AMARO-Box in the airplane.

The actual experiment—the communication with the AMARO-Box—was then conducted by a scientist and a technical assistant on-ground. Equipped with a standard office notebook, they operated the experiment from the user's side in the airfield's restaurant, which provided a stable internet connection. We want to mention that these users could have resided anywhere on Earth and could have used any device, as long as an internet connection was present.

Figure 7. Mosaic of thermal images over the mouth of the Elbe during AMARO's experimental core time.

3.2. Performance Communication

3.2.1. Iridium Signal Quality

During operation time, the signal strength of the connection to the Iridium satellite network was measured and logged in the database *signal.db*. An evaluation of the database revealed an excellent overall reception quality for the whole flight. However, from about 10:30 p.m. to 11:00 p.m., no messages were received nor sent from the on-board AMARO-Box. In the evening, we received a notification from the Iridium SBD service informing us about unplanned intermittent outages which had taken place between 10:42 a.m. and 03:28 p.m. Since issues with the Iridium communication service also impact the AMARO performance in general, potential outages have to be taken into account when analyzing the performance. Nevertheless, it is worth noting that during the 15 months of using the Iridium SBD service, we received a total of five unplanned outage notifications, one of them just on the day of the experimental flight. The percentage distribution of the signal strength can be seen in Table 3. Figure 8a shows the distribution of the signal strength over time.

Table 3. Distribution of Iridium's signal strength over time in [%] measured on-board.

Signal Strength	Signal Strength	Distribution During Operation Time [%]	Distribution During Core Time [%]
0	no signal	0%	0%
1	very low	0%	0%
2	low	0%	0%
3	medium	0.63%	0.35%
4	strong	1.65%	0.70%
5	very strong	97.72%	98.94%

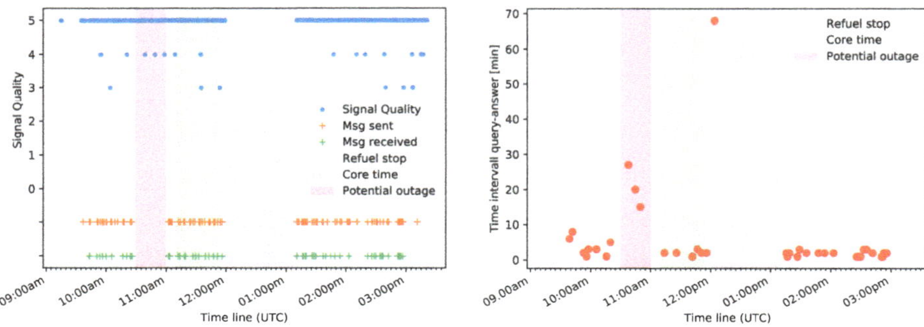

(**a**) Iridium SBD signal quality measured on-board.　　(**b**) Time between query and answer measured on-ground.

Figure 8. Iridium Short Burst Data (SBD) signal and response measurements.

3.2.2. Message Exchange

The first part of the operating time was taken by the flight to the experimental site at the North Sea. During that time, several messages were exchanged to establish and check the connection and to set up push queries.

In total, 56 messages were sent from ground to AMARO, while 169 messages were received from AMARO by the on-ground operator. From these, 13 and 34 messages are contemporary with the experimental core time, respectively.

Tables 4 and 5 show the amount and type of messages sent from ground to AMARO and vice-versa. The push queries contained information about start time, expiry time, and period, i.e., the time interval in which the query should be executed by AMARO. One-time queries were executed as soon as possible after reception by the AMARO-Box on board. The possibility to exchange chat messages between the on-board and on-ground operators was set up in order to facilitate communication between on-board and on-ground operators during the flight. Empty downlink messages occurred due to technical reasons within the Iridium service, as described in ([38], Section 7.1.3).

Here, we give some examples for the message exchange during the experimental core time:

- Via a one-time query, AMARO was instructed to send the five latest log messages;
- via a push query, AMARO was instructed to send the number of hitherto acquired datatakes every 10 min;
- via a push query, AMARO was instructed to send the coordinates of the airplane's current position every 12 min;
- via a push query, AMARO was instructed to send information about the latest 20 detected ships every five minutes;
- via one-time queries, AMARO was instructed to send information about objects with an area greater than 900 pixels and greater than 1800 pixels, respectively;
- via one-time queries, AMARO was instructed to send small quicklook images for several ships;
- via a one-time query, AMARO was instructed to send the MMSI for all ships inside a quadrilateral defined by latitudes and longitudes of its corners;
- via one-time queries, AMARO was instructed to send details for several MMSIs.

To all queries, a category is assigned. The answers are branded with the same category, such that the operator on-ground is able to match them with the corresponding queries.

3.2.3. Query–Response Time Interval

Figure 8b shows the distribution of the time intervals between sending a query and receiving the corresponding answer during the experimental flight. It can be seen that these results coincide with the measurements of the SBD signal strength. Furthermore, Table 6 lists the number of message pairs (query/answer) with the time span between sending the query and receiving the answer. As the messages are sent and received by email, the time between the outgoing of the query and the incoming of the corresponding answer is measured with a temporal resolution of one minute. For time intervals larger than five minutes, the uplink time of the query and the downlink time of the corresponding answer were also analyzed. Note that the computation time is negligible, because the processing times of queries involved only database accesses.

Table 4. Number of uplinked messages.

Message Type	Operating Time	Core Time
One-time queries	43	9
Push queries	6	2
Delete instructions	5	1
Chat messages	2	0
Total	56	13

Table 5. Number of downlinked messages. In M1 messages, the answer fits into one single packet, while for >M1, it had to be split into several parts.

Message Type	Operating Time	Core Time
Answers to one-time queries M1	26	3
Answers to one-time queries >M1	23	3
Answers to push queries	55	16
Empty messages	50	8
Chat messages	15	4
Total	169	34

For 7 out of the 46 query messages, we received no answer at all for different kinds of comprehensible reasons, e.g., due to an incorrect SQL syntax or a preceding delete-query where an answer is not expected. These messages are not taken into account hereafter. Apart from this, one message was answered with a delay of 68 min, where uplinking the query took 67 min and downlinking the answer took 1 min. As the query was sent during the refuel stop of the airplane, during which the AMARO system was deactivated, it is also not taken into account.

For 32 out of the remaining 38 messages, i.e., around 84%, the time span between query and answer was below five minutes, with an average of 1.87 min.

The response time for three messages (8%) was between 5 to 10 min, with an average of 6 min.

Another three messages (8%) were answered between 10 and 30 min. The average delay in this range was 20 min, with an average uplink delay of 18 min and an average downlink delay of 2 min. The most likely explanation for the high delays is the Iridium outage mentioned in Section 3.2.1, as the three queries in question were sent subsequently during the beginning of this time frame.

Table 6. Query–Response time: Time interval between sending a query and receiving the corresponding answer (email to email). Gray values were not taken into account for further analysis.

Time Span [min]	Average Uplink/Downlink Time [min]	Contacts
1	-	10
2	-	16
3	-	6
5	1/4	1
6	4/2	1
8	2/6	1
15	13/2	1
20	18/2	1
27	25/2	1
68	67/1	1 (refuel stop)
-	-	7 (no answer expectable)

3.3. Performance AIS

During the operating time, 303,986 AIS messages were received by the system, with 275,144 AIS messages of types 1/2/3 and 7660 of type 5. A further 13,082 unsupported messages were received. A detailed overview is presented in Table 7.

Table 7. Overview of AIS messages received on-board. For a detailed description, see [26].

Type	Count	Percentage	Description
raw	303,986		AIS data frames
all	295,886	100%	Supported and unsupported messages
supported	282,804	96%	
123	275,144	93%	Position report
5	7660	3%	Static and voyage related data
not supported	13,082	4%	
4	3737	1%	Base station support
6	1	<1%	Binary addressed message
7	9	<1%	Binary acknowledgement
8	5161	2%	Binary broadcast message
9	9	<1%	Standard SAR aircraft position report
10	2	<1%	UTC/date inquiry
11	135	<1%	UTC/date response
15	971	<1%	Interrogation
17	31	<1%	DGNSS broadcast binary message
18	1223	<1%	Standard Class B equipment position report
20	129	<1%	Data link management message
21	886	<1%	Aids-to-navigation report
23	42	<1%	Group assignment command
24	732	<1%	Static data report
27	14	<1%	Position report for long range applications

In Figure 9, the aggregation over time of received AIS messages is displayed. AIS data were received during the complete operating time, except during the refuel stop, during which the AMARO-Box was not activated. All AIS messages were stored on-board in the AIS database, which was queried several times on the return flight. However, to match them with the results of the image processing is left for the next stage of expansion.

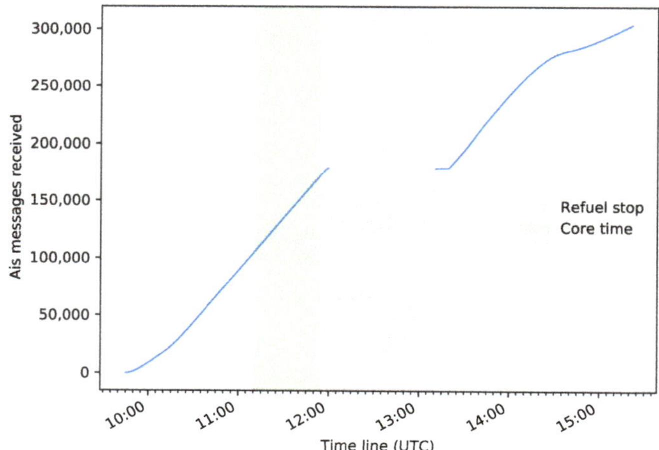

Figure 9. Aggregation of received AIS messages during the operating time.

3.4. Performance Image Processing

As mentioned in Section 2.3.7, image processing was carried out on the thermal channel only. We abstained from creating a mature algorithm in terms of state-of-the-art remote sensing and Earth observation, since our main focus was to demonstrate a prototype for a globally deployable real-time information system. Nevertheless, the algorithm performed quite well. Apart from this, our service also includes the possibility of downlinking a quicklook of the object, such that an operator can double-check the result by visual inspection. An example set of quicklook images is displayed in Figure 10.

Due to the limited communication bandwidth, the maximal data volume of a quicklook was very limited. With a combination of a small image size, reduction of the color depth to one-bit monochrome, the use of a standard run length compression schema, and splitting up of the images into several parts, it was possible to fit the images in one to three SBD messages, each with a size of around 300 bytes.

During the whole experiment, 13,928 thermal images were acquired by the MACS sensor, while 13,607 images were processed by AMARO. Hence, 321 either got lost during transfer or were missed by AMARO because the processing channels were already busy. During the experimental core time, approximately 2570 thermal images were acquired, of which 25 were not processed. All results from the image processing thread were stored on-board in an SQlite database file. The results of the post-flight analysis are summarized in Table 8.

Table 8. Analysis of the results produced by the ship detection thread.

		Full Flight Time	Experimental Core Time
Objects	total	144,988	12,860
	ship-like	2339	324
	assigned to ship	721	294
Ships	total	188	47
	category active ship	68	26
	category initialized ship	120	12

Actually, since the algorithm was designed for objects that are surrounded by water, the results during the flight over land are not meaningful. Therefore, the verification of the algorithm's performance is done for the core time only. From the 26 results that were marked by AMARO as ships, we could verify by visual inspection that 23 were truly ships. Out of these, 13 were assigned one-to-one, i.e., AMARO detected one ship where we also see one ship in the images. An example of the visual inspection of one ship observation is shown in Figure 11.

It happened three times that AMARO detected two distinct ships in a time series of subsequent images, where only one and the same was present. In one case, AMARO detected two ships where there were two ships, but mixed up the results. Apart from this, AMARO re-detected three ships, i.e., these ships were overflown two times (while overflying the mouth of the Elbe forward and backward), and AMARO recognized them as one and the same object, which may be wanted or not, depending on the definition. If this effect is undesired, the time span for identifying "similar" objects could be narrowed further on. No ships were missed by AMARO compared to the visual inspection. For a quicker overview, these results are summarized in Table 9.

Even though the design of the algorithm was not our main focus, development efforts were kept comparatively low and only the thermal channel was used; the results are perfectly satisfactory. However, a thorough comparison with other ship detection algorithms would go beyond the scope of the present paper.

Table 9. Comparison of the results from AMARO with visual inspection.

Description	Number
detected by AMARO	26
unambiguously identifiable by human eye	23
results from AMARO and visual inspection assignable one-to-one	13
AMARO detected two ships where only one was present	3
AMARO mixed two distinct ships	1
AMARO re-detected ships on the return flight	3

Figure 10. Example of quicklook images of potential ship objects.

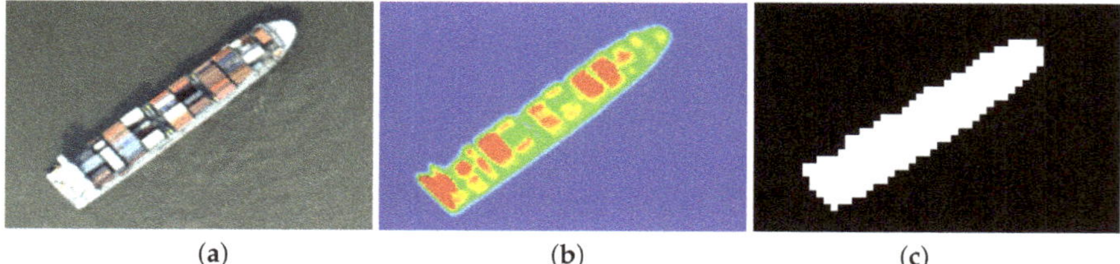

(**a**)　　　　　　(**b**)　　　　　　(**c**)

Figure 11. Ship detected by AMARO, 12th April 2018, 11:45 am UTC, Mouth of the Elbe, Hamburg, Germany: (**a**) RGB image (**b**) thermal image (**c**) quicklook which was sent to ground from AMARO.

4. Discussion

We developed a comprehensive prototype system called AMARO for future real-time ship detection on-board satellites and other Earth observing vehicles. It includes on-board image processing, real-time communication via a satellite network, and a user-driven message exchange. To test the concept, the AMARO-box was built as prototype hardware, and the system was tested within a flight campaign over the North Sea.

4.1. Communication

Most special focus was put on the user-driven near-real-time information capability facilitated by using a satellite communication service. It was successfully demonstrated within our flight campaign, in which the Iridium SBD service was used for message exchange. More than 84% of the user queries were answered in less than five minutes with an average of less than two minutes.

For EO satellites, an information flow within minutes is not possible with the current approach of downloading the sensor data to ground stations and processing them on-ground. In contrast to conventional remote sensing missions, our system does not rely on any direct link to a ground station.

By using satellite communication services, as demonstrated with AMARO, product information can be communicated to any device on the ground with connection to the internet, independent of the localities of both the carrier platform and the user. The system is therefore flexibly deployable at varying monitoring sites and especially suitable for the surveillance of remote areas without ground connection; for example, over the open sea. Especially for micro- and nano-platforms, this can be a feasible approach for enabling real-time capability, as it can be used worldwide, 24/7, and no ground infrastructure is required. Apart from this, the operational costs are affordable, even for smaller missions.

Apart from this, with AMARO, users are not drowned with an unmanageable amount data. They can control the flow of information by interactively interchanging messages with the on-board system. They can configure the automatic notification service during operation to get custom-tailored information about events of their interest. Finally, they can request further details by querying the on-board databases.

Being able to get information about ships within few minutes after observation, as we demonstrated, is beneficial in various situations. For example, it can support maritime safety agencies to take action against smuggling, illegal fishing, and sea pollution or support sea rescue services.

Nevertheless, regarding the communication procedure, some aspects were deemed to be in need of improvement. As described in Section 2.2.1, the queries in the SQL language were recorded in text documents and sent to AMARO as email attachments. The AMARO system replied the same way. It turned out that this procedure was uncomfortable to handle, even for the experienced operator. Requests and their corresponding answers always started with the same ID for an easier matching, but nevertheless, it was difficult to oversee which answers were already received, which were wrong, and which were empty or not present at all.

One of our priorities regarding further development is therefore the design of a graphical user interface. In principle, the interface should handle user-defined requests to a database via the internet. In the upcoming stage of expansion, every authorized user should be able to retrieve the information of their personal interest via a web application, using the device of their choice (smartphone, tablet, laptop, etc.). Apart from this, the limited bandwidth of around 300 B per message was a bottleneck in the communication flow. Sophisticated programming and workarounds were necessary in order to transmit a reasonable amount of information. The quicklook images could only be sent as highly compressed binary shapes. However, here we are sure that our approach will be augmented by ongoing and future development, which will continuously allow higher transmission rates. For example, with their next-generation satellites launched in the recent years, Iridium SBD can now transmit packages of around 2 kB in message size, compared to the previous 300 B. Furthermore, there may evolve even more possibilities with globe-spanning satellite-borne internet systems, such as OneWeb or StarLink.

Regarding the deployment on EO satellites, further investigation is necessary to examine the potential of the existing real-time communication services in LEO orbits. Satellite communication networks are usually designed for operating services on-ground and, hence, provide continuous coverage within their operational area on the Earth's surface. Since EO satellites typically fly in an altitude of approximately 200 to 2000 km, at that height, coverage may be rather discontinuous. Apart from this, depending on the relative orbits of EO and communication satellites, a loss of connection may be encountered due to the amplified Doppler effect [39,40]. However, some on-board experiments were already conducted and yielded apparently promising results [41,42].

4.2. Onboard Data Analysis

Within AMARO, image data are processed directly on-board in order to extract the relevant and rather small-sized product data. In combination with using satellite communication, on-board data reduction is the prerequisite that enables real-time information.

Although the designed algorithm uses the thermal infrared channel only and is altogether kept relatively simple, the results were definitely competitive. More than 88% of the detected objects could be identified as ships. No ships, which were identified by eye, were missed.

We want to mention that this ship detection algorithm was primarily developed to be able to demonstrate the concept of a real-time onboard ship detection system in general. Only limited resources were available for the development and the validation of the ship detection algorithm. For a future version of the system however, we are planning to cooperate with remote sensing experts to integrate a mature, validated, state-of-the-art ship detection and classification processor.

Currently, we are part of the project ScOSA (Scalable On-Board Computing for Space Avionics), which has the goal of developing a high-performance on-board computer for satellite platforms [31]. The ScOSA system consists of multiple hardware nodes, uses a distributed computing approach, and can be dynamically reconfigured during runtime to remove faulty nodes and shift applications to healthy ones. We contribute to this project by porting AMARO to the ScOSA platform in order to stress the overall system and demonstrate its computing capacity [30].

It was not part of our experiment to synchronize the signals from the AIS receiver with the results from the image processing. However, the fusion of AIS and image data would bring a significant benefit. In particular, ships without signals could thus be identified. In the scientific community, there are several ongoing projects engaged in the fusion of AIS and image data [43]. Hence, we are establishing cooperations to rely on profound experience for the future improvement of our application. At this stage, we would like to mention that our system is not limited to optical data and AIS. Other sources of signals, e.g., an SAR camera (Synthetic Aperture Radar) or a pager for mobile phones, can be added without modifying the existing concept or the software structure.

4.3. System Design

Several publications about the individual subsystems exist, e.g., on-board image processing or real-time communication. However, our investigation and development is aimed at designing an operable system as a whole. We designed a comprehensive modular system for on-board data analysis and real-time information. It detects vessels and sends the results to the interested user within minutes after sighting. Our system is not designed as a monolithic block, but is flexibly expandable and deployable. It is modeled similarly to modern internet searching engines, consisting of a big database and several services that request and modify the database. The software system is therefore easy to expand, to adapt, and to maintain. AMARO is not set up as a simple one-way processing chain, i.e., getting images, extracting information, sending results. In fact, it is an autonomously working entity respondent to the user's needs.

4.4. General Limitations of the System

By now, the main benefit of the system is achieved by using optical image data. Therefore, usability of the system heavily depends on the weather and lighting conditions. Operation at night is not supported, and during the day, heavy cloudiness can seriously limit the surveillance performance of the system. In the future, synthetic aperture radar sensors may be used to noticeably enhance the surveillance usability of the system. By now, this option is not feasible due to the weight and energy usage of available sensors and the high computing performance needed to process the data. Furthermore, with satellites, a permanent surveillance of a specified region is not feasible, as geostationary satellites do not provide a reasonable image resolution. However, in such a scenario, we see the benefit of the system as an additional data source, instead of as a single permanent surveillance solution.

4.5. Expansion of Deployment

A field to be investigated in more detail is that of possible flight devices. High-altitude pseudo-satellites seem to be predestined for this, since they offer the possibility of continuously monitoring an area of interest autonomously and for a longer duration. With the DLR working on the development of a high-altitude platform [44] and commercial systems like the Airbus Zephyr [18] starting to become available, we think that, within the next five years, suitable flight platforms may be a realistic option.

Finally, we are planning to expand our system to be deployable for other time-critical Earth observing scenarios that would benefit from a rapid information system; for example, real-time monitoring of traffic, sea ice, or disasters.

Author Contributions: Conceptualization, K.W. and K.S.; methodology, K.W. and K.S.; software, K.W. and K.S.; validation, K.W., K.S., and J.B.; investigation, K.W. and K.S.; resources, K.W., K.S., and J.B.; data curation, K.W.; writing—original draft preparation, K.W., K.S., and J.B.; writing—review and editing, K.W., K.S., and J.B.; supervision, K.W. and K.S.; project administration, K.W. and K.S. All authors have read and agreed to the published version of the manuscript.

Funding: This research received no external funding.

Acknowledgments: We also thank Christian Mietner for building the AMARO-box and providing technical support, Sebastian Pless for organizing the flight campaign, Daniel Hein for providing software support for the MACS camera system and all persons at the DLR involved in the AMARO project.

Conflicts of Interest: The authors declare no conflict of interest.

Abbreviations

The following abbreviations are used in this manuscript:

AIS	Automatic Identification System
AMARO	Autonomous real-time detection of moving maritime objects
COTS	Commercial Off-The-Shelf
EMSA	European Maritime Safety Agency
GNSS	Global Navigation Satellite System
GSD	Ground Sample Distance
MDPI	Multidisciplinary Digital Publishing Institute
MMSI	Maritime Mobile Service Identity
SBD	Short Burst Data
SQL	Structured Query Language

References

1. United Nations. International Maritime Organization (IMO)—IMO Profile. Available online: https://business.un.org/en/entities/13 (accessed on 23 January 2020).
2. EMSA. European Maritime Safety Agency—SafeSeaNet. Available online: http://www.emsa.europa.eu/ssn-main.html (accessed on 18 February 2020).
3. Chintoan-Uta, M.; Silva, J.R. Global maritime domain awareness: A sustainable development perspective. *WMU J. Marit. Aff.* **2017**, *16*, 37–52. [CrossRef]
4. Tikanmäki, I.; Ruoslahti, H. Increasing Cooperation between the European Maritime Domain Authorities; *IARAS* **2017**. Available online: https://www.iaras.org/iaras/home/caijes/increasing-cooperation-between-the-european-maritime-domain-authorities (accessed on 27 February 2020).
5. Proud, R.; Browning, P.; Kocak, D.M. AIS-based Mobile Satellite Service Expands Opportunities for Affordable Global Ocean Observing and Monitoring. In Proceedings of the OCEANS 2016 MTS/IEEE Monterey, Monterey, CA, USA, 19–23 September 2016; pp. 1–8. [CrossRef]
6. Metcalfe, K.; Bréheret, N.; Chauvet, E.; Collins, T.; Curran, B.K.; Parnell, R.J.; Turner, R.A.; Witt, M.J.; Godley, B.J. Using Satellite AIS to Improve our Understanding of Shipping and Fill Gaps in Ocean Observation Data to Support Marine Spatial Planning. *J. Appl. Ecol.* **2018**, *55*, 1834–1845. [CrossRef]
7. Traffic, M. Maritime Traffic—Provider of Ship Tracking and Maritime Intelligence. Available online: https://www.marinetraffic.com (accessed on 21 January 2020).

8. Reggiannini, M.; Righi, M.; Tampucci, M.; Lo Duca, A.; Bacciu, C.; Bedini, L.; D'Errico, A.; Di Paola, C.; Marchetti, A.; Martinelli, M. Remote Sensing for Maritime Prompt Monitoring. *J. Mar. Sci. Eng.* **2019**, *7*. [CrossRef]

9. Nie, T.; Han, X.; He, B.; Li, X.; Liu, H.; Bi, G. Ship Detection in Panchromatic Optical Remote Sensing Images Based on Visual Saliency and Multi-Dimensional Feature Description. *Remote Sens.* **2020**, *12*, 152. [CrossRef]

10. Chang, Y.L.; Anagaw, A.; Chang, L.; Wang, Y.C.; Hsiao, C.Y.; Lee, W.H. Ship Detection Based on YOLOv2 for SAR Imagery. *Remote Sens.* **2019**, *11*, 786. [CrossRef]

11. Krause, D.; Schwarz, E.; Voinov, S.; Damerow, H.; Tomecki, D. Sentinel-1 Near Real-time Application for Maritime Situational Awareness. *CEAS Space J.* **2019**, *11*, 45–53. [CrossRef]

12. EMSA. European Maritime Safety Agency—Copernicus Maritime Surveillance. Available online: http://emsa.europa.eu/publications/information-leaflets-and-brochures/item/2880-copernicus-maritime-surveillance-service-overview.html (accessed on 23 January 2020).

13. EMSA. European Maritime Safety Agency—Earth Observation Services. Available online: http://www.emsa.europa.eu/operations/earthobservationservices.html (accessed on 21 January 2020).

14. Dawood, A.S.; Visser, S.; Williams, J.A. Reconfigurable FPGAs for Real Time Image Processing in Space. In Proceedings of the 2002 14th International Conference on Digital Signal Processing, DSP 2002 (Cat. No. 02TH8628), Santorini, Greece, 1–3 July 2002; Volume 2, pp. 845–848.

15. Nadoushan, M.J.; Assadian, N. Repeat Ground Track Orbit Design With Desired Revisit Time and Optimal Tilt. *Aerosp. Sci. Technol.* **2015**, *40*, 200–208. [CrossRef]

16. Klimkowska, A.; Lee, I.; Choi, K. Possibilities of UAS for Maritime Monitoring. *Int. Arch. Photogramm. Remote Sens. Spat. Inf. Sci.* **2016**, *41*, 885. [CrossRef]

17. EMSA. European Maritime Safety Agency—Remotely Piloted Aircraft Systems (RPAS). Available online: http://emsa.europa.eu/operations/rpas.html (accessed on 22 January 2020).

18. Online, A. Airbus Zephyr S HAPS Sets New World Record. Available online: https://aeronauticsonline.com/airbus-zephyr-s-haps-sets-new-world-record/ (accessed on 23 January 2020).

19. Prior-Jones, M. *Satellite Communications Systems Buyers' Guide*; British Antarctic Survey: Cambridge, UK, 2008.

20. Yao, Y.; Jiang, Z.; Zhang, H.; Zhou, Y. On-Board Ship Detection in Micro-Nano Satellite Based on Deep Learning and COTS Component. *Remote Sens.* **2019**, *11*, 762. [CrossRef]

21. Ji-yang, Y.; Dan, H.; Lu-yuan, W.; Jian, G.; Yan-hua, W. A Real-time On-board Ship Targets Detection Method for Optical Remote Sensing Satellite. In Proceedings of the 2016 IEEE 13th International Conference on Signal Processing (ICSP), Chengdu, China, 6–10 November 2016; pp. 204–208. [CrossRef]

22. Iridium Communications Inc.. Iridium Short Burst Data (SBD). Available online: https://www.iridium.com/services/details/iridium-sbd (accessed on 12 December 2019).

23. Iridium Communications Inc. Iridium 9603. Available online: https://www.iridium.com/products/iridium-9603/ (accessed on 23 January 2020).

24. Wireless Innovation Inc. MiChroBurst—Out of the Box Ready SBD Modem. Available online: https://www.wireless-innovation.co.uk/product/wiltd-michroburst/ (accessed on 24 January 2020).

25. AeroAntenna Technology Inc. Iridium Antenna AT2775-110GA. Available online: https://www.aeroantenna.com/PDF/AT2775-110GA_C.pdf (accessed on 30 January 2020).

26. U.S. Department of Homeland Security, U.D. Navigation Center of Excellence—AIS Messages. Available online: https://www.navcen.uscg.gov/?pageName=AISMessages (accessed on 27 January 2020).

27. Alltek Marine Electronics Corp. AMEC Cypho-150 AIS Receiver User Manual. Available online: http://www.alltekmarine.com/upload/d4fs111712114611153100.pdf (accessed on 27 February 2020).

28. Schwehr, K. C++ decoder for Automatic Identification System for Tracking Ships and Decoding Maritime Information. Available online: https://github.com/schwehr/libais (accessed on 23 January 2020).

29. Procom, A. Portabelantennen HX 2/...-FME—Helixantenne mit universellem FME-Adaptersystem für tragbare Geräte im 2 m Band. Available online: https://amphenolprocom.com/de/produkte/portable-antennas-de/produkter/380-hx-2-fme (accessed on 24 January 2020).

30. Schwenk, K.; Ulmer, M.; Peng, T. ScOSA: Application Development for a High-Performance Space Qualified Onboard Computing Platform. In Proceedings of the SPIE 10792, High-Performance Computing in Geoscience and Remote Sensing VIII, Berlin, Germany, 10–13 September 2018.

31. Treudler, C.J.; Benninghoff, H.; Borchers, K.; Brunner, B.; Cremer, J.; Dumke, M.; Gärtner, T.; Höflinger, K.J.; Lüdtke, D.; Peng, T.; et al. ScOSA-scalable On-board Computing for Space Avionics. In *Proceedings of the International Astronautical Congress*; IAC: New York, NY, USA, 2018.

32. SQLite-Team. Offical SQLite project page. Available online: https://www.sqlite.org/ (accessed on 13 December 2019).

33. Olson, E. Particle Shape Factors and Their Use in Image Analysis Part 1: Theory. *J. GXP Compliance* **2011**, *15*, 85.

34. Schwenk, K. Methoden zur Bildsegmentation und zum Connected Component Labeling auf einem FPGA für eine zukünftige On-Board Bilddatenverarbeitung. Ph.D. Thesis, Raumflugbetrieb und Astronautentraining, Universität der Bundeswehr München, Neubiberg, Germany, 2016

35. Nischwitz, A.; Fischer, M.; Haberäcker, P.; Socher, G. *Bildverarbeitung: Band II des Standardwerks Computergrafik und Bildverarbeitung*; Springer: Berlin, Germany, 2020.

36. Brauchle, J.; Bayer, S.; Hein, D.; Berger, R.; Pless, S. MACS-Mar: A Real-time Remote Sensing System for Maritime Security Applications. *CEAS Space J.* **2019**, *11*, 35–44. [CrossRef]

37. Brauchle, J.; Bayer, S.; Berger, R. Automatic Ship Detection on Multispectral and Thermal Infrared Aerial Images Using MACS-Mar Remote Sensing Platform. In *Pacific-Rim Symposium on Image and Video Technology*; Springer: Berlin, Germany, 2017; pp. 382–395.

38. Iridium Communications Inc. *Iridium Short Burst Data Service Developers Guide, Release 3.0*; Iridium Communications Inc.: McLean, VA, USA, 2012.

39. Horan, S. The Potential for Using LEO Telecommunications Constellations to Support Nanosatellite Formation Flying. *Int. J. Satell. Commun.* **2002**, *20*, 347–361. [CrossRef]

40. Schönegger, S. Investigation of LEO Communication Networks for Space Telemetry and Commanding. Ph.D. Thesis, University of Applied Sciences TECHNIKUM WIEN Department of Electronics, Vienna, Austria, 2004.

41. Kalnins, I. RUBIN Microsatellites for Advanced Space Technology Demonstration. In Proceedings of the IAF Abstracts, 34th COSPAR Scientific Assembly, Houston, TX, USA, October, 10–19 October 2002.

42. Marcus, M.; Ali Guarneros-Luna, R.A.; Wheless, J. Communications for the TechEdSat/PhoneSat Missions. In Proceedings of the Presentation to Small Satellite Pre-Conference Workshop, Campus Utah State University, Logan, UT, USA, 2017. Available online: https://digitalcommons.usu.edu/smallsat/2016/S7Comm/6/ (accessed on 27 February 2020).

43. Andersson, M.; Johansson, R. Multiple Sensor Fusion for Effective Abnormal Behaviour Detection in Counter-Piracy Operations. In Proceedings of the 2010 International WaterSide Security Conference, Carrara, Italy, 3–5 November 2010; pp. 1–7.

44. DLR. Uncrewed High-Altitute Platform (HAP). Available online: https://www.dlr.de/content/en/articles/digitalisation/digitalisation-project-hap.html (accessed on 18 February 2020).

Article

Adaptive Intrawell Matched Stochastic Resonance with a Potential Constraint Aided Line Enhancer for Passive Sonars

Haitao Dong [1,2,†], **Ke He** [1,2,†], **Xiaohong Shen** [1,2], **Shilei Ma** [1,2], **Haiyan Wang** [1,3,*] and **Changcheng Qiao** [4,5]

[1]   School of Marine Science and Technology, Northwestern Polytechnical University, Xi'an 710072, China; haitaodong@mail.nwpu.edu.cn (H.D.); hk@nwpu.edu.cn (K.H.); xhshen@nwpu.edu.cn (X.S.); slma@mail.nwpu.edu.cn (S.M.)

[2]   Key Laboratory of Ocean Acoustics and Sensing (Northwestern Polytechnical University), Ministry of Industry and Information Technology, Xi'an 710072, China

[3]   School of Electronic Information and Artificial Intelligence, Shaanxi University of Science and Technology, Xi'an 710021, China

[4]   The 27th Research Institude of China Electronic Technology Group Corporation, Zhengzhou 450047, China; ccqiao1227@sina.com

[5]   Zhengzhou Key Laboratory of Underwater Information System Technology, Zhengzhou 450000, China

*   Correspondence: hywang@sust.edu.cn

†   These authors contributed equally to this work.

Received: 1 May 2020; Accepted: 1 June 2020; Published: 8 June 2020

Abstract: Remote passive sonar detection and classification are challenging problems that require the user to extract signatures under low signal-to-noise (SNR) ratio conditions. Adaptive line enhancers (ALEs) have been widely utilized in passive sonars for enhancing narrowband discrete components, but the performance is limited. In this paper, we propose an adaptive intrawell matched stochastic resonance (AIMSR) method, aiming to break through the limitation of the conventional ALE by nonlinear filtering effects. To make it practically applicable, we addressed two problems: (1) the parameterized implementation of stochastic resonance (SR) under the low sampling rate condition and (2) the feasibility of realization in an embedded system with low computational complexity. For the first problem, the framework of intrawell matched stochastic resonance with potential constraint is implemented with three distinct merits: (a) it can ease the insufficient time-scale matching constraint so as to weaken the uncertain affect on potential parameter tuning; (b) the inaccurate noise intensity estimation can be eased; (c) it can release the limitation on system response which allows a higher input frequency in breaking through the large sampling rate limitation. For the second problem, we assumed a particular case to ease the potential parameter $a_{opt} = 1$. As a result, the computation complexity is greatly reduced, and the extremely large parameter limitation is relaxed simultaneously. Simulation analyses are conducted with a discrete line signature and harmonic related line signature that reflect the superior filtering performance with limited sampling rate conditions; without loss of generality of detection, we considered two circumstances corresponding to H_1 (periodic signal with noise) and H_0 (pure noise) hypotheses, respectively, which indicates the detection performance fairly well. Application verification was experimentally conducted in a reservoir with an autonomous underwater vehicle (AUV) to validate the feasibility and efficiency of the proposed method. The results indicate that the proposed method surpasses the conventional ALE method in lower frequency contexts, where there is about 10 dB improvement for the fundamental frequency in the sense of power spectrum density (PSD).

Keywords: adaptive stochastic resonance (ASR); matched intrawell response; nonlinear filter; line enhancer; autonomous underwater vehicles (AUVs)

1. Introduction

Passive sonars have been proven to have practical efficiency in detecting and recognising selfemitting underwater targets such as ships, submarines, and autonomous underwater vehicle (AUVs), etc. [1–3]. Generally, narrowband discrete components of ship-radiated noise, known simply as lines or tonals, have largely prevailed so far [4–6]. From the view of rotatory machinery, ship-radiated line signatures that are relevant to the engine and shaft/propeller rotation attract lots of attention, of which the long-term challenging problem is to address the weak signatures from heavy noise background in the far field scenario [3,7,8]. In this way, increasing the signal-to-noise ratio (SNR) is expected as a tenet to detection performance for applications. To enhance the narrowband discrete components, passive sonars usually employ an adaptive line enhancer (ALE) as a pre-processing step [9–12]. However, performance of the conventional ALE is limited, and advanced signal processing techniques dedicated to better denoising performance are of highly practical importance.

In general, more noise in the system often leads to worse detection performance and degrades the estimation accuracy. However, despite its disruptive character in nature, noise does play an important constructive role under certain conditions [13]. This phenomenon so called stochastic resonance (SR), in which the output periodic signals can be enhanced by the cooperative effect between "signal", "noise" and certain "nonlinear systems" [14]. Since proposed by Benzi et al. in 1981 [15], it has become a hot research topic in the field of nonlinear science. Previous studies have addressed a lot of research progress, both theoretical and experimental, in achieving some noteworthy contributions in practical applications [8,16–21]. As a result, taking SR for weak signal detection is regarded as a potential novel technique for weak signal detection, especially under low SNR conditions.

Classical SR theories generally refer to a noise enhanced phenomenon by means of adding an appropriate amount of noise [14]. This is restricted to applications of how to remove proper noise, especially under low SNR circumstances. Rather than adjusting the input noise levels, Xu et al. [22] proposed a parameter-induced stochastic resonance (PSR) which highlighted the effect of the nonlinear systems and promoted the flexibility of SR utilization in designing systems to deal with the noise. In view of this, utilizing SR in weak signal processing could be considered as a special nonlinear filter, which can achieve superior performance compared with the traditional noise-suppression-based filters [8,23–25]. Among the publications on this point, the simplest first-order overdamped Langevin equation (LE) model is largely adopted, while the filtering performance is limited [8,26,27]. To get a cleaner filtered signal with higher signal-to-noise ratio (SNR), the second order Duffing equation with an underdamped system is used for a secondary filtering effect, which reflects a superior filtering performance compared with the traditional state-of-the-art methods [23,24,28]. For the purpose of better improvements, some authors reported cascaded SR systems [29,30], coupled SR system [31,32], etc. with bistable or multi-stable potentials. As a matter of fact, superior performance generally required a high sampling frequency condition to fully take advantage of the nonlinear property by concentrating most of the noise energy into the low-frequency region [33]. Several efforts have addressed the problem of large parameter stochastic resonance (LPSR) by tuning the signal structure and (or) the system parameters [24,27,34]. However, all the methods of LPSR are based on the classical SR part, and the classical SR generally requires a large sampling frequency that is more than 50 times the driving frequency. All these processing approaches are essential to deal with the periodic signal individually, and hence, flexible limited and lack of computation efficiency. An open problem for practical applications is "Can the SR be realized under limited sampling rate?"

To address this problem, limited studies on intrawell SR can give us inspirations. Actually, intrawell SR can exist for a periodically driven noisy signal in terms of underdamped bistable nonlinear dynamic systems [14]. Alfonsi et al. [35] gave a phemonological nonadiabatic description of intrawell SR by adding Lorentz' colored noise. On the basis of this, Li et al. [36] further demonstrate that the intrawell SR phenomenon usually occurs in a system with optimal system parameters. Since the system response speed of intrawell motion is much larger than that of interwell, it can release the limitation on system response speed which allows a higher input frequency. In this way, the large sampling condition can be eased in properly controlling the escape time scale (characterizing interwell jumps) together with the relaxation time of the particle. Back to the view of potential parameter tuning, the adjustable potential parameters can be equivalent to tuning potential depth and well separation [28]. There is lack of flexibility in matching the escape time scale and the relaxation time only with the damping effect. In our previous work [37], the nonlinear filtering effects of intrawell matched stochastic resonance is analyzed, which have shown a superior filtering performance as well as a wider range of frequency response. This can give us a guidance to ease the large sampling frequency limitation.

On the basis of the aforementioned analyses, this paper proposes a novel parameterized filter implementation on adaptive intrawell matched stochastic resonance (AIMSR) with a potential constrained Duffing system. As can be seen in [24,37] matching a high frequency signal requires extremely large system parameters. This refers to a large range of parameter searching interval to limit the computational efficiency and parameterized realization, especially in an embedded system. In this regard, we further eased the potential parameter $a = 1$ which makes it practically realizable. The nonlinear filtering effects are analyzed and evaluated, reflecting a superior performance in enhancing the lines especially under a limited sampling rate. Application verification is further conducted with a set of low sampled data fragment of autonomous underwater vehicles (AUVs). The output performance is further compared with the conventional ALE method, which shows an excellent filtering performance in enhancing the lines, especially for the lower frequency fundamental signature.

The rest of the paper is arranged as follows. After introducing the signal model and measurement in Section 2, in Section 3, the theory and implementation of the AIMSR method are detailed and described with a potential constraint Duffing system. Section 4 verifies the validity of utilizing AIMSR in dealing with the single and muli-harmonic line signature signals. Section 5 verifies the practicability of the proposed method by analyzing a set of low sampled AUVs data fragments. Besides, intensive discussions are made to give an insight into the principal results and inspired future investigations. Finally, concluding remarks are drawn in Section 6.

2. Signal Model and Measurement

Ship radiated noise have been studied for years. Ross and Urick have given an excellent description of the mechanisms of sound generation by large surface ships and submarines [4,5]. They have shown that the radiated noise from a ship is a combination of broadband noise and sinusoidal tonal signals that are generated by many sources. Theoreitical and experimental results have shown signatures related to the rotation of engines, shaft-line dynamics, propeller cavitations are mostly efficient for the purpose of passive detection, tracking, and classification [1,6,38,39]. From the view of rotary machinery of the shaft and propeller, the harmonic signature is typically connected to the running state and can be measured by vibrational and acoustic approaches. For an underwater surveillance system, to address the shaft and propeller signatures, algorithms such as the Detection of Envelope Modulation on Noise (DEMON) [40], cyclic modulation coherence (CMC) [7], etc. are employed to check for the presence of periodic components to the broadband cavitation noise. This generally requires acoustic measurement of broadband noise in a high-frequency range of thousands of Hertz. Additional studies in recent years have shown that the harmonic signature as a propeller rotates can be observed by low frequency acoustic/vibrational measurement as well, which may propagate for a very long distance [41]. Such measurement techniques have been ubiquitously seen in sonobuoys, ocean bottom seismometers (OBS), acoustic vector hydrophone and other small-scale equipment [41–43].

As a matter of fact, these sinusoidal tonals are commonly considered to be the "acoustic fingerprint" of a moving vessel. Table 1 shows the major contributions and relations to the sinusoidal tonal signals from ship engine and propeller. In the situation of practical applications, these sinusoidal components can be modeled as follows,

(i) Discrete line signature

$$r(t) = s(t) + n(t)$$
$$= A_0 cos(2\pi f_0 t + \varphi_0) + n(t) \tag{1}$$

where f_0 represent the character frequency, A_0 and φ_0 are the corresponding amplitude and phase, respectively. $n(t)$ represent the combination of radiated broadband noise and ambient noise.

(ii) Harmonic related line signature

$$r(t) = s(t) + n(t)$$
$$= \sum_{h=1}^{M} A_h cos(2\pi h f_0 t + \varphi_i) + n(t) \tag{2}$$

in which h is the harmonic number, A_h and φ_h are the amplitude and phase of the hth harmonic component, and f_0 is the fundamental frequency of shaft rotation, $n(t)$ represent the combination of radiated broadband noise and ambient noise.

Table 1. Character frequencies of ship engine and propeller.

Engine Rates	Propeller Rates
Cylinder Firing Rate $f_{CF} = f_{CR}/2$	Shaft Rotation Rate $f_{SR} = f_{CR}/\lambda_g$ λ_g: Gear Ratio
Crankshaft Rotation Rate $f_{CR} = RPM/60$ RPM: Engine Speed	Blade Rotation Rate $f_{BR} = N_b f_{SR}$ N_b: Number of Blades
Engine Firing Rate $f_{EF} = N_c f_{CF}$ N_c: Number of Cylinders	

For small targets such as AUVs, speedboats, etc. with rotation marine engine and propeller, the signatures are in a periodic pattern as well. Their acoustic characteristics have been comprehensively studied by measurement tests and modeling [2,39,44], which indicate that the model of radiated noise can be the same with large ships. Since an AUV's prop does not rotate sufficiently fast to cause cavitation, it is hard to be detected with broadband noise in the high-frequency range. As is known, the low frequency harmonics are quite important for remote passive detection, and the experimental study in this work is set up with a very low frequency measurement by ocean bottom seismometer (samapling frequency f_s =200 Hz).

3. Adaptive Intrawell Matched Stochastic Resonance

3.1. Generalized Matched Stochastic Resonance with Duffing Oscillator

The bistable SR phenomenon can be described as a particle driven by periodic force and random force in a quartic double-well system, where the periodic motion can be heightened with the assistance of moderate noise. This can be governed by a two-dimensional Duffing oscillator as below [24],

$$\frac{\mathrm{d}^2 x}{\mathrm{d}t^2} + \gamma \frac{\mathrm{d}x}{\mathrm{d}t} = -\frac{\mathrm{d}V(x)}{\mathrm{d}x} + s(t) + \xi(t) \tag{3}$$

where γ is the damping factor, $s(t)$ represents the input periodic signal, and $n(t)$ stands for the noise item with noise intensity D. In particular, Equation (3) describes the movement of a particle in a quartic double well potential $V(x)$,

$$V(x) = -\frac{a}{2}x^2 + \frac{b}{4}x^4, \qquad a, b > 0 \tag{4}$$

in which a and b are barrier parameters of bistable SR system, and the potential barrier can be calculated by $\Delta V = a^2/(4b)$. As the analytic form of the SR output cannot be easily obtained, it is generally numerically calculated through the discrete fourth-order Runge-Kutta method [23]. Mathematically, let $\dfrac{dx}{dt} = y$, then Equation (3) can be separated into two first-order differential equations as,

$$\begin{aligned} \frac{dx}{dt} &= y \\ \frac{dy}{dt} &= ax - bx^3 - \gamma y + s(t) + \xi(t) \end{aligned} \tag{5}$$

and then the output discrete time series $x[n]$ corresponding to Equation (3) can be calculated by,

$$\left\{ \begin{aligned} y_1 &= y[n] \\ x_1 &= -V'(x[n]) - \gamma y_1 + s[n] + \xi[n] \\ y_2 &= y[n] + x_1 h/2 \\ x_2 &= -V'\left(x[n] + y_1 h/2\right) - \gamma y_2 + s[n] + \xi[n] \\ y_3 &= y[n] + x_2 h/2 \\ x_3 &= -V'\left(x[n] + y_2 h/2\right) - \gamma y_3 + s[n+1] + \xi[n+1] \\ y_4 &= y[n] + x_3 h \\ x_4 &= -V'\left(x[n] + y_3 h\right) - \gamma y_4 + s[n+1] + \xi[n+1] \\ x[n+1] &= x[n] + (y_1 + 2y_2 + 2y_3 + y_4)\, h/6 \\ y[n+1] &= y[n] + (x_1 + 2x_2 + 2x_3 + x_4)\, h/6 \end{aligned} \right. \tag{6}$$

where h is the calculation step of the Runge-Kutta method.

From the perspective of parameter tuning, it can be considered as a special nonlinear filter, and the main issue is to achieve an "matched filter" output [24]. In general, a bistable system response is optimized by maximizing the signal-to-noise ratio improvement (SNRI) on system potential parameters a and b, time scaling parameter h, and damping factor γ. In this way, the nonlinear matched SR filter can be generally modeled with a generalized time-scale matching constraint as,

$$\begin{aligned} \max_{a,b,\gamma,h} \quad & \text{SNRI} \\ s.t. \quad & r_K = \frac{\Omega}{\pi} \\ & \gamma \in (0,1], \ h \in (0,1] \end{aligned} \tag{7}$$

in which $r_K = \dfrac{a}{\sqrt{2}\pi\gamma} \exp(-\dfrac{a^2}{4bD})$ is the famous Kramers rate, $\Omega = 2\pi f_0$ represents the angular frequency of periodic forcing. The constraint of $r_K = \Omega/\pi$ is the time-scale matching condition that represent the statistical synchronization of interwell transition [14]. It is necessary to point out that this generalized condition is not sufficient as other optimized factors are required to judge the occurance of SR.

For sinusoidal signals with additive noise, the SNRI corresponding to Equation (3) could have a maximum at $D_{SR} = \Delta V$, and can be equivalent to $\hat{D} = a_{opt}/4b_{opt}$ in according to [24]. And as a consequence, the mathematic model of SR based nonlinear matched filter can be rewritten as,

$$\left(\gamma^*_{opt}, h^*_{opt}\right) = \underset{\gamma, h}{\operatorname{\mathbf{argmax}}} \ \mathrm{SNRI}$$

$$\text{s.t.} \quad a_{opt} = 2\sqrt{2}\pi f_0 \gamma e$$

$$b_{opt} = a^4_{opt}/4\hat{D}$$

$$\gamma \in \left(0, 16\sqrt{2}\pi f_0 e\right], h \in (0,1] \tag{8}$$

where e is the base of natural logarithms, and the optimal potential parameters a_{opt} and b_{opt} can be obtained with matched relationship that related to the frequency of periodic signal f_0, damping factor γ and the noise intensity estimator $\hat{D}$.

To adaptively adjust these system parameters, the measurement index of input-output SNR improvement (SNRI) is extensively utilized and can be estimated by measuring the power spectral of a time series. This can be further limited with a bandwidth ΔB for the local SNR (SNR_{local}) calculation as it is generally connected to the performance of signal detection [45]. Besides, evaluation indexes such as power spectrum kurtosis (PSK), peak SNR (PSNR), spectral correlation coefficient (SC), structural similarity index (SSI), etc. can be well adopted and fused [20].

3.2. Framework of Intrawell Matched Stochastic Resonance with Potential Constraint

The "classical" description of the SR phenomenon in a bistable system is that a particle can jump across the potential barrier back and forth in the assistance of proper noise. This is called interwell jumping behavior as the random-switching frequency r_K (Kramers rate) is made to agree closely with the periodic forcing angular frequency [14,35]. From the view of parameter tuning, the adjustable potential parameters can be equivalent to tuning the potential depth ΔV and well location (stable focus) $\pm x_m$ [28], where Equation (4) can be transformed into the function of well location and potential depth parameter as,

$$V(x) = -\Delta V \left[2\left(\frac{x}{x_m}\right)^2 - \left(\frac{x}{x_m}\right)^4 \right] \tag{9}$$

In this way, changing the potential parameters a and b is essential to adjust the potential depth ΔV and the separation $2x_m$. A demonstration of particles in a bistable model with different potential parameters is given in Figure 1. It can be seen that a large potential depth can lead to longer system response time for transition, and even intrawell response [35,36].

As is known, the time-scale matching constraint is directly connected to the system response time. The nonlinear filtering effect of matched stochastic resonance can be essentially regarded as a proper control of the particle's response. From this view, it is easy to know that the more the degrees of freedom for parameter tuning, the better the output response. Such results can be found in plenty of SR-related publications with multi-parameter optimization. The classical Duffing equation model with bistable potential utilizes parameter a, b, and γ are used to characterize the system response. It is intuitive to add new degrees of freedom for better characterizing and controlling of the particle's response. Hence, a potential constraint is further adopted here to ease the insufficient time-scale matching constraint [37]. By this means, the mathematical framework of matched stochastic resonance can be further generalized by intuitively adding a constraint of barrier factor K as below,

$$\left(\gamma^*_{opt}, h^*_{opt}, K^*_{opt}\right) = \underset{\gamma, h, K}{\operatorname{\mathbf{argmax}}} \ \mathrm{SNRI}$$

$$\text{s.t.} \quad a_{opt} = 2\sqrt{2}\pi f_0 \gamma e$$

$$b_{opt} = a^4_{opt}/4K\hat{D}$$

$$\gamma \in \left(0, 16\sqrt{2}\pi f_0 e\right], h \in (0,1], K \in (1, K_{max}] \tag{10}$$

in which K should be a postive real number to adjust the particle motion, and can be optimized from a proper searching interval.

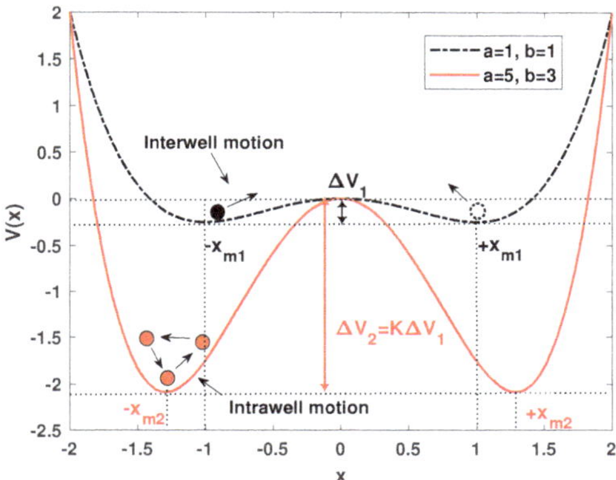

Figure 1. A demonstration of bistable stochastic resonance with interwell and intrawell motion.

The merits of this potential constraint model can be summarized as follows: (1) it can ease the insufficient time-scale matching constraint so as to weaken the uncertain effect on potential parameter tuning; (2) the inaccurate noise intensity estimation can be eased as well; (3) it can release the limitation on a system response which allows us a higher input frequency in breaking through the large sampling rate limitation. All in all, the potential constraint model is anticipated to be superior in nonlinear filtering performance, especially under a low sampled circumstance that is highly concerned for engineering applications.

3.3. Adaptive Strategy for Optimized Implementation

As mentioned in Equation (10), it is a multi-parameter optimization problem, where the constraint of the matched potential relationship is a nonlinearity that can properly be solved by intelligent optimization algorithms such as genetic algorithm (GA), particle swarm optimization (PSO), ant colony optimization (ACO), grey wolf optimization (GWO), et al. The main problem is to determine the proper searching interval of each parameter.

To address this problem, it can be known from [24,37] that in matching high frequency, signals require extremely large system parameters. As for practical engineering fields, the frequency of received signals generally varies from tens to thousands of Hertz, and hence, the order of magnitude for a matched potential parameter should be extremely large, especially for parameter b as there is a relationship with $b \propto a^4$. This problem is the same to the barrier factor K as there is an experimental guidance that K^*_{opt} approximate to f_0^4 [37]. This refers to a large range of parameter searching interval to limit the computational efficiency and parameterized realization. Such a problem will limit the utilization in a real-time online embedded system in confronting data overflow problem (such as commonly utilized 32 bits digital signal processor (DSP)). From this view, the frequency shift to low frequency should be a kind of efficient solution for this problem, while lack of flexibility in dealing with unknown frequency signal or multi-frequency signal. In the consideration of the degree of freedom, the potential constraint of a matched relationship can be relaxed by the new freedom of the barrier factor K. As a consequence, the matched potential relationship can be eased. From the aforementioned analysis, for this problem, one can use the biquadrate relationship between the potential parameter a and b. In this regard, we can assume a particular case here by easing the potential parameter $a_{opt} = 1$. As a result, the computation complexity is greatly deduced, and the extremely large parameter limitation is relaxed simultaneously. The simplified model can be expressed as,

$$
\begin{aligned}
\left(h_{opt}^*, K_{opt}^*\right) \quad &= \underset{\gamma,h,K}{\textbf{argmax}}\ \text{SNRI} \\
\text{s.t.} \quad a_{opt} &= 1 \\
b_{opt} &= 1/(4K\hat{D}) \\
\gamma_{opt} &= (2\sqrt{2}\pi f_0 e)^{-1} \\
h &\in (0,1],\ K \in (0, K_{max}]
\end{aligned}
\tag{11}
$$

where the K_{max} can be further limitted, and the optimal damping factor γ_{opt} is underdamped as well. Here, we need to point out that in this simplification process, the insufficient time-scale matching constraint result in an insufficient relation of $\gamma_{opt} = (2\sqrt{2}\pi f_0 e)^{-1}$. Therefore, this constraint should be eased by a generalized underdamped condition to keep the degrees of freedom. Hence, the optimization model can be further expressed as

$$
\begin{aligned}
\left(\gamma_{opt}^*, h_{opt}^*, K_{opt}^*\right) \quad &= \underset{\gamma,h,K}{\textbf{argmax}}\ \text{SNRI} \\
\text{s.t.} \quad a_{opt} &= 1 \\
b_{opt} &= 1/(4K\hat{D}) \\
\gamma &\in (0,1],\ h \in (0,1],\ K \in (0, K_{max}]
\end{aligned}
\tag{12}
$$

3.4. Implementation of Adaptive Intrawell Matched Stochastic Resonance

In the last subsection, we have the adaptive strategy as an optimization problem with nonlinear constraints. Here, a signal processing strategy is proposed by jointly optimizing the parameters with a classical genetic algorithm (GA) method for the sake of global optimality. Consequently, the detailed implementation steps are summarized as follows,

(1) Signal pretreament. Data normalization and prewhitening are executed on the actual received noisy signals.

(2) Searching range initialization. Initialize the optimization searching range of parameters $\gamma \in [0,1]$, $h \in [0.001, 1]$ and $K \in [0.001, 200]$, respectively.

(3) Parameter optimization. Obtain the optimal parameters γ_{opt}, h_{opt}, and K_{opt} according to the following objective function,

$$
\left(\gamma_{opt}^*, h_{opt}^*, K_{opt}^*\right) \quad = \underset{\gamma,h,K}{\textbf{argmax}}\ \text{SNRI}
\tag{13}
$$

where SNRI corresponds to the fitness criteria of GA method. Here, the input–output SNR improvement (SNRI) is constructed by the superposition of global SNRI (SNRI$_{global}$) and local SNRI (SNRI$_{local}$) to better characterize the filtering performance both in the view of global and local conditions. The SNR calculation for the input and output time series can be defined as,

$$
\text{SNR} = 10\log_{10}\frac{P_s - P_n}{P_n}
\tag{14}
$$

in which $P_s = \sum_{i=f-\Delta B_s/2}^{f+\Delta B_s/2} S_i$ represent the total power around the characteristic frequency f_0 within a small bandwidth of ΔB_s, and $P_n = \frac{1}{\Delta B}\left(\sum_{i=f-\Delta B/2}^{f+\Delta B/2} S_i - P_s\right)$ represent the average power of background noise within ΔB bandwidth around the characteristic frequency f_0. For the global SNR and the local SNR calculation, ΔB is setted by the full bandwidth and $f_s/100$, respectively.

For a multi-harmonic signal circumstance, a modified measurement index of average SNRI is utlized, which can be defined as,

$$\text{SNRI} = \frac{1}{M} \sum_{h=1}^{M} \text{SNRI}(h) \tag{15}$$

where M is the harmonic number of the received signal. The parameter optimization algorithm is summarized in Algorithm 1.

(4) Signal post-treatment. Output optimal waveform and the coresponding optimal fittest value.

Algorithm 1   Parameter Optimization Algorithm.

Parameter Initialization:

$C_r = [\gamma_{start} = 0, h_{start} = 0.001, K_{start} = 0.001; \gamma_{end} = 1, h_{end} = 1, K_{end} = 200]$: the searching intervals;

$N_c = 3$: number of chromosome;

$N_p = 100$: Number of individuals in the population;

$\chi = 0.95$: The fraction to be replaced by crossover in each iteration;

$\mu = 0.01$: The mutation rate;

$M = 10$: The maximal iteration times;

$\lambda_{stop} = 0$: The threshold of stop condition.

Initialize generation 0:

$k:=0$;

$P_k:=$a population of N_p randomly-generated individuals;

Evaluate P_k:

Compute fitness criteria SNRI for each $i \in P_k$;

{

   1: Compute the corresponded MSR output by fourth order Runge–Kutta (RK4) method according to

      Equation (6) and obtain $x[n]\,(n = 1, 2, ..., N)$, where N is the length of the time series;

   2: Compute the SNRI according to Equation (14) and Equation (15);

} **Create generation** k+1:

do

{

   1: **Copy**: Select $(1 - \chi) * n$ members of P_k and insert into P_{k+1};

   2: **Crossover**: Select $\chi * n$ members, pair them up to produce offspring and insert the offspring into

      P_{k+1};

   3: **Mutate**: Select $\mu * n$ members of P_{k+1}, and invert a randomly selected bit;

   4: **Evaluate** P_{k+1};

   5: **if** $P_{k+1} - P_k \le \lambda_{stop}$ **then** break;

   6: **else**

   7:    **Increment:** k:=k+1;

   8: **end if**

}

while $k \le M$;

return the optimal fittest individual from P_M;

---

4. Filtering Performance Analysis and Evaluation

SR can be regarded as a specific kind of nonlinear filter, while generally requires a high sampling rate condition. As is noticed in the Section 3.2, the proposed method can release the limitation on system response so as to allow us a higher input frequency in breaking through the large sampling rate limitation. Hence, here we intuitively exhibit the filtering performance referred to the discrete line signature and harmonic related line signature under a low sampling rate condition.

4.1. Discrete Line Signature Signal Analysis

Without loss of generality of detection, we consider two circumstances corresponding to H_1 and H_0 hypotheses, respectively. The sampling frequency is set to $f_s = 100$ Hz, and the data length $N = 2048$.

For H_1 hypothesis circumstance, the input is composed of a narrowband component with the frequency $f_0 = 10$ Hz, and the ambient noise that simulated by the Gaussian noise. The input SNR is set as -15 dB, and the normalized waveform and power spectrum are shown in Figure 2a. First, we employed the lofargram to qualitatively evaluate the performance. A 2.56 s short-time Fourier transformation (STFT) with the overlapping length of 0.1 s was used to generate the lofargrams. The lofargrams of the original input is illustrated in Figure 2b, where we can see that the narrowband component is not easy to be identified. By utilizing the proposed AIMSR method, the optimal output waveform can be seen in Figure 2c, where it reflects an intrawell response with large amplitude. After a response delay, it is stable and periodic at 10 Hz. The lofargram of the AIMSR output is demonstrated in Figure 2d. It clearly identifies the narrowband component f_0. It is worth noting that there is a second-order harmonic with periodic driving. This can be attribiuted to the resonance phenomenon and anticipate benefiting harmonic signature extraction. A comparison of normalized power spectrum density (PSD) of input and output (Note a highpass filter is adopted with passband frequency 1 Hz to deal with the DC bias, and this is the same in the rest of the paper) with the Welch method is given in Figure 2f, where we can see the superior filtering performance. The frequency response of the AIMSR background noise has the form of Lorentzian distribution by concentrating most of the noise energy into the low-frequency region. This refers to the special property of nonlinear SR phenomenon. The optimal SNRI obtained at each iteration is shown in Figure 2e. It can be seen after three iterations that an optimal result is obtained, which reflects a convergence of the proposed algorithm.

For H_0 hypothesis circumstance, the input is simulated by the pure white Gaussian noise. As is assumed by prior unknown, we conduct the optimization with $f_0 = 10$ Hz for parameters initialization as well. The corresponding results are demonstrated in Figure 3a–f. By utilizing the proposed AIMSR method, the optimal output waveform is shown in Figure 3c, where it reflects an intrawell response as well. After a response delay, the amplitude looks more fluctuant compared with H_1 hypothesis circumstance as illustrated in Figure 2c. The lofargram of the AIMSR output is demonstrated in Figure 3d, where there seems to be a narrowband component near 10 Hz. To further compare the normalized power spectrum density (PSD) of the input and output in Figure 3f, we can see a frequency bias with Δf that does not match the initialized frequency f_0. Such a circumstance does not have a high-order harmonic that is regarded as a non-SR occurrence. The optimal SNRI obtained at each iteration is shown in Figure 3e, where the optimal result is obtained after two iterations. The optimal fitness value is larger than H_1 hypothesis circumstance. This should be affected by the calculation result of input as the input SNR is extremely smaller.

In summary, it is clear to see the superior filtering performance of AIMSR with a limited sampling rate condition. The difference in response between the H_1 and H_0 hypotheses can be identified, which indicates the detection performance fairly well.

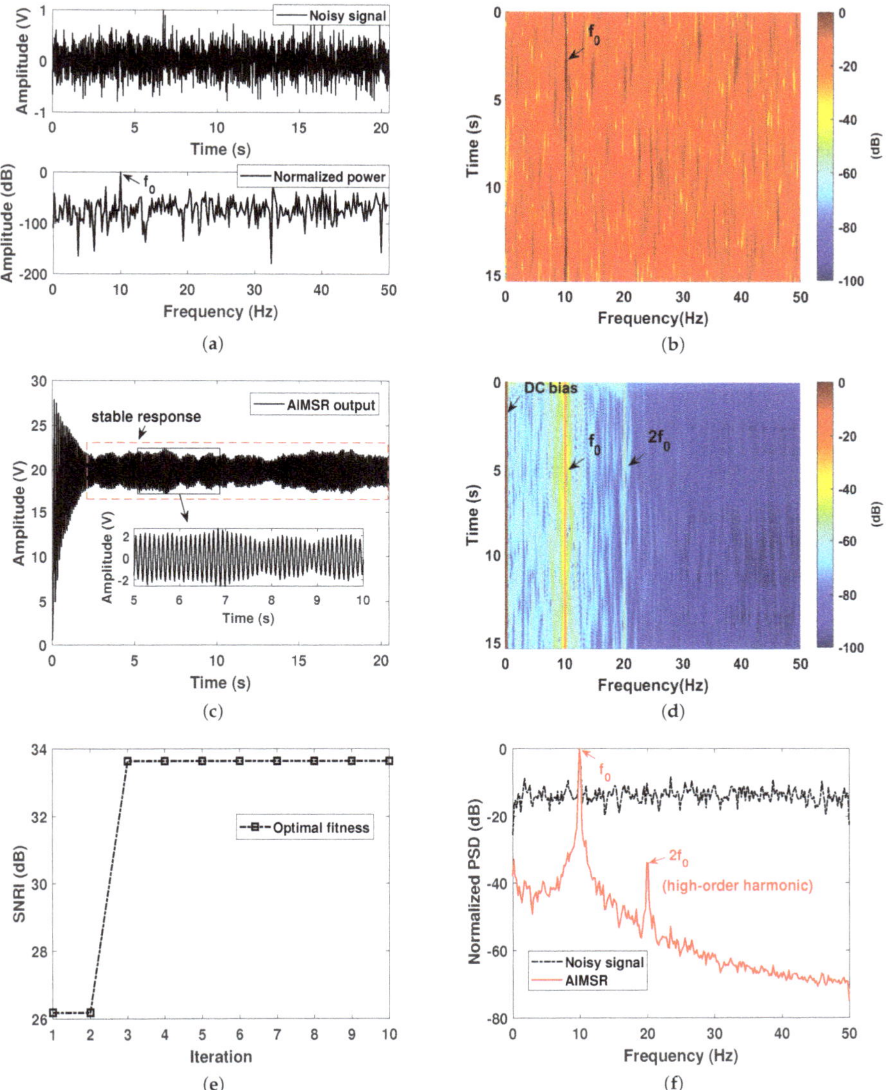

Figure 2. A simulation of discrete line signature signal on H_1 assumption: (**a**) the input waveform and its normalized power spectrum; (**b**) lofargram obtained with the input; (**c**) the adaptive intrawell matched stochastic resonance (AIMSR) output waveform ($\gamma_{opt} = 0.0322$, $K_{opt} = 79.05$, $h_{opt} = 0.4413$); (**d**) lofargram obtained with the AIMSR output; (**e**) the optimal fitness at each iteration with genetic algorithm (GA) method; (**f**) filtering performance comparison with normalized power spectrum density (PSD, Welch method).

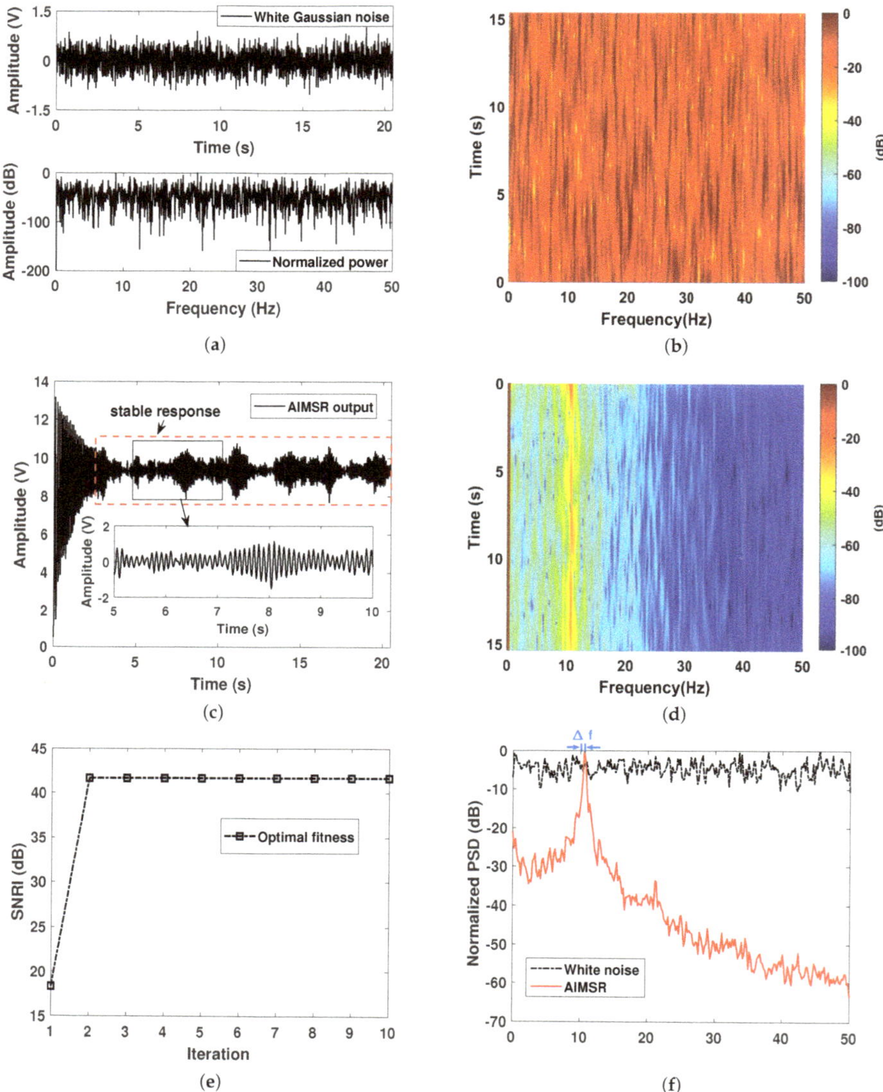

Figure 3. A simulation of discrete line signature signal on H_0 assumption: (**a**) the input waveform and its normalized power spectrum; (**b**) lofargram obtained with the input; (**c**) the AIMSR output waveform ($\gamma_{opt} = 0.2875$, $K_{opt} = 109.4507$, $h_{opt} = 0.4465$); (**d**) lofargram obtained with the AIMSR output; (**e**) the optimal fitness at each iteration with GA; (**f**) filtering performance comparison with normalized PSD (Welch method).

4.2. Harmonic Related Line Signature Signal Analysis

Harmonic vibration is a typical vibration generated by rotary machinery. As is mentioned in Section 2, the harmonic-related signature is commonly referred to in a ship engine and propeller. The sampling frequency is set to be $f_s = 100$ Hz, and the data length $N = 2048$. We consider two circumstances corresponding to H_1 and H_0 hypotheses as well.

For H_1 hypothesis circumstance, the tested harmonic signal is a combination of a fundamental sinusoid and its two high-order harmonics, whose frequencies are all harmonically related to two times and four times the integer multiples of the fundamental frequency f_0. These three sinusoidal signals have the same amplitude of 0.1, and the same Gaussian noise background. The input SNR is set as -15 dB. The lofargram of the original input is illustrated in Figure 4b, where we can see the three harmonic-related narrowband components. By utilizing the proposed AIMSR method, the optimal output waveform still responds well with a response delay as shown in Figure 4c. In comparing with the single periodic driving, the amplitude of stable response is smaller. The corresponding lofargram of the AIMSR output is demonstrated in Figure 4d, where the three harmonic-related narrowband components can be identified. To further evaluate the filtering performance with normalized power spectrum density (PSD) of input and output in Figure 4f, we see the great local denoising performance with the fundamental frequency f_0, which is gradually lost with the 4th order harmonic ($4f_0$). The optimal SNRI obtained after two iterations as shown in Figure 4e, where we can find the optimal value is so small compared with the discrete line signature signal circumstance. There is a negative improvement of global SNRI for the high order harmonics. In this way, the local SNRI is recommended to be the evaluation index of high-order harmonics.

For the H_0 hypothesis, the input is simulated by the pure white Gaussian noise. As is assumed, we conducted the optimization with $f_0 = 10$ Hz for parameter initialization as well. The corresponding results are demonstrated in Figure 5a–f. By utilizing the proposed AIMSR method, the optimal output waveform is shown in Figure 3c. The output is an intrawell response with a larger response delay, and the amplitude of stable response reflects larger effects. The lofargram of the AIMSR output is demonstrated in Figure 5d, where we can see a broadband harmonic energy in the first 5 seconds, and then the energy is focused to a periodic mode with a frequency bias Δf that does not match the initialized frequency f_0 as shown in Figure 5f. This means no SR occurs as the noisy input can not be matched to the nonlinear system. The optimal fitness value is extremely large which is consistent with the H_0 hypothesis of the discrete line signature signal case. This indicates the detectability as well.

In summary, the proposed AIMSR method can be utilized to enhance the harmonic-related signature, especially for the fundamental frequency. Since the fundamental frequency estimation is a topic that spans many disciplines including passive sonars [39], speech recognition [46], biomedical signal processing [47], musical pitch estimation [48], and etc., this method is anticipated to be a potential new technique for the future. The difference in responses between the H_1 and H_0 hypotheses can be identified with proper measurement index to establish a detector. The computation is efficient and generally converges to an optimum within five iterations, which can be realized in the embedded system. This work will be a topic for further study in the future.

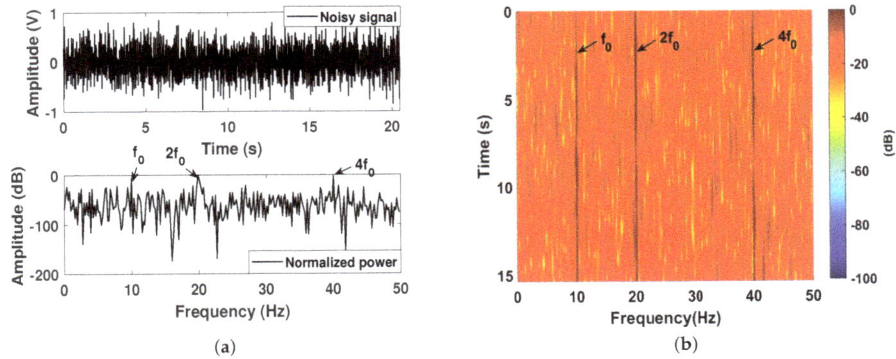

Figure 4. *Cont.*

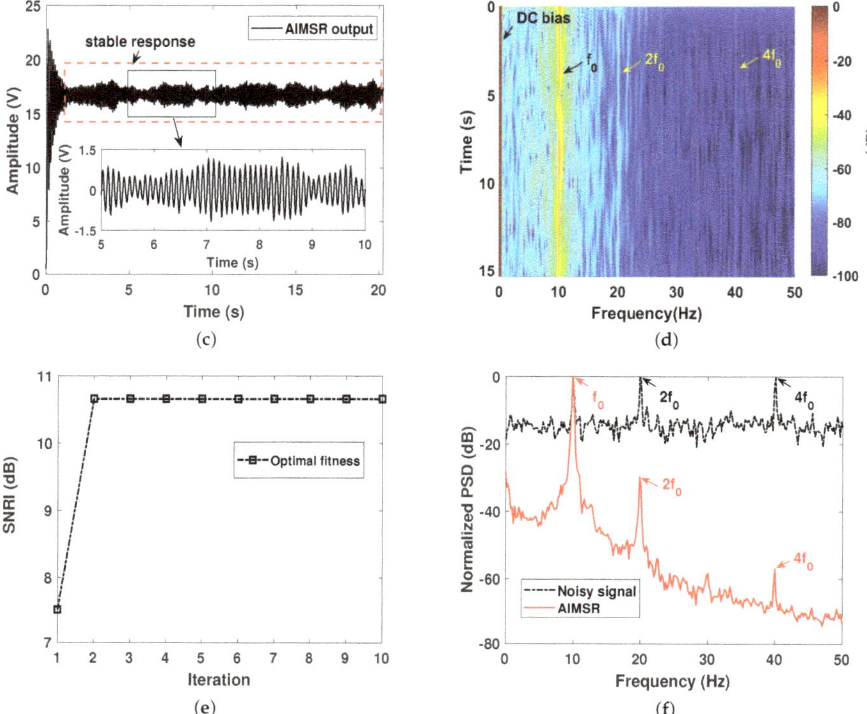

Figure 4. A simulation of harmonic-related line signature signals on H_1 assumption: (**a**) the input waveform and its normalized power spectrum; (**b**) lofargram obtained with the input; (**c**) the AIMSR output waveform ($\gamma_{opt} = 0.1132$, $K_{opt} = 70.4507$, $h_{opt} = 0.4429$); (**d**) lofargram obtained with the AIMSR output; (**e**) the optimal fitness at each iteration with GA; (**f**) filtering performance comparison with normalized PSD (Welch method).

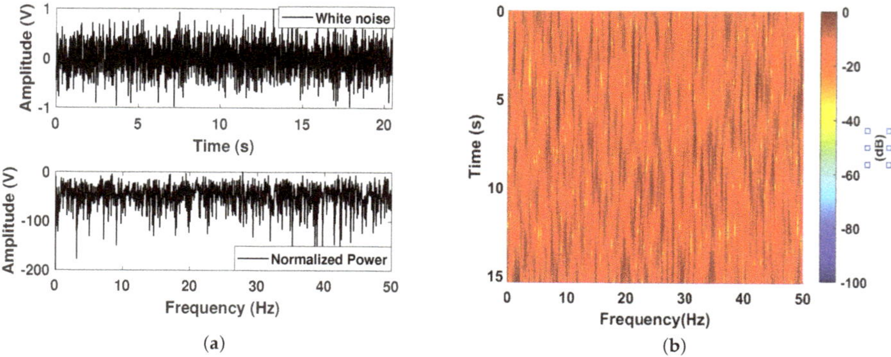

Figure 5. *Cont.*

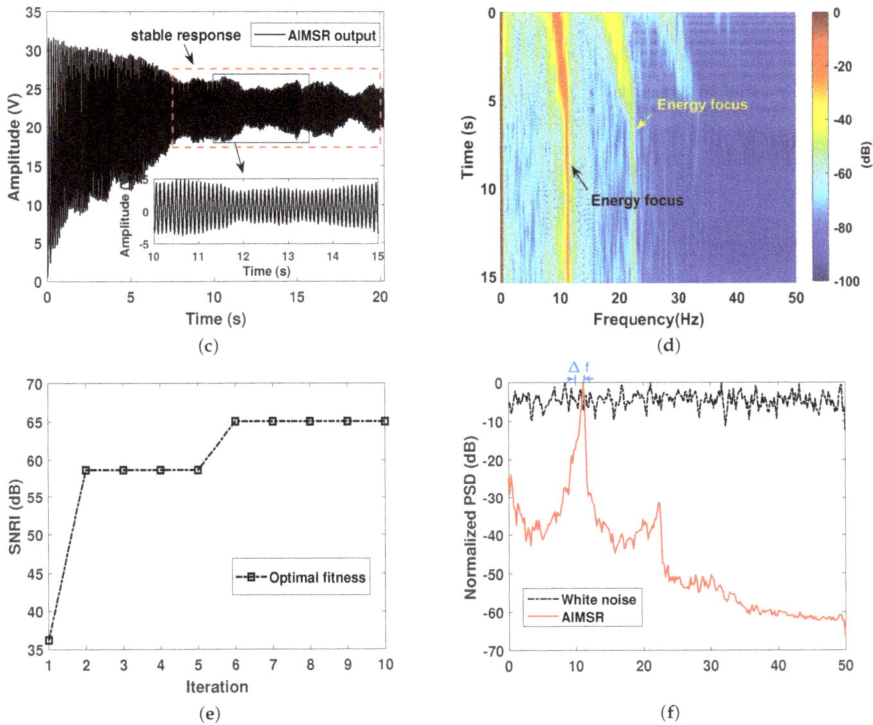

Figure 5. A simulation of a harmonic-related line signature signal on H_1 assumption: (**a**) the input waveform and its normalized power spectrum; (**b**) lofargram obtained with the input; (**c**) the AIMSR output waveform ($\gamma_{opt} = 0.0462$, $K_{opt} = 159.2903$, $h_{opt} = 0.5763$); (**d**) lofargram obtained with the AIMSR output; (**e**) the optimal fitness at each iteration with GA; (**f**) filtering performance comparison with normalized PSD (Welch method).

5. Application Verification and Discussion

5.1. Verfication on AUV's Low Frequency Propeller Harmonic Tonals

To validate the feasibility and efficiency of the proposed method, an experiment was conducted with an AUV in the Zhanghe reservoir. The AUV moved away in a trajectory of straight line, and the radiated noise was measured by a low frequency seismometer with the sampling frequency $f_s = 200$ Hz. Figure 6a shows the lofargram of the received signal, where we can see three harmonic lines. The conventional ALE that employs the least mean square (LMS) algorithm is conducted to enhance the line signatures as shown in Figure 6b, where the harmonic lines are more clearly identified. By utilizing the proposed AIMSR method, the optimal output is the intrawell as well. The corresponding lofagrams of the AIMSR output and the direct current (DC) offset filtered output are illustrated in Figure 6c,d, respectively. It can be seen that the background noise of the AIMSR output is much lower than that of the conventional ALE. A comparison of normalized power spectrum denisity (PSD) is shown in Figure 6f, where there is about 10 dB improvement for the fundamental frequency compared to that of the conventional ALE. For high order harmonics, the conventional ALE should be better. This indicates that a combination of the two methods gives a better output.

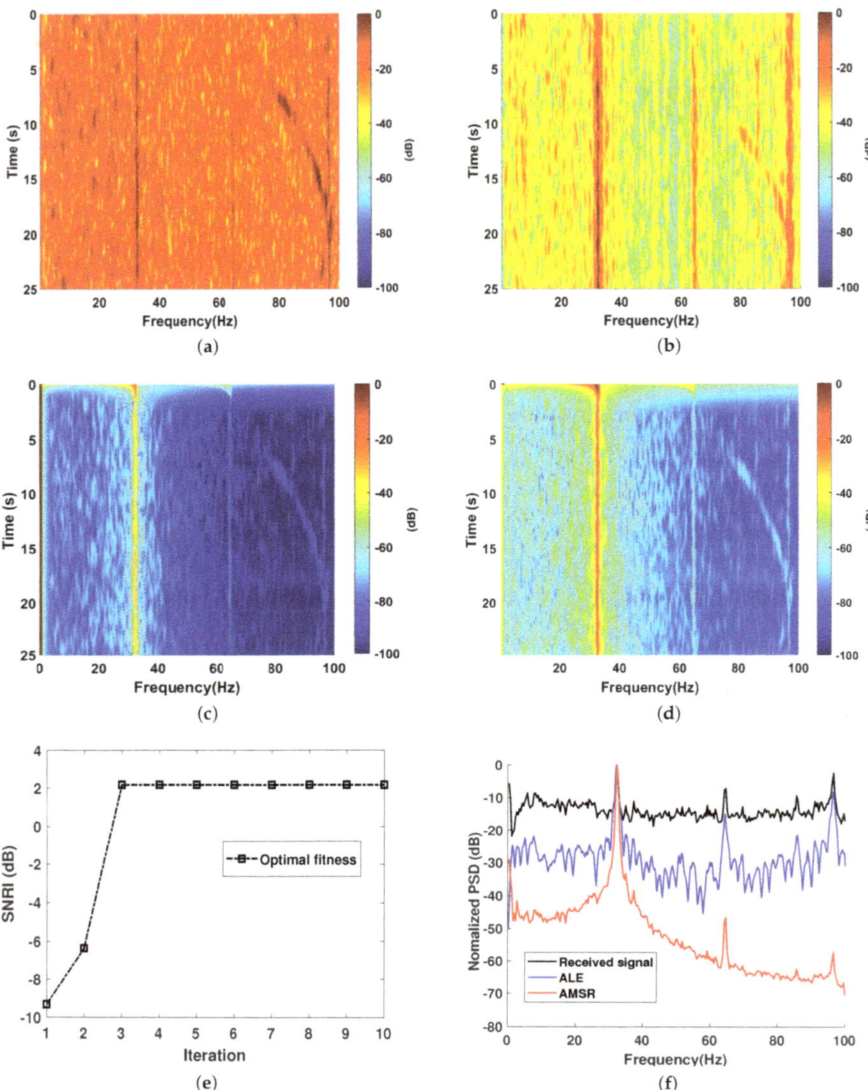

Figure 6. Filtering performance verification: (**a**) lofagram of the original received signal of an autonomous underwater vehicle (AUV); (**b**) lofagram of the adaptive line enhancer (ALE)output (LMS); (**c**) lofagram of the AMSR output ($\gamma_{opt} = 0.1227$, $K_{opt} = 85.6350$, $h_{opt} = 0.4707$); (**d**) lofagram of the AMSR output with direct current (DC) offset filter; (**e**) the optimal fitness at each iteration with GA; (**f**) normalized power spectrum denisty (PSD) comparison.

5.2. Discussion

(1) From the aforementioned numerical and practical analyses on AIMSR with potential constrainted bistable model, the SR is implemented under a low sampling rate condition. Essentially, it is to add a new degree of freedom to the bistable potential which can have an effect on the limitation of the system response. As we consider the system parameter tuning SR as a special nonlinear filter,

it is a tenet that the more the tuning parameters (system complexity), the better output response. Similar results can also be found like multi-parameters tuning [49], improved potentials [30,50,51], coupled systems [31,32], etc. Hence, one of the contribution of this paper is to explain and lead the further explore of nonlinear filter. In addition, from the view of system nonlinearity, there seems to be a connection to deep neural networks for the system complexity. This might lead to an inspiration and possibility in training the deep SR network for applications, and gives us a guidance to better understand the innate character of deep learning from the view of nonlinearity as well.

(2) The evaluation index of SNRI in this paper generally requires a prior knowledge of signal frequency. However, for engineering applications, it is commonly unknown. As is known, the SR will always force the energy focus to a periodic mode. It is hard to distinguish whether the correctness of output response is simply achieved by frequency searching. Although the difference of response between the H_1 and H_0 hypotheses can be fairly well identified, the problem of proper selection of the measurement index in the sense of detection is still a problem for further studies.

(3) The superior filtering performance is validated in this paper, which means the advantage of AIMSR is a potential. However, its intrinsic Lorentzian property leads to a preference in enhancing the low-frequency signals, and as a result, to a loss of the performance for higher frequencies. As is mentioned in the last subsection, a conventional ALE should be better in dealing with the high-order harmonics. Proper combination on these two methods is needed for better outputs. This work will probably be studied in the future.

(4) The proposed AIMSR method is of low computational complexity which enables implementation in an embedded system. This is an interesting and highly important engineering problem. However, related work can rarely be found. To further improve the applicablity of the SR in the engineering fields, we think this work is necessary for future studies.

6. Conclusions

In this paper, we proposed an adaptive intrawell matched stochastic resonance (AIMSR) method to break through the limitation of the conventional ALE by a nonlinear filtering effect. The problem of parameterized implementation of SR under a low sampling rate condition is firstly addressed by implementing a framework of intrawell matched stochastic resonance with potential constraint. To further promote its practicality in an embedded system, the large parameter limitation of matched relationship is eased to deduce the computation complexity. Simulation analyses are conducted with a discrete line signature and harmonic-related line signature that reflect the superior filtering performance as well as the computational efficiency fairy well. Besides, two hypotheses corresponding to H_1 (periodic signal with noise) and H_0 (pure noise) circumstances are further considered to reveal the feasibility of detection. Application verification was experimentally conducted in a reservoir with an AUV to validate the proposed AIMSR method compared with the conventional ALE method. Additional intensive discussions have been made to give an insight into the principal results and inspire future investigations. In the end, we anticipate that the proposed method can be a potential new technique for passive sonar detection in the future.

Author Contributions: Conceptualization, H.D.; Formal analysis, H.D. and S.M.; Funding acquisition, X.S. and H.W.; Investigation, S.M. and C.Q.; Resources, C.Q.; Methodology, H.D. and K.H.; Validation, H.D., X.S. and S.M.; Writing—original draft, H.D.; Writing—review & editing, K.H., X.S. and H.W. All authors have read and agreed to the published version of the manuscript.

Funding: This work was funded by the National Natural Science Foundation of China (Grant No. 61571365, 61671386, 61771401, 61571367) and National Key Research and Development Program of China (Grant No. 2016YFC1400200

Conflicts of Interest: The authors declare no conflict of interest. The funding sponsors have the role in the design of the study and in the decision to publish the results.

References

1. Carter, G.C. Passive sonar signal processing. In *Underwater Acoustics and Signal Processing*; Springer: Berlin, Germany, 1981; pp. 499–508.
2. Zimmerman, R.; D'Spain, G.; Chadwell, C.D. Decreasing the radiated acoustic and vibration noise of a mid-size AUV. *IEEE J. Ocean. Eng.* **2005**, *30*, 179–187. [CrossRef]
3. Etter, P.C. Advanced applications for underwater acoustic modeling. *Adv. Acoust. Vib.* **2012**, *2012*. [CrossRef]
4. Ross, D. *Mechanics of Underwater Noise*; Pergamon Press: Los Altos, CA, USA, 1976.
5. Urick, R.J. *Principles of Underwater Sound*; McGraw-Hill: New York, NY, USA, 1983.
6. Waite, A.D.; Waite, A. *Sonar for Practising Engineers*; Wiley: New York, NY, USA, 2002; Volume 3.
7. Antoni, J.; Hanson, D. Detection of surface ships from interception of cyclostationary signature with the cyclic modulation coherence. *IEEE J. Ocean. Eng.* **2012**, *37*, 478–493. [CrossRef]
8. Dong, H.; Wang, H.; Shen, X.; He, K. Parameter matched stochastic resonance with damping for passive sonar detection. *J. Sound Vib.* **2019**, *458*, 479–496. [CrossRef]
9. Hollmann, L.J.; Stevenson, R.L. Adaptive whitening of ambient ocean noise with narrowband signal preservation. *J. Acoust. Soc. Am.* **2016**, *139*, 3122–3133. [CrossRef]
10. Guo, Y.; Zhao, J.; Chen, H. A novel algorithm for underwater moving-target dynamic line enhancement. *Appl. Acoust.* **2003**, *64*, 1159–1169. [CrossRef]
11. Bershad, N.J.; Eweda, E.; Bermudez, J.C. Stochastic analysis of an adaptive line enhancer/canceler with a cyclostationary input. *IEEE Trans. Signal Process.* **2015**, *64*, 104–119. [CrossRef]
12. Hao, Y.; Chi, C.; Qiu, L.; Liang, G. Sparsity-based adaptive line enhancer for passive sonars. *IET Radar Sonar Navig.* **2019**, *13*, 1796–1804. [CrossRef]
13. Kay, S. Can detectability be improved by adding noise? *IEEE Signal Process. Lett.* **2000**, *7*, 8–10. [CrossRef]
14. Gammaitoni, L.; Hänggi, P.; Jung, P.; Marchesoni, F. Stochastic resonance. *Rev. Mod. Phys.* **1998**, *70*, 223–286. [CrossRef]
15. Benzi, R.; Sutera, A.; Vulpiani, A. The mechanism of stochastic resonance. *J. Phys. A Math. Gen.* **1981**, *14*, L453. [CrossRef]
16. Harmer, G.P.; Davis, B.R.; Abbott, D. A review of stochastic resonance: Circuits and measurement. *IEEE Trans. Instrum. Meas.* **2002**, *51*, 299–309. [CrossRef]
17. Moss, F.; Ward, L.M.; Sannita, W.G. Stochastic resonance and sensory information processing: a tutorial and review of application. *Clin. Neurophysiol.* **2004**, *115*, 267–281. [CrossRef]
18. McDonnell, M.D.; Abbott, D. What is stochastic resonance? Definitions, misconceptions, debates, and its relevance to biology. *PLoS Comput. Biol.* **2009**, *5*, e1000348. [CrossRef]
19. Dylov, D.V.; Fleischer, J.W. Nonlinear self-filtering of noisy images via dynamical stochastic resonance. *Nat. Photonics* **2010**, *4*, 323–328. [CrossRef]
20. Lu, S.; He, Q.; Wang, J. A review of stochastic resonance in rotating machine fault detection. *Mech. Syst. Signal Process.* **2019**, *116*, 230–260. [CrossRef]
21. Qiao, Z.; Lei, Y.; Li, N. Applications of stochastic resonance to machinery fault detection: A review and tutorial. *Mech. Syst. Signal Process.* **2019**, *122*, 502–536. [CrossRef]
22. Xu, B.; Duan, F.; Bao, R.; Li, J. Stochastic resonance with tuning system parameters: The application of bistable systems in signal processing. *Chaos Solitons Fractals* **2002**, *13*, 633–644. [CrossRef]
23. Lu, S.; He, Q.; Kong, F. Effects of underdamped step-varying second-order stochastic resonance for weak signal detection. *Digit. Signal Process.* **2015**, *36*, 93–103. [CrossRef]
24. Dong, H.; Wang, H.; Shen, X.; Jiang, Z. Effects of Second-Order Matched Stochastic Resonance for Weak Signal Detection. *IEEE Access* **2018**, *6*, 46505–46515. [CrossRef]
25. Lu, S.; He, Q.; Yuan, T.; Kong, F. Online fault diagnosis of motor bearing via stochastic-resonance-based adaptive filter in an embedded system. *IEEE Trans. Syst. Man Cybern. Syst.* **2017**, *47*, 1111–1122. [CrossRef]
26. Hu, N.Q. *Theory and Methods for Weak Characteristic Signal Detection with Stochastic Resonance*; National Defense Industry Press: Beijing, China, 2012.
27. Leng, Y.G.; Wang, T.Y.; Guo, Y.; Xu, Y.G.; Fan, S.B. Engineering signal processing based on bistable stochastic resonance. *Mech. Syst. Signal Process.* **2007**, *21*, 138–150. [CrossRef]
28. Rebolledo-Herrera, L.F.; FV, G.E. Quartic double-well system modulation for under-damped stochastic resonance tuning. *Digit. Signal Process.* **2016**, *52*, 55–63. [CrossRef]

29. Li, J.; Zhang, Y.; Xie, P. A new adaptive cascaded stochastic resonance method for impact features extraction in gear fault diagnosis. *Measurement* **2016**, *91*, 499–508. [CrossRef]

30. Shi, P.; An, S.; Li, P.; Han, D. Signal feature extraction based on cascaded multi-stable stochastic resonance denoising and EMD method. *Measurement* **2016**, *90*, 318–328. [CrossRef]

31. Li, J.; Zhang, J.; Li, M.; Zhang, Y. A novel adaptive stochastic resonance method based on coupled bistable systems and its application in rolling bearing fault diagnosis. *Mech. Syst. Signal Process.* **2019**, *114*, 128–145. [CrossRef]

32. Xiao, L.; Tang, J.; Zhang, X.; Xia, T. Weak fault detection in rotating machineries by using vibrational resonance and coupled varying-stable nonlinear systems. *J. Sound Vib.* **2020**, *478*, 115355. [CrossRef]

33. Andò, B.; Graziani, S. *Stochastic Resonance: Theory and Applications*; Springer Science & Business Media: Berlin, Germany, 2012.

34. Tan, J.; Chen, X.; Wang, J.; Chen, H.; Cao, H.; Zi, Y.; He, Z. Study of frequency-shifted and re-scaling stochastic resonance and its application to fault diagnosis. *Mech. Syst. Signal Process.* **2009**, *23*, 811–822. [CrossRef]

35. Alfonsi, L.; Gammaitoni, L.; Santucci, S.; Bulsara, A. Intrawell stochastic resonance versus interwell stochastic resonance in underdamped bistable systems. *Phys. Rev. E* **2000**, *62*, 299. [CrossRef]

36. Li, H.; Bao, R.; Xu, B.; Zheng, J. Intrawell stochastic resonance of bistable system. *J. Sound Vib.* **2004**, *272*, 155–167. [CrossRef]

37. Dong, H.; Wang, H.; Shen, X.; Yong, X. Nonlinear filtering effects of intrawell matched stochastic resonance with barrier constrainted Duffing system for ship radiated line signature extraction. *J. Sound Vib.* **2020**, submitted.

38. Arveson, P.T.; Vendittis, D.J. Radiated noise characteristics of a modern cargo ship. *J. Acoust. Soc. Am.* **2000**, *107*, 118–129. [CrossRef] [PubMed]

39. Ogden, G.L.; Zurk, L.M.; Jones, M.E.; Peterson, M.E. Extraction of small boat harmonic signatures from passive sonar. *J. Acoust. Soc. Am.* **2011**, *129*, 3768–3776. [CrossRef] [PubMed]

40. Bao, F.; Wang, X.; Tao, Z.; Wang, Q.; Du, S. Adaptive extraction of modulation for cavitation noise. *J. Acoust. Soc. Am.* **2009**, *126*, 3106–3113. [CrossRef] [PubMed]

41. Dong, H.; Wang, H.; Shen, X.; Ma, S. Adaptive Matched Stochastic Resonance Enhanced Speedboat Seismoacoustic Signature under Heavy Background noise. In Proceedings of the OCEANS'2018 MTS/IEEE Charleston, Charleston, SC, USA, 22–25 October 2018; IEEE: Piscataway, NJ, USA, 2018; pp. 1–5.

42. Zhang, R.; Zhou, S.; Qi, Y.; Liang, Y.; Sui, Y. Characteristics of very-low-frequency pulse acoustic fields measured by vector sensor and ocean bottom seismometer in shallow water. *J. Acoust. Soc. Am.* **2018**, *144*, 1916–1916. [CrossRef]

43. Silvia, M.T.; Richards, R.T. A theoretical and experimental investigation of low-frequency acoustic vector sensors. In Proceedings of the OCEANS'02 MTS/IEEE, Biloxi, MI, USA, 29–31 October 2002; IEEE: Piscataway, NJ, USA, 2002; pp. 1886–1897.

44. Lepper, P.A.; D'Spain, G.L. Measurement and modeling of the acoustic field near an underwater vehicle and implications for acoustic source localization. *J. Acoust. Soc. Am.* **2007**, *122*, 892–905. [CrossRef]

45. Zozor, S.; Amblard, P.O. On the use of stochastic resonance in sine detection. *Signal Process.* **2002**, *82*, 353–367. [CrossRef]

46. Medan, Y.; Yair, E.; Chazan, D. Super resolution pitch determination of speech signals. *IEEE Trans. Signal Process.* **1991**, *39*, 40–48. [CrossRef]

47. Akay, M. *Detection and Estimation Methods for Biomedical Signals*; Academic Press: San Diego, CA, USA, 1996.

48. Christensen, M.G.; Stoica, P.; Jakobsson, A.; Jensen, S.H. Multi-pitch estimation. *Signal Process.* **2008**, *88*, 972–983. [CrossRef]

49. Lai, Z.h.; Leng, Y.g. Weak-signal detection based on the stochastic resonance of bistable Duffing oscillator and its application in incipient fault diagnosis. *Mech. Syst. Signal Process.* **2016**, *81*, 60–74. [CrossRef]

50. Lu, S.; He, Q.; Kong, F. Stochastic resonance with Woods–Saxon potential for rolling element bearing fault diagnosis. *Mech. Syst. Signal Process.* **2014**, *45*, 488–503. [CrossRef]

51. López, C.; Zhong, W.; Lu, S.; Cong, F.; Cortese, I. Stochastic resonance in an underdamped system with FitzHug-Nagumo potential for weak signal detection. *J. Sound Vib.* **2017**, *411*, 34–46. [CrossRef]

Article

Extremely Robust Remote-Target Detection Based on Carbon Dioxide-Double Spikes in Midwave Spectral Imaging

Sungho Kim [1,*]**, Jungsub Shin** [2]**, Joonmo Ahn** [2] **and Sunho Kim** [2]

[1] Department of Electronic Engineering, Yeungnam University, 280 Daehak-Ro, Gyeongsan,
     Gyeongbuk 38541, Korea
[2] Agency for Defense Development, P.O. Box 35, Daejeon 34186, Korea; jss@add.re.kr (J.S.);
     ahnjm@add.re.kr (J.A.); edl423@add.re.kr (S.K.)
*   Correspondence: sunghokim@ynu.ac.kr; Tel.: +82-53-810-3530

Received: 6 April 2020; Accepted: 15 May 2020; Published: 20 May 2020

Abstract: Infrared ship-target detection for sea surveillance from the coast is very challenging because of strong background clutter, such as cloud and sea glint. Conventional approaches utilize either spatial or temporal information to reduce false positives. This paper proposes a completely different approach, called carbon dioxide-double spike (CO_2-DS) detection in midwave spectral imaging. The proposed CO_2-DS is based on the spectral feature where a hot CO_2 emission band is broader than that which is absorbed by normal atmospheric CO_2, which generates CO_2-double spikes. A directional-mean subtraction filter (D-MSF) detects each CO_2 spike, and final targets are detected by joint analysis of both types of detection. The most important property of CO_2-DS detection is that it generates an extremely low number of false positive caused by background clutter. Only the hot CO_2 spike of a ship plume can penetrate atmosphere, and furthermore, there are only ship CO_2 plume signatures in the double spikes of different spectral bands. Experimental results using midwave Fourier transform infrared (FTIR) in a remote sea environment validate the extreme robustness of the proposed ship-target detection.

Keywords: ship detection; false alarm; farbon dioxide peaks; midwave infrared; FTIR

1. Introduction

Remote-ship detection is important in various applications, such as maritime navigation [1], coast guard searches [2], and homeland security procedures [3]. Infrared images are frequently adopted, especially due to their operational capability, day or night.

The key issue is how to reduce the high number of false detections caused by cloud clutter and sea surface glint while maintaining an acceptable detection rate [4,5]. Cloud edges produce false detections, and sea glint is similar to small infrared targets, both of which degrade detection performance. Since the 1990s, a lot of methods have been proposed to reduce the false positives by using either spatial information or temporal information from infrared images.

Spatial information processing, such as background subtraction, can be a feasible approach to reducing background clutter. Background images are estimated by spatial filters, such as the mean filter [6], the least mean square filter [7,8], the median filter [9], and the morphological filter [10]. These filters use neighboring pixels to estimate background pixel values, but with different strategies. In particular, mean subtraction filter (MSF)-based target detection is the most simple, but is weak when it comes to cloud edges. Non-linear filters, such as max-median, morphology, and data-fitting, show better cloud clutter suppression capabilities [11,12]. Local directional Laplacian-of-Gaussian filtering can remove false positives from cloud edges [13]. In the spatial information approach, classification with a spatial shape feature can discriminate clutter. Hysteresis thresholding [14], statistics-based adaptive thresholding [15], the Bayesian classifier [16], and support vector machines [17] are well-known methods. Voting by various classifiers can enhance dim-target detection rates [18]. Spatial frequency information can be used to remove low-frequency clutter, such as the three-dimensional fast Fourier transform (3D-FFT) spectrum [19]. The wavelet transform is effective against sea glint [20], and low pass filter can also deal with sea glint [21]. An adaptive high-pass filter can reduce cloud and sea glint clutter [22], while spatial target-background contextual information enhances the target signature and reduces background clutter, which is effective for sea glint reduction [23]. Spatial multi-feature fusion can increase the clutter discrimination capability [24].

Temporal information processing, such as track-before-detection, can remove slowly moving cloud clutter and fast-blinking sea glint [25]. Temporal profiles are useful to discriminate between a fast-moving target and slow-moving cloud [26–28]. The temporal contrast filter is suitable for detecting small, supersonic, infrared targets [29], and a 3D matched filter for wide-to-exact searches can remove cloud clutter very fast [30]. A power-law detector for image frames is effective for target detection in dense clutter [31]. Since a previous frame can be regarded as background, a weighted autocorrelation matrix update using the recursive technique can be useful in order to eliminate sea glint [32]. An advanced adaptive spatial-temporal filter can achieve tremendous gain [33], and principal component analysis in multi-frames can alleviate false positives from sun glint [34].

Synthetic aperture radar (SAR) with deep learning can be a useful approach for remote ship detection application [35–37]. Faster R-CNN [35], DRBox [36], and Contextual CNN [37] can provide improved detection performance but these approaches require huge number of training dataset to reduce false positives.

As mentioned above, the conventional approaches in the spatial- or temporal-image domains have their own pros and cons, depending on the situations and conditions. The spatial filter based approach relies on target shapes and intensity distributions, which is ambiguous if targets are far away. The temporal filter based approach assumes fast target motion with stationary sensors. If this assumption is not satisfied, it will generate many false positives in the maritime environment.

This paper proposes a novel spectral–spatial signature analysis-based ship detection method in midwave hyperspectral images, instead of the conventional spatial or temporal methods. Based on the combustion process, CO_2 emissions show a double-spike signature at around 4.16 μm (2400 cm^{-1}) and 4.34 μm (2300 cm^{-1}). The hot CO_2 emission band is broader than that which is absorbed by normal atmospheric CO_2, which generates CO_2-double spikes. The directional-mean subtraction filter (D-MSF) detects each CO_2 spike, and final targets are detected by joint analysis of the detection of both spikes. So, we call the proposed method carbon dioxide-double spike (CO_2-DS) detection.

The contributions from this paper can be summarized as follows. First, the phenomenon of the CO_2-double spike emission in the midwave band is analyzed systematically. Second, a novel spectral–spatial analysis is proposed to detect remote ships in the maritime environment based on this analysis. Third, CO_2-DS detection can suppress background clutter (cloud, sea glints) completely, which leads to an extremely low false positive rate in remote-ship detection. Finally, the performance of CO_2-DS detection is demonstrated in a real sea environment with real ships.

The remainder of this paper is organized as follows. Section 2 explains the proposed CO_2-DS method, including analysis of carbon dioxide-double spike radiation. Section 3 evaluates the clutter suppression performance of the CO_2-DS method in the maritime environment. The paper concludes in Section 4.

2. Proposed Ship CO$_2$ Plume Detection

2.1. Signature Analysis of Carbon Dioxide-Double Spikes

Before explaining the details of the proposed CO_2-DS, the double-spike phenomenon in midwave spectral bands should be introduced [38,39]. Figure 1 summarizes the overall mechanism of

CO_2-double spikes in received spectral radiance from ship plumes. Hot CO_2 emits thermal energy in wider spectral band than that of normal air CO_2, as shown in Figure 1a. Atmospheric spectral transmittance shows strong absorption at 2300–2380 cm^{-1} by normal CO_2 in the atmosphere, as shown in Figure 1b. Therefore, the received spectral radiance in a TELOPS Fourier transform infrared (FTIR) camera [40] shows a double-spike signature: the first spike is around 2276 cm^{-1} (4.39 μm), and the second spike is around 2393 cm^{-1} (4.18 μm).

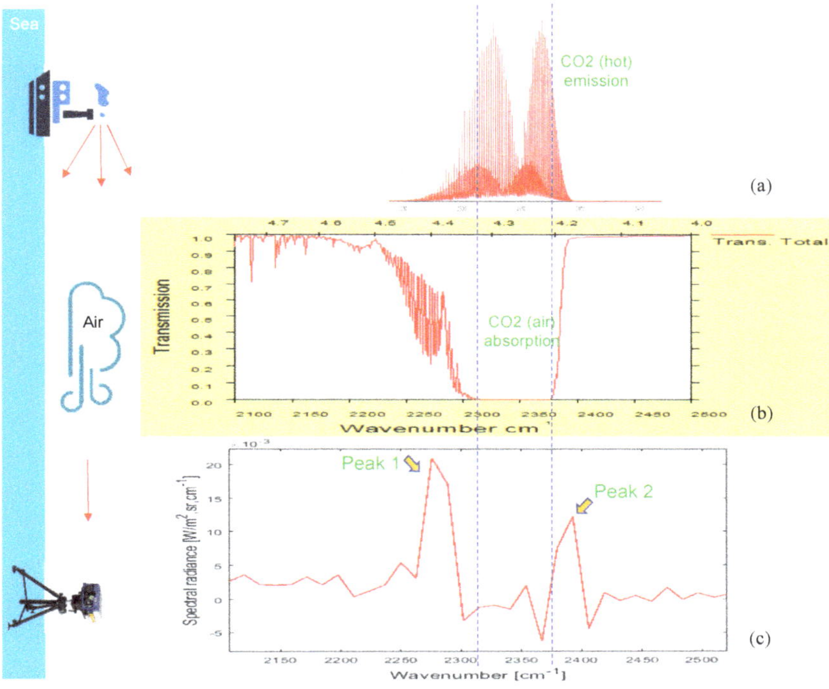

Figure 1. Operational concept of remote ship detection and the double spike-signature generation mechanism of hot carbon dioxide: (**a**) spectrum of a hot CO_2 emission by a ship, (**b**) spectral atmospheric transmittance, and (**c**) received spectral radiance of hot CO_2 from a ship's plume.

The measurement of CO_2-double spikes can be derived from radiative transfer calculated with Equation (1). We adopted the radiative-transfer equation used in the Moderate-Resolution Atmospheric Radiance and Transmittance (MODTRAN) [41]. In general, at-sensor received radiance in the midwave infrared (MWIR) region consists of CO_2-emitted radiance, transmitted background radiance, and total atmospheric path radiance (thermal+solar components).

$$L_{target}(\lambda) = \tau(\lambda)\left[\varepsilon(\lambda)B_{CO_2}(\lambda, T_{CO_2}) + (1 - \varepsilon(\lambda))L_{bg}(\lambda)\right] + L_s^\uparrow(\lambda) + L_{atm}^\uparrow(\lambda) \tag{1}$$

$L_{target}(\lambda)$ is the observed at-sensor target radiance; λ is wavelenth; $\varepsilon(\lambda)$ is spectral CO_2 gas emissivity; $B_{CO_2}(\lambda, T_{CO_2})$ is the spectral radiance of the hot CO_2 gas, assuming a blackbody in the Planck function with the combusted CO_2 gas temperature (T_{CO_2}); $\tau(\lambda)$ is the spectral atmospheric transmittance, and $L_s^\uparrow(\lambda)$ and $L_{atm}^\uparrow(\lambda)$ are the spectral upwelling solar and thermal path radiance, respectively, reaching the sensor.

According to the spectral emissivity of combusted CO_2 gas [38], $\varepsilon(\lambda)$ is relatively high. Therefore, the term L_{bg} can be removed. According to MWIR radiometric characteristics [42], the contribution of solar radiance, $L_s^{\uparrow}(\lambda_{CO_2})$, from air scattering is very small, even for very dry conditions (less than 2% at 5 μm) [42]. Ignoring the solar term, we can simplify Equation (1) to Equation (2):

$$L_{target}(\lambda) = \tau(\lambda)\left[\varepsilon(\lambda)B_{CO_2}(\lambda, T_{CO_2})\right] + (1 - \tau(\lambda))B_{atm}(\lambda, T_{atm}) \tag{2}$$

where the definition of thermal upwelling, $L_{atm}^{\uparrow}(\lambda)$, is replaced by the multiplication of blackbody radiation of the atmosphere, $B_{atm}(\lambda, T_{atm})$ and $1 - \tau(\lambda)$. The observed target spectral radiance, L_{target}, directly depends on both emissivity $\varepsilon(\lambda)$ of hot CO_2, and atmospheric transmittance, $\tau(\lambda)$, that should be analyzed specifically.

In addition, neighboring pixels to target hot CO_2 has only atmospheric spectral path radiance at wavelenth λ. Therefore, the received background signal can be expressed with Equation (3):

$$L_{neighbor}(\lambda) = (1 - \tau(\lambda))B_{atm}(\lambda, T_{atm}) \tag{3}$$

If we apply center-surround difference by subtracting Equation (3) from Equation (2), which is the same as a mean subtraction filter (MSF) operation for a specific band image, the target-only signal can be extracted with Equation (4). This physical property is used in the D-MSF to extract the target (hot CO_2) signature:

$$L_{targetOnly}(\lambda) = L_{target}(\lambda) - L_{neighbor}(\lambda) = \tau(\lambda)\left[\varepsilon(\lambda)B_{CO_2}(\lambda, T_{CO_2})\right] \tag{4}$$

Conventional ships are powered by diesel engines, and the corresponding hydrocarbon combustion produces water vapor, H_2O) and CO_2 as expressed in Equation (5) [39,43]:

$$C_nH_{2n+2} + (3n + 1)O_2 \rightarrow (n + 1)H_2O + nCO_2 \tag{5}$$

The first important parameter is spectral emissivity $\varepsilon(\lambda)$ in Equation (2). The normal temperature of combusted CO_2 gas from ships is approximately 600 K by infrared signature suppression (IRSS) [44]. The emission band-width of CO_2 gas is proportional to the temperature of the CO_2 gas [39]. Figure 2 demonstrates this phenomenon by changing CO_2 temperature using the high-resolution transmission molecular absorption database (HITRAN) simulation on the web (http://hitran.iao.ru/molecule/). As the CO_2 temperature increases, the absorption band range is broader. The absorption band range of normal atmospheric temperature (300 K) is 2300–2380 cm^{-1} (or 4.20–4.35 μm), where the inverse acts as atmospheric transmittance. The spectral unit can be either the wavelength (λ) [μm] or wavenumber (k) [cm^{-1}] by unit selection ($k = 10,000/\lambda$). For hot CO_2, the spectral absorption coefficient is the same as the spectral emission coefficient , $\varepsilon(\lambda)$, in Equation (2).

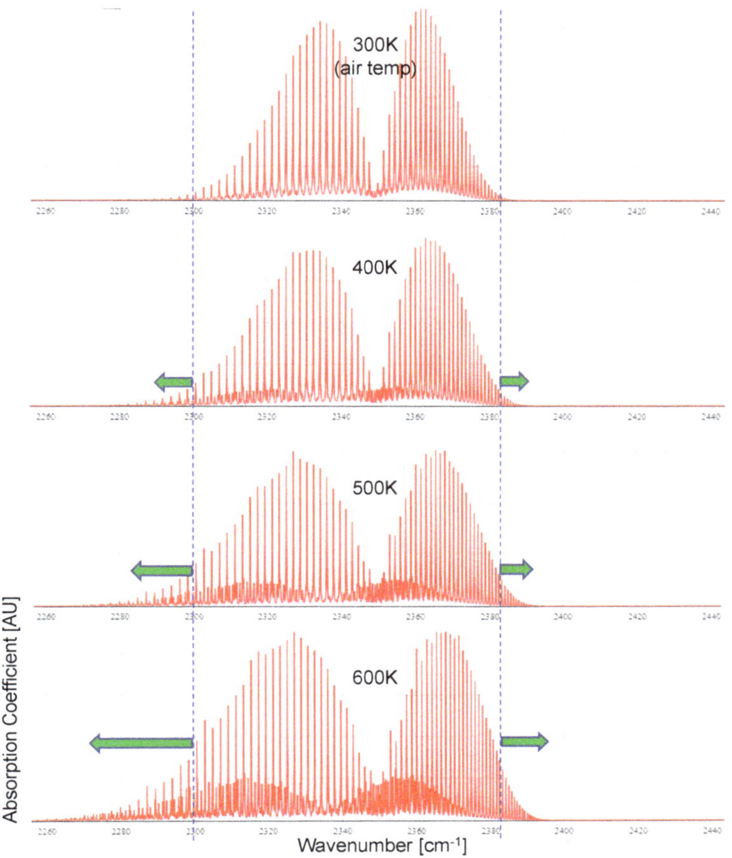

Figure 2. Spectral absorption coefficient variations of carbon dioxide according to different gas temperatures.

The second important parameter is spectral atmospheric transmittance $\tau(\lambda)$ in Equation (2). In the MODTRAN simulation in the MWIR band, the spectral transmittance of the CO_2 band (2320–2375 cm^{-1} or 4.21–4.31 μm) decreases abruptly with distance, as shown in Figure 3. The average transmittance in the CO_2 band is 0.13, 0.005, and 0 at 5, 20, and 100 m, respectively. Note that the atmospheric transmittance band of zero value coincides with the spectral absorption coefficient of CO_2 at 300 K in the top part of Figure 2. In addition, it remains only the atmospheric path radiance in Equation (2) in the CO_2 absorption band, $\tau(\lambda) = 0$), where there is no target or background signals.

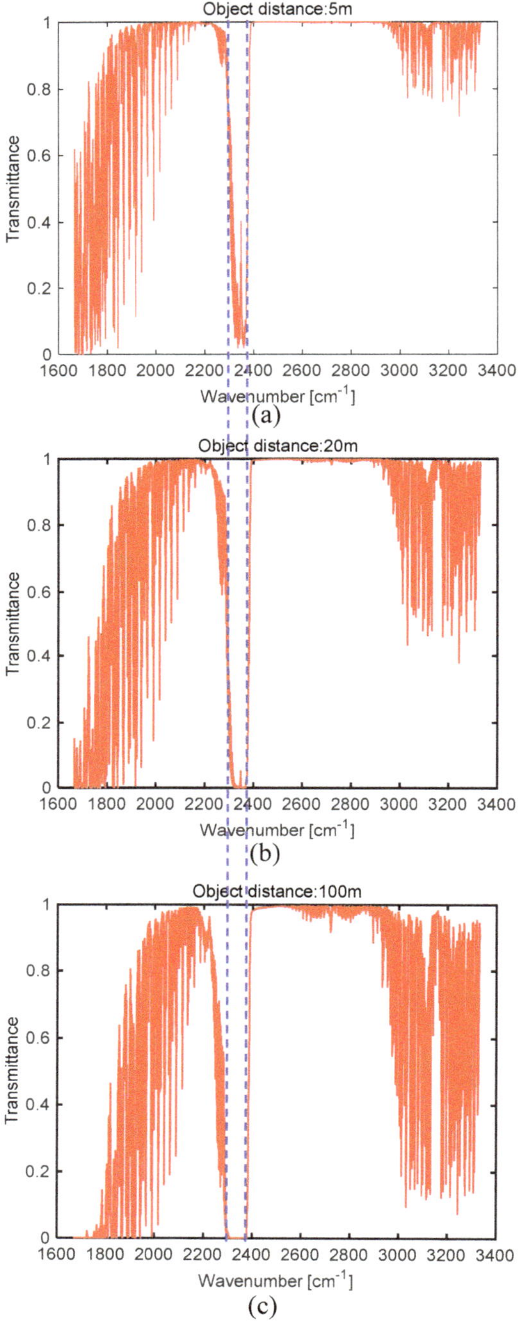

Figure 3. Spectral transmittance according to ship-camera distance in the MWIR band: (**a**) 5, (**b**) 20, and (**c**) 100 m.

Multiplication of the spectral emissivity of hot CO_2 and atmospheric transmittance produces double spikes, as shown in the bottom portion of Figure 1. The spectral feature where a hot CO_2 emission band is broader than that which is absorbed by normal atmospheric CO_2 generates CO_2 double spikes. The midwave spectral information is acquired via TELOPS MWIR hyperspectral camera [40]. It can provide calibrated spectral radiance images with a high spatial and spectral resolution from a Michelson interferometer in the short-wave to midwave bands (1.5–5.6 μm).

2.2. CO_2-DS-Based Ship Plume Detection

Motivated by the double-spike phenomenon of hot CO_2, a novel ship-detection method is proposed, as shown in Figure 4. The proposed method is called CO_2-DS because it is based on the carbon dioxide-double spike feature. The proposed CO_2-DS approach consists of three steps: spectral band selection, spatial small-target detection, and the final ship detection. The spectral band selection sets the search range in the spectral domain focusing on the CO_2 absorption band. Then, each CO_2 spike is detected by a carefully designed spatial filter (the directional-mean subtraction filter). Finally, the target ship is confirmed using a joint detection operation.

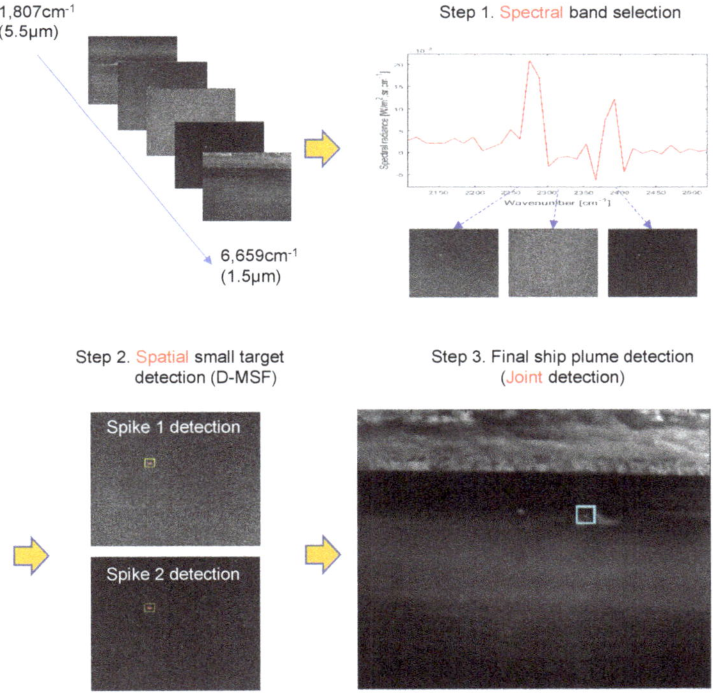

Figure 4. Overall processing flow of remote-ship detection using spectral–spatial information of hot CO_2 plumes.

In Step 1, the specific spectral band range should be defined to attain successful remote-ship detection with extremely few false positives. The proposed CO_2-DS method is based on the atmospheric CO_2 absorption band (2320–2375 cm⁻¹ or 4.21–4.31 μm) that should be included. In addition, double-spike bands emitted by hot CO_2 should be included. The first spike range is below 2300 cm⁻¹, and the second spike range is above 2380 cm⁻¹.

The details of Step 2 and Step 3 in Figure 4 are described in Figure 5. Each spike signal is detected by a parallel search algorithm: the first spike detection and the second spike detection.

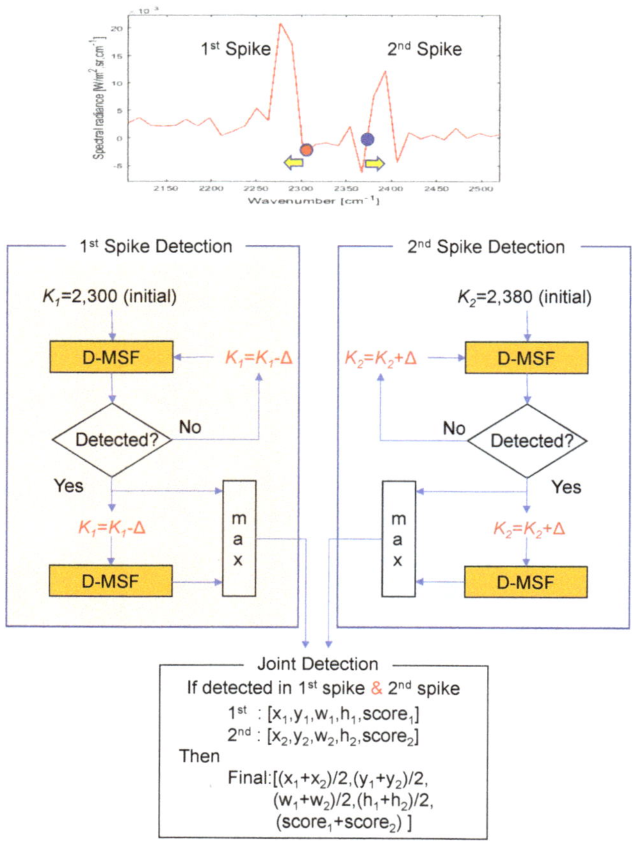

Figure 5. Details of the proposed CO_2-DS: first spike detection, second spike detection, and joint detection.

The starting wavenumbers are 2300 cm^{-1} and 2380 cm^{-1} because there is no signal at all except for atmospheric path radiance. Given a wavenumber, a test image of this band is probed by D-MSF-based small infrared target detection [4]. D-MSF is adopted in this paper because it is a well-proven method and robust against thermal noise in the maritime environment. Figure 6 summarizes the detailed flow of the D-MSF graphically. Given a hypothesized spectral band image, $I_\lambda(x,y)$, a mean subtraction filter, $h(x,y)$, is used to enhance the signal-to-clutter ratio (1). Then, directional local median $I_\lambda^D(x,y)$ is used to estimate row directional background (2) from the MSF result, $I_\lambda^{MSF}(x,y)$. Subtracting (2) from (1) in Figure 6 produces a clearer signature image. The region of interest (ROI) is generated by an initial thresholding ($th1$) and eight-nearest neighbor-based clustering. Final detection is achieved by applying adaptive thresholding $th2$ in constant false alarm rate (CFAR) detector to the subtracted image with the given ROI. It can provide a signal-to-clutter ratio (SCR) as well as the ROI of the target.

If there is no detection, the wavenumber (k_1) of the first spike is decreased by Δ, and that of the second spike is increased by Δ, the value of which is 13 cm^{-1} in this paper. If there is a detection, each neighboring band image is probed additionally to obtain a maximum signal score.

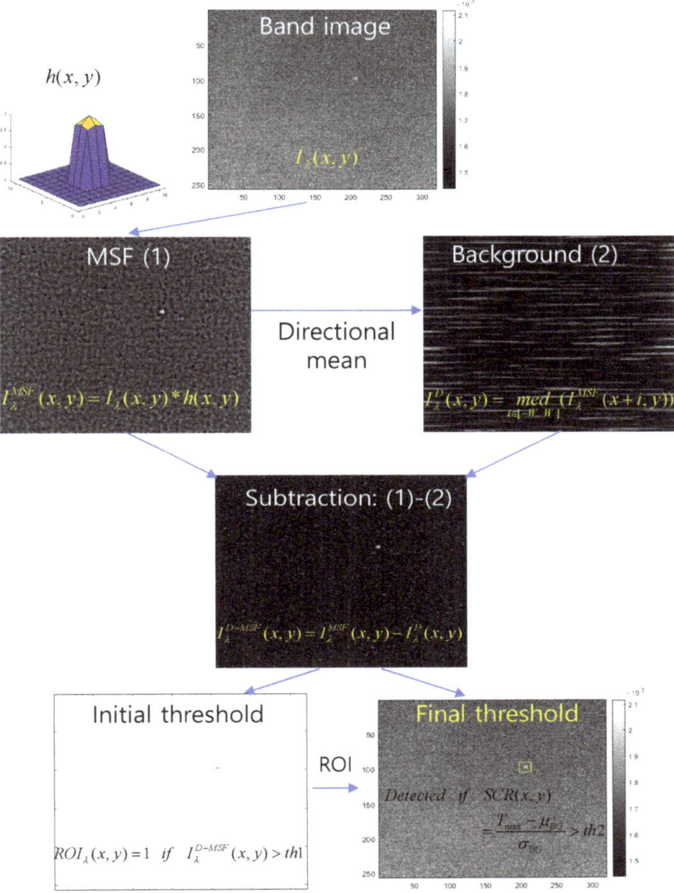

Figure 6. Operational flow of the directional-mean subtraction filter (D-MSF) for small-target detection in a band image.

Finally, in Step 3, the joint detection process is performed from the first and second spike detections: $[x_1, y_1, w_1, h_1, SCR_1]$, $[x_2, y_2, w_2, h_2, SCR_2]$, where $[\cdot]$ denotes [column, row, width, height, SCR] of the detected target. If two detections exist, then the final target information is merged with Equation (6). If the merged SCR is larger than a predefined threshold ($th3$), then it is declared a final target.

$$Final\ target\ info = \left[\frac{x_1 + x_2}{2}, \frac{y_1 + y_2}{2}, \frac{w_1 + w_2}{2}, \frac{h_1 + h_2}{2}, (SCR_1 + SCR_2)\right] \tag{6}$$

Figure 7 shows the spectral profile of a remote ship's plume and an enlarged view focusing on the atmospheric CO_2 absorption band. The numbers (1), (2), (3), and (4) represent the probing band locations of the proposed CO_2-DS method; (1) and (2) belong to the first spike band, and (3) and (4) belong to the second spike band. Note that the first spike is not clear in remote cases but it does not matter in our CO_2-DS method since D-MSF is used. Figure 8 shows the corresponding band images, and the max SCR is used to select a band image for each spike: (2) and (3) are selected. The bottom image in Figure 8 represents the result of joint detection from the detection results of the first spike and the second spike.

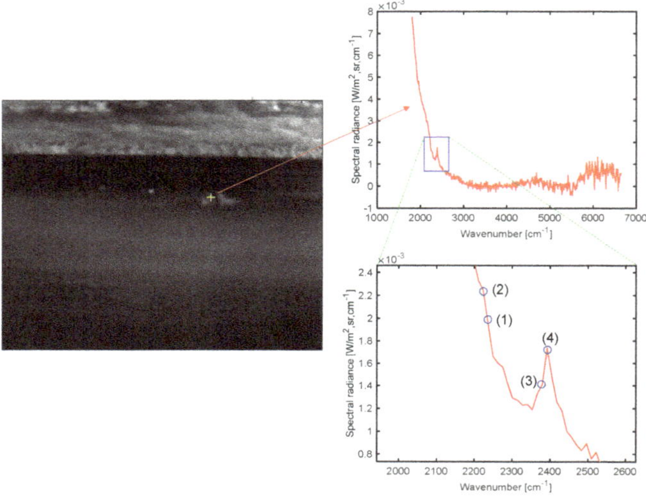

Figure 7. Spectral profile of a remote ship's plume with the probed band positions: first spike: (1) and (2); second spike: (3) and(4).

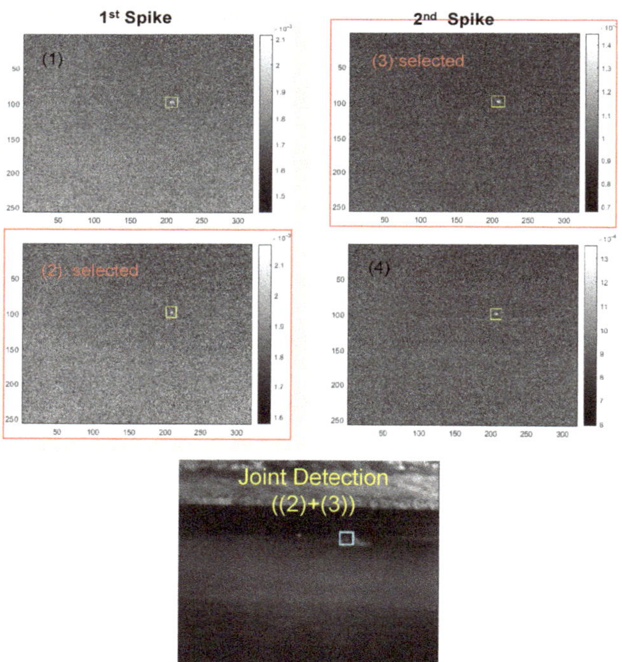

Figure 8. Probed band images of first and second spikes corresponding to (1) and (2), and (3) and (4), respectively. The final target is detected via joint detection from the selected max signature for each spike.

3. Experimental Results

3.1. Experiment 1: Analysis of Signature Variation

MWIR hyperspectral images of remote ships were acquired with the open path TELOPS Hyper-Cam Mid-Wave Extended (MWE) model [40]. It could provide calibrated spectral radiance images with a high spatial and spectral resolution from a Michelson interferometer in the shortwave to midwave band (1.5–5.6 μm). The spatial image resolution was 320×256, and the spectral resolution was up to 0.25 cm^{-1}. The noise-equivalent spectral radiance (NESR) was 7 $[\mathrm{nW}/(\mathrm{cm}^2 \cdot \mathrm{sr} \cdot \mathrm{cm}^{-1})]$, and the radiometric accuracy was approximately 2 K. The field of view was 6.5×5.1 degrees.

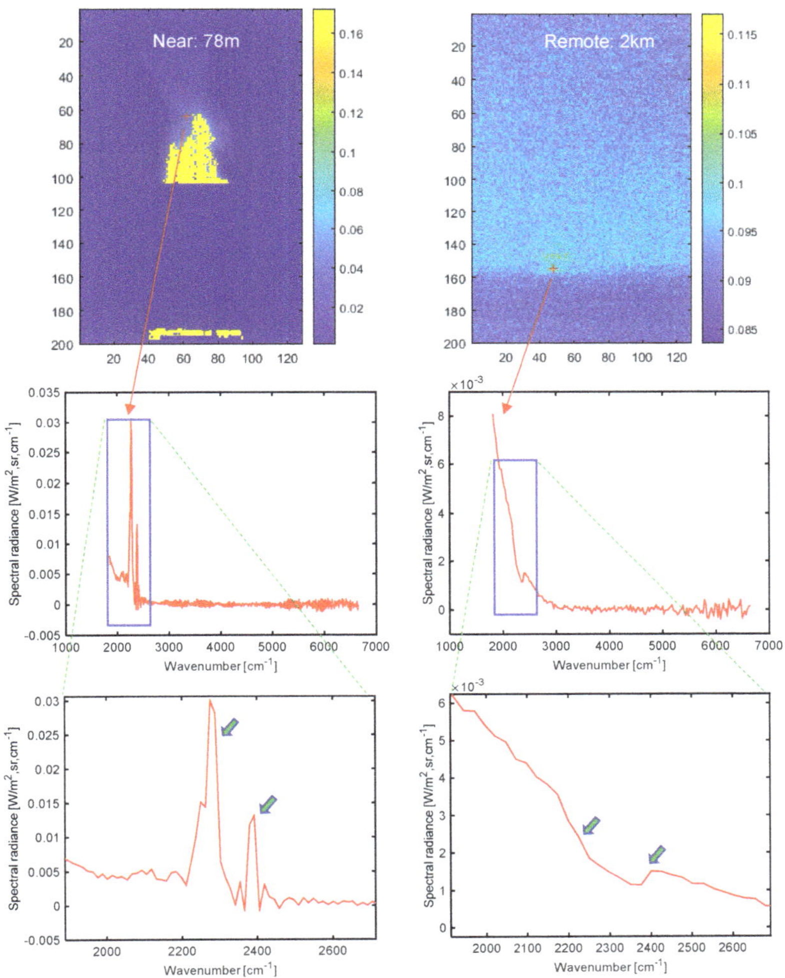

Figure 9. Comparison of CO_2 signatures in terms of distance: (left) near CO_2 (78 m), (right) remote CO_2 (2 km).

In the first experiment, the effect of signature variation was compared, depending on target distance. Figure 9 shows the spectral profiles of hot CO_2 plumes, both near (78 m) and remote (2 km).

A strong double-spike phenomenon can be observed for near CO_2, as shown in Figure 9 (left side) due to relatively weak signal attenuation. However, a remote target does not show such a distinctive double-spike pattern as seen in Figure 9 (right side), which makes it clear that only the spectral profile-based approach was not effective at detecting targets. This conclusion was confirmed through a real experiment applying the well-known spectral angle mapper-based detector [45] to the spectral profiles. Figure 10 shows the spectral profile-based detection results. The left column is the training region; the middle column is a representative spectral profile after training (just the mean of spectral profiles); and the right column is SAM-based detection results indicated by red dots. For the near target, the hot CO_2 plume region could be detected correctly with a similarity threshold of 0.8, as shown in Figure 10a. On the other hand, the remote target signature was unclear, which led to false positive detections with a similarity threshold of 0.98, as shown in Figure 10b.

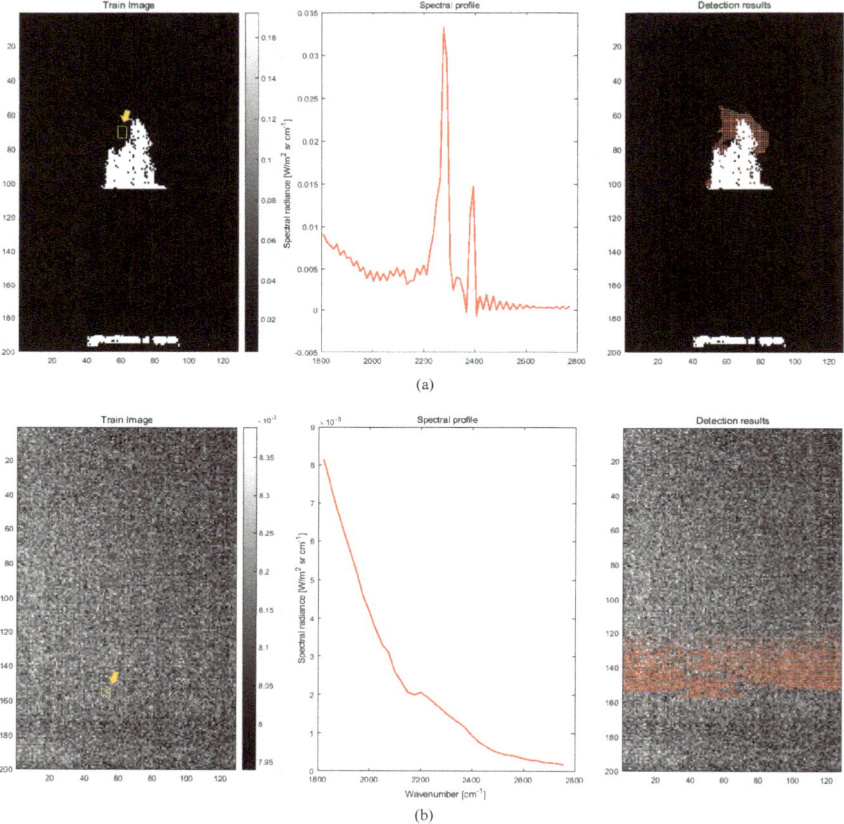

Figure 10. Spectral profile-based target detection results: (**a**) near CO_2, (**b**) remote CO_2.

3.2. Experiment 2: Detection Parameter Analysis

In the second experiment, key parameters of the proposed CO_2-DS method were analyzed and evaluated. There are four parameters; spectral band range for probing, initial threshold (th1) and CFAR threshold (th2) in the D-MSF, and final threshold (th3) from joint detection. The spectral band range for robust ship detection should be determined analytically based on Figure 11. Basically, the atmospheric CO_2 absorption band should be included. Since the air temperature is usually 300 K, this band is fixed at [2320–2380 cm^{-1}] as shown in the bottom part of Figure 11. The first-spike band starts at 2300 cm^{-1} and ends at 2180 cm^{-1} considering band offset. The second-spike band starts at 2380 cm^{-1} and ends at 2450 cm^{-1} considering band offset. The band images below 2180 cm^{-1} or above 2450 cm^{-1} show background clutter with target information that should be excluded to ensure extremely few false positives.

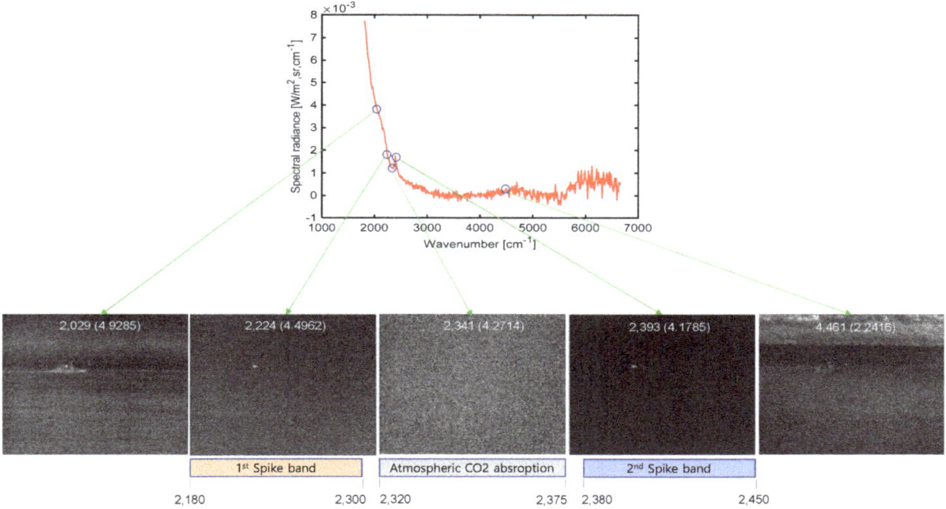

Figure 11. Analysis of spectral band ranges for ship detection by visualizing corresponding band images.

The selection of threshold1 (*th1*) is used to determine an ROI in order to investigate whether it is a target or not, as shown in Figure 6. Figure 12 shows the image for $I_\lambda^{MSF}(x, y)$, where the bright blob represents the hot CO_2 target, and others are background pixels. The 3D surf views of the target patch and background patch show the signal levels, and the lower left graph of Figure 6 displays the signal intensity profile corresponding to the dotted red line in the image. Note that the max signal is $20 \times 10^{-5}[\text{W/m}^2,\text{sr,cm}^{-1}]$ and the max background signal is $5 \times 10^{-5}[\text{W/m}^2,\text{sr,cm}^{-1}]$. Considering the safety margin of factor 2, *th1* was set at $1.0 \times 10^{-4}[\text{W/m}^2,\text{sr,cm}^{-1}]$. The threshold 2 parameter (*th2*) can be set by analyzing the SCR values for different targets, as shown in Figure 13. Most SCRs of a spike band image are larger than 12.8, so, *th2* can be set at 10.0 for a safe detection capability. The last threshold (*th3*) was varied to calculate the receiver operating characteristic (ROC) curve in order to compare different target-detection methods.

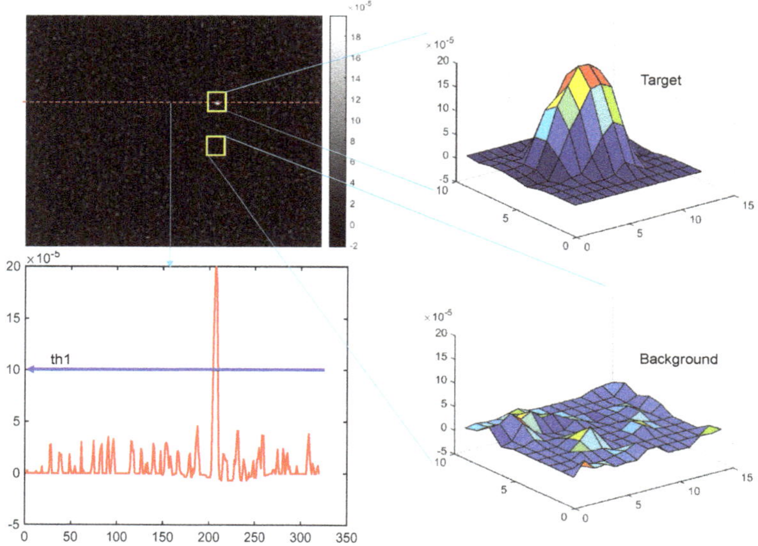

Figure 12. Parameter analysis of threshold 1 (*th*1) from directional mean subtraction image.

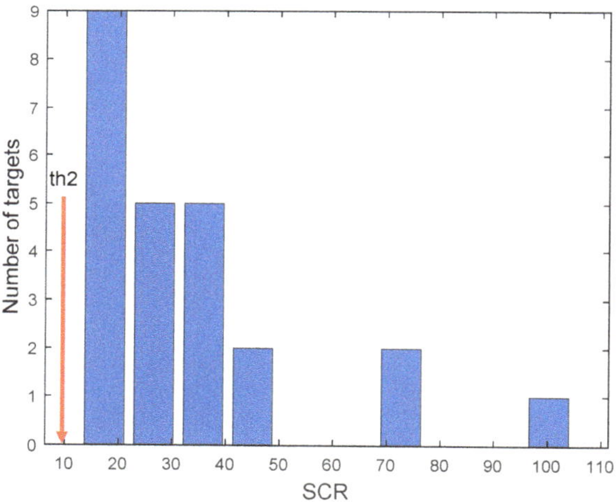

Figure 13. Parameter analysis of threshold 2 (*th*2) from statistics of SCR values of different targets.

3.3. Experimental 3: Performance Evaluation

Based on these parameter analyses, the final detection performance was evaluated by comparing the proposed CO$_2$-DS, D-MSF [4], and the high-boost multi-scale contrast measure (MLCM) [46]. MLCM is based on the human visual system, and greatly improves the detection rate for small infrared targets. The test images consist of 53 hypercubes acquired by a TELOPS MWIR hyperspectral camera. Three different ships with distance ranges between 1.5 km and 4 km were considered. For the baseline methods, test images were prepared by making broad band images in the spectral range of 3.5–5.6 μm. The ROC metric was used for a fair detection performance based on false positives per image (FPPI). Figure 14 summarizes detection performance. The proposed method (the red solid line) showed an ideal detection curve. MLCM ranked second, and D-MSF showed the worst result. Visual inspection was conducted at the indicated FPPI (0.65/image). With a normal homogeneous background, the three methods showed reasonable working performance, as seen in Figure 15. Magenta circles represent ground truth, cyan rectangles are detection results from the proposed CO_2-DS, and yellow rectangles represent baseline detection results. If there was strong cloud clutter, D-MSF generated many false alarms, as shown in Figure 16. In a strong sea-clutter environment, MLCM detected a lot of sea-glints, as shown in Figure 17. Note that the proposed CO_2-DS showed stable detection results without any false positives from sky or sea clutter. In addition, the ship was located at 4 km (near the horizon) and was detected successfully by our proposed method.

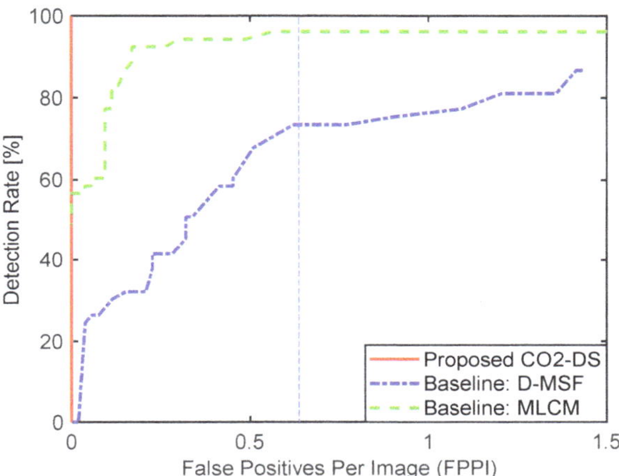

Figure 14. Receiver operating characteristic (ROC)-based comparison of detection performance.

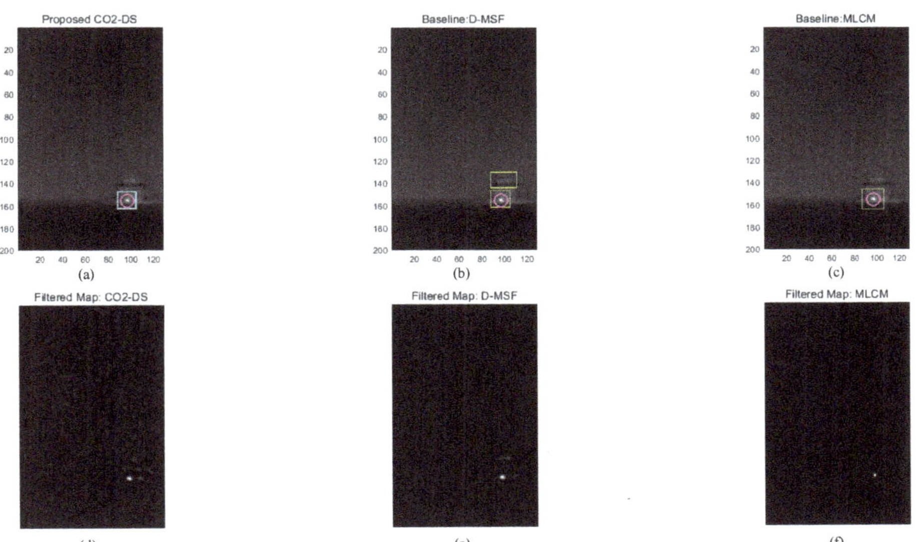

Figure 15. Comparative detection examples for a homogeneous background: (**a–c**) detection results, (**d–f**) filtered maps.

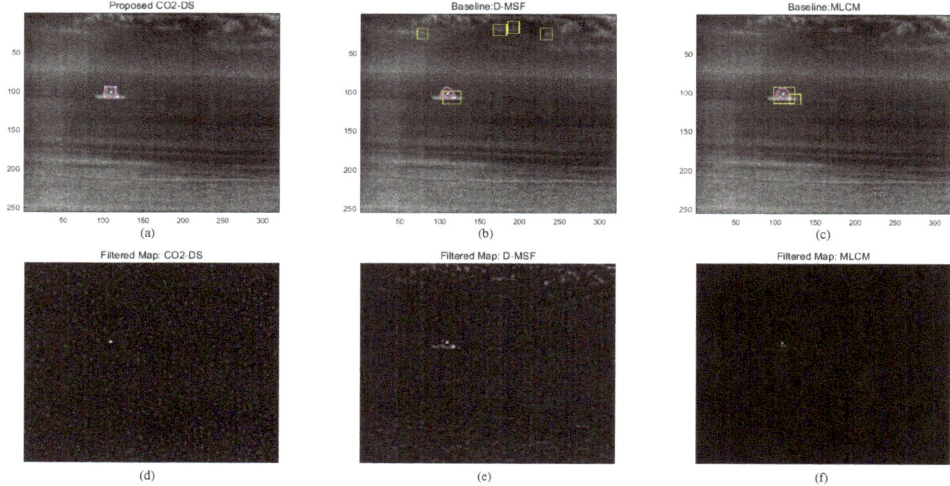

Figure 16. Comparative detection examples for cloud-clutter backgrounds: (**a–c**) detection results, (**d–f**) filtered maps.

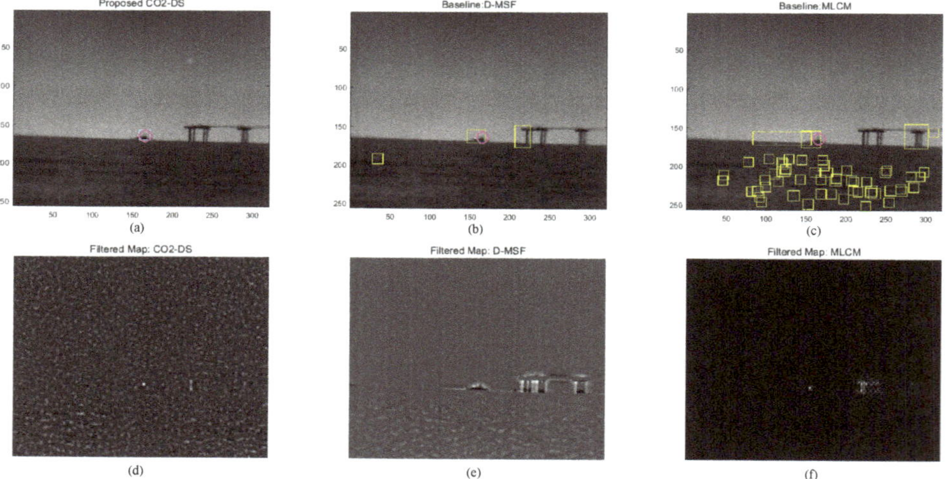

Figure 17. Comparative detection examples for sea-clutter backgrounds: (**a–c**) detection results, (**d–f**) filtered maps.

3.4. Experiment 4: Detection Performance Analysis

It can be useful to analyze the effect of weather condition and target distance to the detection performance. The proposed CO_2-DS method is based on Equation (4), especially atmospheric transmittance ($\tau(\lambda)$). Therefore, if we use the MODTRAN-based atmospheric transmittance calculation and measured background noise, we can conduct the Monte Carlo simulation. Figure 18 represents the MODTRAN-based atmospheric transmittance calculation by changing distance at a specific wavenumber (2390.16 cm^{-1}) or wavelength (4.184 μm). The atmosphere model was selected either

"MidLatitude Summer" or "MidLatitude Winter" to consider the effect of humidity and temperature. The distance length changed from 0.1 km to 6.0 km with 0.1 km interval.

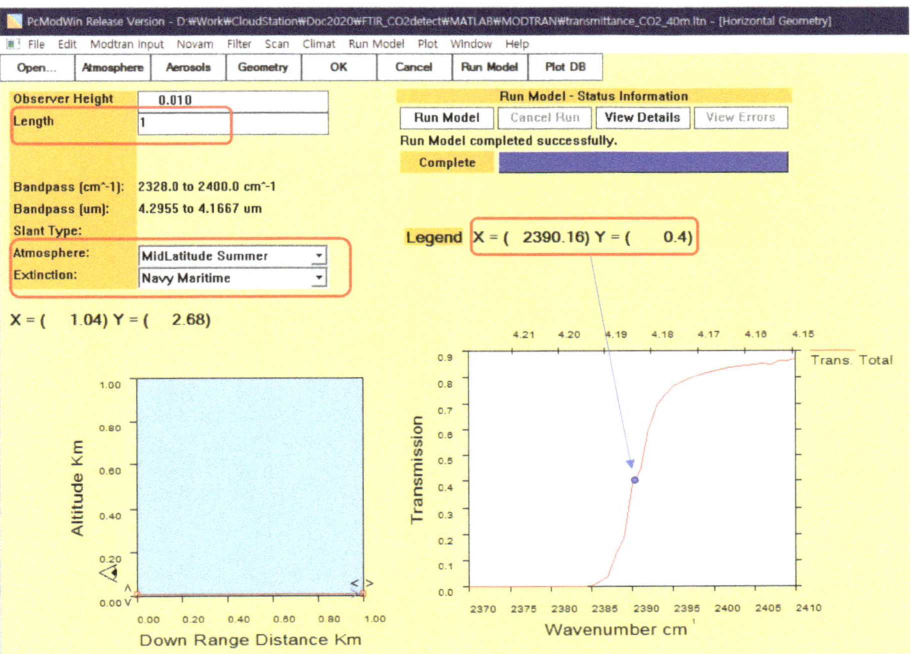

Figure 18. Moderate-Resolution Atmospheric Radiance and Transmittance (MODTRAN) environment for atmospheric transmittance calculation.

The proposed CO_2-MS detector uses target-background difference method (mean subtraction filter). Therefore, the path radiance term can be removed. Figure 19 shows a filtered map for the spectral band image at 2393 cm^{-1}. The maximum target signal-background at 2 km is 6.1742×10^{-4} [W/m$^2 \cdot$ sr $\cdot$ cm^{-1}]. If one uses transmittance of 0.18 at 2 km, the original signal-background is 3.4×10^{-3} [W/m$^2 \cdot$ sr $\cdot$ cm^{-1}]. Background noise can be modeled as the rectified Gaussian noise model with 0 mean and standard deviation of 1.3964×10^{-5} [W/m$^2 \cdot$ sr $\cdot$ cm^{-1}]. Th2 (SCR) is set as 10 in this simulation. Figure 20 shows the atmospheric transmittance and Monte Carlo simulation-based detection performance according to the target distance and weather conditions. Normally, the transmittance in mid-latitude summer is lower than that in mid-latitude winter because of higher water vapor contents (higher humidity), which leads to shorten the detection range as Figure 20b. If we fix target distance at 5 km, the detection rate is reduced from 95.4% to 5.2%.

The proposed CO_2-MS detection method shows extremely low false positives in ship detection. In addition, CO_2-MS is based on the thresholding and clustering. So, it can detect multiple targets. Because the CO_2-MS can use both the spectral and spatial information, the size of the ship (fishing boat, military ship, or cargo ship) can be identified if each ship has different type of fuel, engine, gas volume, and temperature. However, it has several limitations such as expensive spectral measurement device, weak to weather condition such as high humidity.

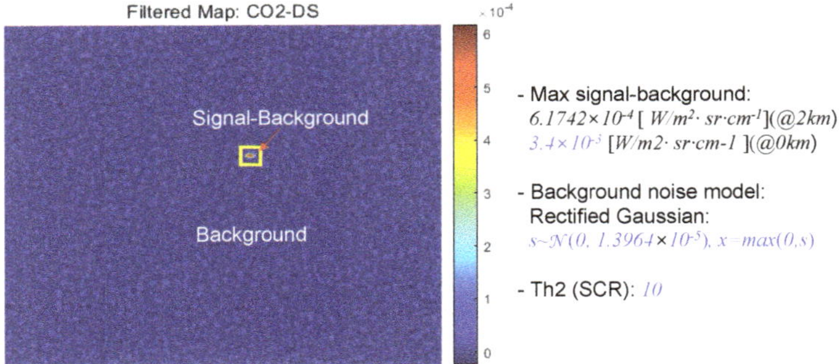

Figure 19. MODTRAN environment for atmospheric transmittance calculation.

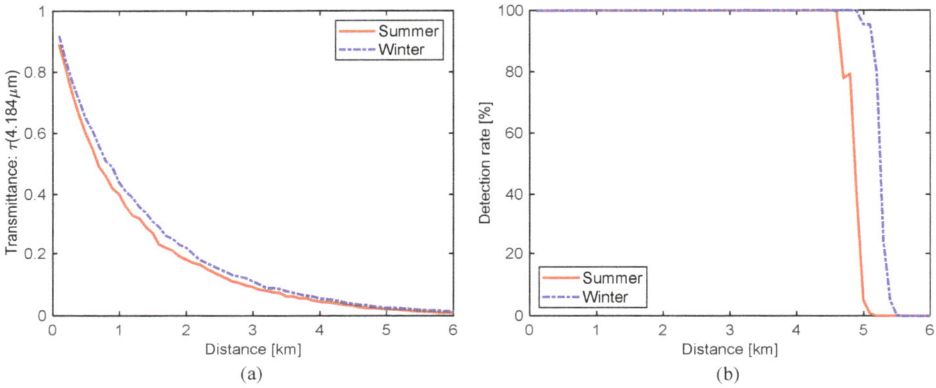

Figure 20. (**a**) Atmospheric transmittance vs. distance, (**b**) Monte Carlo simulation-based detection results.

4. Conclusions

Remote-ship detection is important in various applications, such as navigation and surveillance. Midwave infrared image-based ship detection is deployed due to its long-range detection capability in the maritime environment. However, MWIR images show strong response to cloud clutter and sea glint, which generate many false positives, even in state-of-the-art methods. This paper proposed a completely different approach by fusing spectral and spatial information in the MWIR carbon dioxide absorption band with the double-spike phenomenon. The atmospheric CO_2 absorption band (2320–2375 cm^{-1}) can block all background radiance, and remote hot CO_2 can penetrate through the spectral double-spike band, which is detected by the proposed carbon dioxide-double spike algorithm. Experimental results in a real cluttered sea environment with remote ships showed excellent detection performance with extremely few false positives if the ships' diesel engines are working.

Author Contributions: The contributions were distributed between authors as follows: S.K. (Sungho Kim) wrote the text of the manuscript and programmed the hyperspectral detection method. J.S., J.A. and S.K. (Sunho Kim) provided the midwave infrared hyperspectral database, operational scenario, performed the in-depth discussion of the related literature, and confirmed the accuracy experiments that are exclusive to this paper. All authors have read and agree to the published version of the manuscript.

Funding: This research was funded by ADD grant number UE191095FD, 2020 Yeungnam University Research Grants, NRF (NRF-2018R1D1A3B07049069) and the APC was funded by NRF.

Acknowledgments: This study was supported by the Agency for Defense Development (UE191095FD). This work was supported by the 2020 Yeungnam University Research Grants. This research was also supported by Basic

Science Research Program through the National Research Foundation of Korea(NRF) funded by the Ministry of Education (NRF-2018R1D1A3B07049069).

Conflicts of Interest: The authors declare no conflict of interest.

References

1. Willburger, K.; Schwenk, K.; Brauchle, J. AMARO—An On-Board Ship Detection and Real-Time Information System. *Sensors* **2020**, *20*, 1324. [CrossRef] [PubMed]
2. Brauchle, J.; Bayer, S.; Berger, R. Automatic Ship Detection on Multispectral and Thermal Infrared Aerial Images Using MACS-Mar Remote Sensing Platform. In *Image and Video Technology*; Satoh, S., Ed.; Springer International Publishing: Cham, Switzerland, 2018; pp. 382–395.
3. Li, Y.; Li, Z.; Zhu, Y.; Li, B.; Xiong, W.; Huang, Y. Thermal Infrared Small Ship Detection in Sea Clutter Based on Morphological Reconstruction and Multi-Feature Analysis. *Appl. Sci.* **2019**, *9*, 3786. [CrossRef]
4. Kim, S.; Lee, J. Small Infrared Target Detection by Region-Adaptive Clutter Rejection for Sea-Based Infrared Search and Track. *Sensors* **2014**, *14*, 13210–13242. [CrossRef]
5. Xu, F.; Liu, J.; Sun, M.; Zeng, D.; Wang, X. A Hierarchical Maritime Target Detection Method for Optical Remote Sensing Imagery. *Remote Sens.* **2017**, *9*, 280. [CrossRef]
6. Warren, R.C. Detection of Distant Airborne Targets in Cluttered Backgrounds in Infrared Image Sequences. Ph.D. Thesis, University of South Australia, Sydney, Australia, 2002.
7. Soni, T.; Zeidler, J.R.; Ku, W.H. Performance Evaluation of 2-D Adaptive Prediction Filters for Detection of Small Objects in Image Data. *IEEE Trans. Image Process.* **1993**, *2*, 327–340. [CrossRef]
8. Sang, H.; Shen, X.; Chen, C. Architecture of a configurable 2-D adaptive filter used for small object detection and digital image processing. *Opical Eng.* **2003**, *48*, 2182–2189. [CrossRef]
9. Sang, N.; Zhang, T.; Shi, W. Detection of Sea Surface Small Targets in Infrared Images based on Multi-level Filters. *Proc. SPIE* **1998**, *3373*, 123–129.
10. Wang, Y.L.; Dai, J.M.; Sun, X.G.; Wang, Q. An efficient method of small targets detection in low SNR. *J. Phys. Conf. Ser.* **2006**, *48*, 427–430. [CrossRef]
11. Despande, S.D.; Er, M.H.; Venkateswarlu, R.; Chan, P. Max-Mean and Max-Median Filters for Detection of Small-targets. *Proc. SPIE* **1999**, *3809*, 74–83.
12. Van den Broek, S.P.; Bakker, E.J.; de Lange, D.J.; Theil, A. Detection and Classification of Infrared Decoys and Small Targets in a Sea Background. *Proc. SPIE* **2000**, *4029*, 70–80.
13. Kim, S. Min-local-LoG filter for detecting small targets in cluttered background. *Electron. Lett* **2011**, *47*, 105–106. [CrossRef]
14. De Lange, H.B.D.J.J.; van den Broek, S.P.; Kemp, R.A.W.; Schwering, P.B.W. Automatic Detection of Small Surface Targets with Electro-Optical Sensors in a Harbor Environment. Proc. SPIE **2008**, *7114*, 9–16.
15. Crosby, F. Signature Adaptive Target Detection and Threshold Selection for Constant False Alarm Rate. *J. Electron. Imaging* **2005**, *14*, 1–10. [CrossRef]
16. Hubbard, W.A.; Page, G.A.; Carroll, B.D.; Manson, D.C. Feature Measurement Augmentation for a Dynamic Programming based IR Target Detection Algorithm in the Naval Environment. *Proc. SPIE* **1999**, *2698*, 2–9.
17. Shirvaikar, M.V.; Trivedi, M.M. A Neural Network Filter to Detect Small Targets in High Clutter Backgrounds. *IEEE Trans. Neural Netw.* **1995**, *6*, 252–257. [CrossRef] [PubMed]
18. Lim, E.T.; Shue, L.; Ronda, V. Multi-mode Fusion Algorithm for Robust Dim Point-like Target Detection. Proc. SPIE **2003**, *5082*, 94–102.
19. Kojima, A.; Sakurai, N.; Kishigami, J.I. Motion detection using 3D-FFT spectrum. In Proceedings of the IEEE International Conference on Acoustics, Speech, and Signal Processing (ICASSP), Minneapolis, MN, USA, 27–30 April 1993; Volume 5, pp. 213–216.
20. Ye, Z.; Wang, J.; Yu, R.; Jiang, Y.; Zou, Y. Infrared clutter rejection in detection of point targets. *Proc. SPIE* **2002**, *4077*, 533–537.
21. Zuo, Z.; Zhang, T. Detection of Sea Surface Small Targets in Infrared Images based on Multi-level Filters. *Proc. SPIE* **1999**, *3544*, 372–377.
22. Yang, L.; Yang, J.; Yang, K. Adaptive Detection for Infrared Small Target under Sea-sky Complex Background. *Electron. Lett.* **2004**, *40*, 1083–1085. [CrossRef]

23. Chen, Z.; Wang, G.; Liu, J.; Liu, C. Small Target Detection Algorithm based on Average Absolute Difference Maximum and Background Forecast. *Int. J. Infrared Milli. Waves* **2007**, *28*, 87–97. [CrossRef]
24. Chi, J.N.; Fu, P.; Wang, D.S.; Xu, X.H. A Detection Method of Infrared Image Small Target based on Order Morphology Transformation and Image Entropy Difference. In Proceedings of the 2005 International Conference on Machine Learning and Cybernetics, Guangzhou, China, 18–21 August 2005; Volume 8, pp. 5111–5116.
25. Rozovskii, B.; Petrov, A. Optimal Nonlinear Filtering for Track-before-Detect in IR Image Sequences. *Proc. SPIE* **1999**, *3809*, 152–163.
26. Silverman, J.; Mooney, J.M.; Caefer, C.E. Tracking Point Targets in Cloud Clutterr. *Proc. SPIE* **1997**, *3061*, 496–507.
27. Tzannes, A.P.; Brooks, D.H. Point Target Detection in IR Image Sequences: A Hypothesis-Testing Approach based on Target and Clutter Temporal Profile Modeling. *Opt. Eng.* **2000**, *39*, 2270–2278.
28. Thiam, E.; Shue, L.; Venkateswarlu, R. Adaptive Mean and Variance Filter for Detection of Dim Point-Like Targets. *Proc. SPIE* **2002**, *4728*, 492–502.
29. Kim, S.; Sun, S.G.; Kim, K.T. Highly efficient supersonic small infrared target detection using temporal contrast filter. *Electron. Lett* **2014**, *50*, 81–83. [CrossRef]
30. Zhang, B.Z.T.; Cao, Z.; Zhang, K. Fast New Small-target Detection Algorithm based on a Modified Partial Differential Equation in Infrared Clutter. *Opt. Eng.* **2007**, *46*, 106401 [CrossRef]
31. Wu, B.; Ji, H.B. Improved power-law-detector-based moving small dim target detection in infrared images. *Opt. Eng.* **2008**, *47*, 010503. [CrossRef]
32. Soni, T.; Zeidler, R.; Ku, W.H. Recursive estimation techniques for detection of small objects in infrared image data. In Proceedings of the 1992 IEEE International Conference on Acoustics, Speech, and Signal Processing, San Francisco, CA, USA, 23–26 March 1992; Volume 3, pp. 581–584.
33. Tartakovsky, A.; Blazek, R. Effective adaptive spatial-temporal technique for clutter rejection in IRST. *Proc. SPIE* **2000**, *4048*, 85–95.
34. Lopez-Alonso, J.M.; Alda, J. Characterization of Dynamic Sea Scenarios with Infrared Imagers. *Infrared Phy. Technol.* **2005**, *46*, 355–363. [CrossRef]
35. Lin, Z.; Ji, K.; Leng, X.; Kuang, G. Squeeze and Excitation Rank Faster R-CNN for Ship Detection in SAR Images. *IEEE Geosci. Remote Sens. Lett.* **2019**, *16*, 751–755. [CrossRef]
36. An, Q.; Pan, Z.; Liu, L.; You, H. DRBox-v2: An Improved Detector with Rotatable Boxes for Target Detection in SAR Images. *IEEE Trans. Geosci. Remote Sens.* **2019**, *57*, 8333–8349. [CrossRef]
37. Kang, M.; Ji, K.L.X.L.Z. Contextual Region-Based Convolutional Neural Network with Multilayer Fusion for SAR Ship Detection. *Remote Sens.* **2017**, *9*, 860. [CrossRef]
38. Yuan, H.; rui Wang, X.; Yuan, Y.; Li, K.; Zhang, C.; shun Zhao, Z. Space-based full chain multi-spectral imaging features accurate prediction and analysis for aircraft plume under sea/cloud background. *Opt. Express* **2019**, *27*, 26027–26043. [CrossRef]
39. Gagnon, M.; Gagnon, J.; Tremblay, P.; Savary, S.; Farley, V.; Guyot, E.; Lagueux, P.; Chamberland, M. Standoff midwave infrared hyperspectral imaging of ship plumes. In Proceedings of the 2015 7th Workshop on Hyperspectral Image and Signal Processing: Evolution in Remote Sensing (WHISPERS), Tokyo, Japan, 2–5 June 2015; pp. 1–4.
40. Gagnon, M.A.; Gagnon, J.P.; Tremblay, P.; Savary, S.; Farley, V.; Guyot, É.; Lagueux, P.; Chamberland, M. Standoff midwave infrared hyperspectral imaging of ship plumes. *Proc. SPIE* **2016**, *9988*, 998806. [CrossRef]
41. Romaniello, V.; Spinetti, C.; Silvestri, M.; Buongiorno, M.F. A Sensitivity Study of the 4.8 μm Carbon Dioxide Absorption Band in the MWIR Spectral Range. *Remote Sens.* **2020**, *12*, 172. [CrossRef]
42. Griffin, M.K.; Hua, K. Burke, H.; Kerekes, J.P. Understanding radiative transfer in the midwave infrared: A precursor to full-spectrum atmospheric compensation. *Proc. SPIE* **2004**, *5425*, 348–356. [CrossRef]
43. Gagnon, M.A.; Gagnon, J.P.; Tremblay, P.; Savary, S.; Farley, V.; Guyot, E.; Lagueux, P.; Chamberland, M. Standoff midwave infrared hyperspectral imaging of ship plumes. In *Electro-Optical Remote Sensing X*; Kamerman, G., Steinvall, O., Eds.; International Society for Optics and Photonics, SPIE: Bellingham, WA, USA, 2016; Volume 9988, pp. 40–48. [CrossRef]

44. Schoemaker, R.; Schleijpen, R. Evaluation tools for the effectiveness of infrared countermeasures and signature reduction for ships. In *Infrared Imaging Systems: Design, Analysis, Modeling, and Testing XXI*; Holst, G.C., Krapels, K.A., Eds.; International Society for Optics and Photonics, SPIE: Bellingham, WA, USA, 2010; Volume 7662, pp. 248–257. [CrossRef]

45. Ibraheem, I. Comparative study of maximum likelihood and spectral angle mapper algorithms used for automated detection of melanoma. *Skin Res. Technol.* **2014**, *21*. [CrossRef]

46. Shi, Y.; Wei, Y.; Yao, H.; Pan, D.; Xiao, G. High-Boost-Based Multiscale Local Contrast Measure for Infrared Small Target Detection. *IEEE Geosci. Remote Sens. Lett.* **2018**, *15*, 33–37. [CrossRef]

Article

Aircraft and Ship Velocity Determination in Sentinel-2 Multispectral Images

Henning Heiselberg

National Space Institute, Technical University of Denmark, 2800 Kongens Lyngby, Denmark; hh@space.dtu.dk; Tel.: +45-4525-9760

Received: 31 May 2019; Accepted: 26 June 2019; Published: 28 June 2019

Abstract: The Sentinel-2 satellites in the Copernicus program provide high resolution multispectral images, which are recorded with temporal offsets up to 2.6 s. Moving aircrafts and ships are therefore observed at different positions due to the multispectral band offsets, from which velocities can be determined. We describe an algorithm for detecting aircrafts and ships, and determining their speed, heading, position, length, etc. Aircraft velocities are also affected by the parallax effect and jet streams, and we show how the altitude and the jet stream speed can be determined from the geometry of the aircraft and/or contrail heading. Ship speeds are more difficult to determine as wakes affect the average ship positions differently in the various multispectral bands, and more advanced corrections methods are shown to improve the velocity determination.

Keywords: Sentinel-2; multispectral; temporal offsets; ship; aircraft; velocity; altitude; parallax; jet stream

---

1. Introduction

Surveillance for marine and air space situation awareness is essential for monitoring and controlling traffic safety, piracy, smuggling, fishing, irregular migration, trespassing, spying, icebergs, shipwrecks, the environment (oil spill or pollution), etc. Dark ships and aircrafts are non-cooperative vessels with non-functioning transponder systems such as the automatic identification system (AIS) for ships or automatic dependent surveillance (ADS-B) for aircrafts. Their transmission may be jammed, spoofed, sometimes experience erroneous returns, or simply turned off deliberately or by accident. Furthermore, AIS and ADS-B land based and satellite coverage is sparse at sea and at high latitudes. Therefore, other non-cooperative surveillance systems as satellite or airborne systems are required.

The Sentinel-2 satellites under the Copernicus program [1–3] carry multispectral imaging (MSI) instruments that provide excellent and freely available imagery with pixel resolutions down to 10 m. The orbital period is 5 days between the Sentinel-2 (S2) satellites A + B, but as the swaths from different satellite orbits overlap at higher latitudes, the typical revisit period for each satellite is two or three days in Europe and almost daily in the Arctic. S2 MSI has the potential to greatly improve the marine and airspace situational awareness, especially for non-cooperative ships and aircrafts.

Ship detection, recognition, and identification in optical satellite imagery has been studied in a number of papers with good results [4–10]. The resolution and sensitivity are generally better and the number of multispectral bands is larger, but clouds reduce the continuous coverage. Ship positions, and their length, breadth, form and heading can be determined accurately and the multispectral reflections can be fingerprints for ID. Ship speeds have only been determined from satellite imagery in a few cases where Kelvin wakes are observed [7].

Detection, tracking and speed determination of vehicles on ground has been studied in video sequences and images recorded with time intervals fx. by change detection. Heights of tall buildings or altitudes of clouds [11–13] and other static or slow moving objects have been determined by shadow lengths or parallax methods, when the images are recorded from a flying platform as an aircraft, drone or satellite with time delay imaging or from two platforms.

Recently, parallax effects were observed for aircrafts and their condensation trails (contrails) in Sentinel-2 color images, referred to as "plainbows" [14]. The work presented here is novel when it comes to exploiting the temporal offsets in Sentinel-2 MSI, and specifically for determining aircraft velocities and altitudes, and ship speeds. We consider this work as the first analysis of such effects due to temporal offsets, as we have not been able to find any studies with scientific analyses or applications. In this work, we outline the basic physics for moving objects in satellite multispectral images with temporal offsets, the parallax effects and influence of jet streams. We primarily consider aircrafts and ships, but the analysis also applies in principle to all kinds of moving vehicles including cars and helicopters, and also clouds, auroras, etc. The basic formulas are derived, and as proof of principle, we show a number of representative examples for both aircrafts and ships. From Sentinel-2 multispectral images with known temporal offsets we calculate the resulting speed, heading, altitudes, etc. Subsequently, we test our results by comparing to data from the navigation systems ADS-B or AIS.

In Section 2 we analyze how the temporal offset affects moving objects in S2 multispectral images and include the parallax effect for aircrafts as well as the effect of jet streams. After a description of the S2 MSI offset times and resulting apparent velocities, we calculate aircraft velocities and altitudes from S2 multispectral images in Section 3. In Section 4 we turn to ship velocities, which are not affected by parallax but are slower and more difficult to determine accurately due to long wakes. We describe a simple but effective correction method, which improves the speed calculations considerably.

In the summary and outlook we suggest ways to improve the model by better position determination and understanding of spectral dependence of object reflections [15]. In addition, calculations for a large number of ships and aircrafts are required for better statistics and improving the model by fine-tuning the parameters. This could also lead to an annotated database useful for machine learning methods.

2. Satellite Images and Method of Analysis

The S2 multispectral images were analyzed using dedicated software developed specifically to detect ships and aircrafts in large images with different pixel resolutions and determine their precise position and orientation as described in [7,8]. As detection is not the focus of this paper, we mainly describe the MSI temporal offsets and how they affect the multispectral images for velocity calculations. A flow chart for the algorithm is shown in Figure 1.

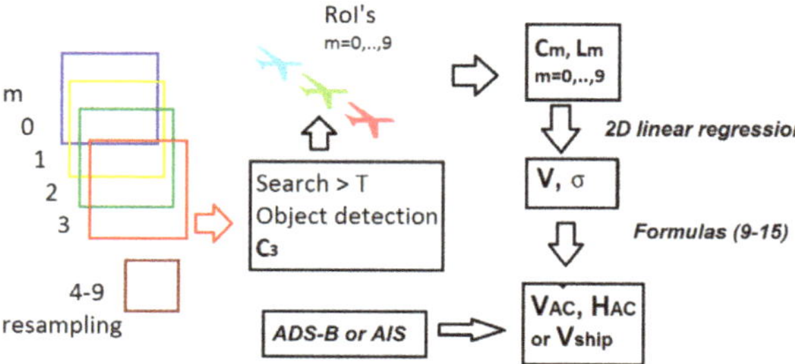

Figure 1. Flow chart of algorithm: For a given scene the m = 0, ... , 9 high resolution S2 multispectral images are selected. Search and detection of objects above a threshold T is done for the red band m = 3. For each object, a region of interest (RoI) around the object center $\vec{c}_3$ is chosen for all ten images, where m = 4–9, are resampled to 10 m pixel size. The center $\vec{c}_m$ and length $\vec{L}_m$ are calculated for m = 0, ..., 9. From 2D linear regression (3) of the variance σ, the apparent velocity $\vec{V}$ is found and inserted in (9–15), whereby the aircraft velocity and altitude or ship velocity is found. By comparing to ADS-B or AIS, the velocities, altitudes and positions are validated.

2.1. Sentinel-2 Multispectral Images

S2 carries the Multispectral Sensor Imager [1–3] that records images in 13 multispectral bands (see Table 1) with different resolutions and time offsets. As we are interested in small object detection and tracking, we focused on analyzing the high resolution images, the 4 bands with 10 m and the 6 bands with 20 m pixel resolution. These are mega- to giga-pixel images with 16 bit grey levels.

Table 1. Spectral bands for the Sentinel-2B Multispectral Imager. The 10 high resolution bands $m = 0$, ..., 9 are ordered according to temporal offset [1].

S2 Spectral Band	Temporal Order (m)	Temporal Offset t_m (s)	Central Wavelength (nm)	Bandwidth (nm)	Spatial Resolution (m)
1	-	2.314	442.2	21	60
2	0	0	492.1	66	10
3	2	0.527	559.0	36	10
4	3	1.005	664.9	31	10
5	4	1.269	703.8	16	20
6	6	1.525	739.1	15	20
7	7	1.790	779.7	20	20
8	1	0.263	832.9	106	10
8a	8	2.055	864.0	22	20
9	-	2.586	943.2	21	60
10	-	0.851	1376.9	30	60
11	5	1.468	1610.4	94	20
12	9	2.085	2185.7	185	20

We analyzed several S2 level 2A processed images from 2019 covering Copenhagen airport in Denmark, see Figure 2, and the Channel between Amsterdam and London. These images are convenient because several aircrafts are usually present after takeoff or before landing. In addition, there are a larger number of ships present in the strait of Øresund surrounding the airport and in the Channel.

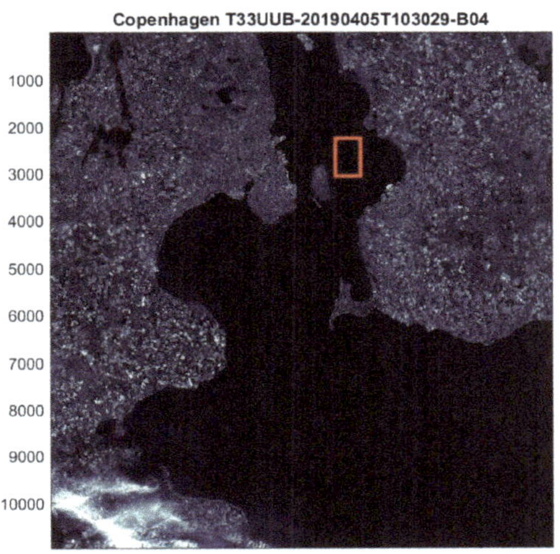

Figure 2. East part of Denmark in the red band B4 (m = 3) from 5 April 2019, 10:30 a.m. UTC. The red box is a ROI in the strait of Øresund surrounding Copenhagen airport.

In the S2 multispectral images $I_m(i, j)$, the spatial coordinates $\vec{r} = (x, y)$ are the pixel coordinates (i,j) multiplied by the pixel resolution $l = 10$ m, 20 m or 60 m as given in Table 1 for the 13 bands. The 10 high resolution multispectral images with pixel size 10 m or 20 m are ordered according to temporal offset t_m, $m = 0, 1, ..., 9$. As shown in Table 1, they range from 0 s to 2.085 s in temporal offset. Due to the odd and even detector array in MSI, the offsets are either delayed or advanced, respectively. The imaging sequence is such that the offsets are reversed in stripes along track within the image [13].

2.2. Object Detection and Position Determination

To detect an object, its reflection must deviate from the background. For proving the principle of velocity determination, we chose for simplicity a region of interest with sea background, which is usually darker than the objects and therefore makes detection easier. The multispectral variant background over land will require a more elaborate detection algorithm, but has the potential of determining velocities of driving vehicles as well.

When the sea covers more than half of the image after land removal, the median reflection value provides an accurate and robust value for the average background. For detection, we chose the red band $m = 3$ (see Figure 2) because it has high resolution and average time offset such that temporal offset objects will appear around the red center (see Figures 3–5). In addition, solar reflections from ships and aircrafts generally have high contrast in red with respect to the sea background. For each object a small region, e.g., 100×100 pixels or smaller, is extracted around the central object coordinate, such that it covers the object extent including movement, wakes or contrails. The same 1 km $\times$ 1 km region is then extracted for the 10 high resolution bands $m = 0, 1, 2, 3$ with spatial resolution 10 m and $m = 4, ..., 9$ with 20 m. The latter are corrected for the different resolutions. For each band the median value is chosen as the background.

The pan-sharpening technique [7,16] for increasing the resolution in lower resolution images can only be applied to static images. As moving objects change pixel position in the multispectral images due to the temporal offset, we could not apply pan-sharpening in this analysis.

For each multispectral image, the object is defined spatially by the pixels with reflections above the background value plus a threshold T, which depends on target type as will be discussed below. The central object position $\vec{c}_m = (x, y)$, length L_m, breadth B_m and orientation/heading angle θ_m are calculated in each band $m = 0, ..., 9$ by weighting the object pixels with their reflection $I_m(i, j)$ and calculating the first moments in i and j, as described in detail in [7].

Unfortunately, ship wakes and aircraft contrails can corrupt the position determination considerably. Both generally move the central position backwards with respect to vessel direction by an amount that varies with the band. At the same time, the object length is extended. We corrected for this effect to the first order by adding the distance from the average center position to the ship front, which is half of the object length L_m, in the vessel heading direction, i.e., the vector $\vec{L}_m = L_m(\cos\theta_m, \sin\theta_m)$.

$$\vec{r}_m = \vec{c}_m + \frac{1}{2}\vec{L}_m \tag{1}$$

This position is now at the front of the object in each band m as shown in the images below.

2.3. Multispectral Temporal Offsets and Velocity Determination

We define the apparent velocity as the change in position as observed in the multispectral images divided by the band dependent time delay. An object moving with apparent velocity $\vec{V}$ will ideally be recorded in band m at position

$$\vec{r}_m \sim \vec{r}_V + \vec{V} \cdot t_m \tag{2}$$

here, $\vec{r}_V$ is the vessel position at zero temporal offset—ideally the blue band $m = 0$ for which $t_0 = 0$. For ships, the apparent velocity $\vec{V}$ is simply the vessel speed and direction, whereas for aircrafts, the parallax effect due to satellite motion must be included as will described below. Currents and jet streams also influence V.

For example, the aircrafts shown in Figures 4 and 5 fly with apparent speeds V ~ 200 m/s, and move a distance of ~400 m or 40 pixels in the time interval of 2.085 s. Consequently, the aircrafts show up as ten "pearls on a string" when the high resolution multispectral bands are plotted all together in a (false) color image.

The object positions $\vec{r}_m$ that are calculated for each multispectral image $m = 0, ..., 9$ will generally scatter around the linear prediction of Equation (2). We define the variance as the mean square average of distance deviations

$$\sigma^2 = \frac{1}{10} \sum_{m=0}^{9} \left(\vec{r}_m - \vec{r}_V - \vec{V} \cdot t_m \right)^2 \tag{3}$$

By minimizing this variance, which is equivalent to a two-dimensional linear regression, we obtain the best fit values for vessel position $\vec{r}_V$ and the apparent velocity vector $\vec{V} = V(\cos \theta_V, \sin \theta_V)$. An estimate for the uncertainty in apparent velocity is the lowest standard deviation in distance divided by the temporal interval

$$\sigma_V = \sigma / t_9 \tag{4}$$

Typically, the positions are accurate up to a pixel size l or less in each multispectral image. The two-dimensional linear regression fit of $\vec{r}_m$ vs. t_m therefore has a standard deviation less than $\sigma \sim l / \sqrt{10}$. Dividing by the temporal offset interval $t_9 = 2.085$ s, we obtain the uncertainty for the apparent velocity of $\sigma_V = l / t_9 \sqrt{10}$ ~5–10 kph. This is comparable to speeds of slow ships, whereas typical aircraft cruise velocities are 800–1000 kph, and gives a relatively accurate aircraft velocity determination.

2.4. Comparison to AIS and ADS-B

By matching positions of aircrafts from ADS-B and ships from AIS at the same time and positions, we can identify the vessels and compare size, heading, velocities and altitudes. Unfortunately, the positioning systems sometimes lag or the updating is delayed or is infrequent. Therefore, we find that ships and aircraft do not always match precisely at the correct position and time. In addtion, the S2 overflight time included in the file name is not the local image recording time. In Figure 2, the Copenhagen regions are recorded five minutes later than the time 10 h 30 m 29 s given in the file name, and the local recording time is 14 s later from north to south for the descending track.

3. Aircraft Velocities

For low flying aircrafts and ships the parallax effect is negligible and the apparent velocity $\vec{V}$ is just the aircraft velocity $\vec{V}_{AC}$. For high altitude aircraft, we need to consider the satellite orbit, velocity and viewing angle in order to correct for the parallax effect. In addition, the jet stream must be considered as it affects the contrails.

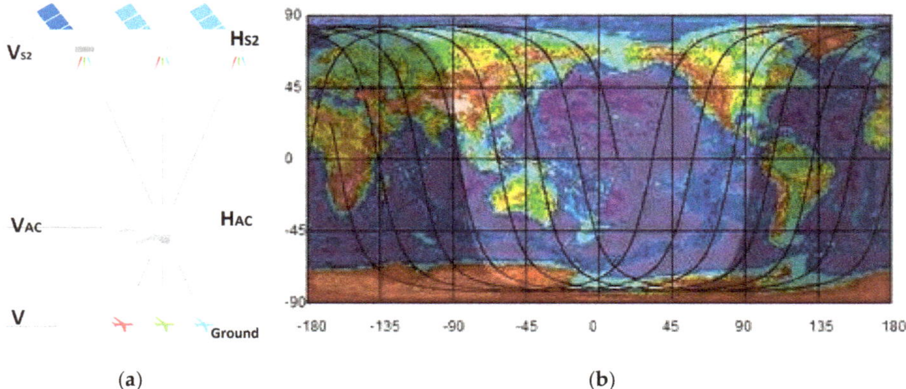

(**a**)   (**b**)

Figure 3. (**a**) Illustration of the satellite-aircraft-ground parallax effect. (**b**) Earth map with projected satellite orbits. The orbits cross the vertical lines (latitudes ϕ) at angles θ_{S2}.

3.1. Satellite Direction w.r.t. Ground

We used standard mathematical and celestial convention where angles are measured from the equator counter-clockwise. In navigation, angles are measured from the North Pole clockwise and thus differ by 90° and angular direction.

The S2 satellites fly in a sun-synchronous orbit at mean altitude H_{S2} = 786 km with speed V_{S2} = 7.44 km/s. Their polar orbit is slightly retrograde, descending with inclination angle i = −98.62° on the dayside. Due to Earth's curvature, the orientation of the satellite track θ_{S2} with respect to latitude ϕ is (see Figure 3b)

$$\cos\theta_{S2} = \cos i / \cos\phi \tag{5}$$

At the equator, $\phi = 0°$ and $\theta_{S2} = i$, but at maximum polar S2 latitude $\phi = 180° - i = 81.38°$, the S2 satellite flies straight west, i.e., $\theta_{S2} = \pm 180°$. For the images around Copenhagen, $\phi \simeq 55°$ and we find $\theta_{S2} \simeq -105°$. At Amsterdam, $\phi \simeq 52,5°$ and $\theta_{S2} \simeq -104°$. The resulting satellite velocity is $\vec{V}_{S2} = V_{S2}(\cos\theta_{S2}, \sin\theta_{S2})$, w.r.t. ground.

3.2. Parallax Effect

The satellite movement during the multispectral temporal delays causes a parallax effect (see Figure 3a) that moves objects at an altitude H northeast-wards in direction $-\theta_{S2}$ with respect to the ground. Stationary objects such as tall buildings, clouds, balloons, stalling aircrafts, etc. will be moved by an apparent velocity $\vec{V} = -\vec{V}_{S2} \cdot H/H_{S2}$ with respect to ground due to their parallax. Determining V from a linear regression of Equation (3), we find the object altitude

$$H = \frac{V}{V_{S2}} \cdot H_{S2} \tag{6}$$

The parallax has recently been exploited for determining altitudes and movement of, for example, volcanic plumes [11,12].

The parallax effect moves the flight paths opposite to the satellite direction, and separates each multispectral band such that an aircraft (with its contrails) appears as a multispectral rainbow, when plotted in a false color image as shown in Figures 4 and 5. Previously [14], the RGB contrails were named "plainbows". We named our ten multispectral contrails as a "planebows". Contrails are usually observed at high altitudes 7.5–12 km.

3.3. Moving Objects

When the object moves, its velocity must be added to the apparent velocity. In the absence of wind, a moving object, such as an aircraft at altitude H_{AC} with velocity $\vec{V}_{AC} = V_{AC}(\cos\theta_{AC}, \sin\theta_{AC})$, will appear to have velocity

$$\vec{V} = \vec{V}_{AC} - \vec{V}_{S2} \cdot \frac{H_{AC}}{H_{S2}} \tag{7}$$

or

$$\begin{pmatrix} \cos\theta_{AC} \\ \sin\theta_{AC} \end{pmatrix} V_{AC} = \begin{pmatrix} \cos\theta_V \\ \sin\theta_V \end{pmatrix} V + \begin{pmatrix} \cos\theta_{S2} \\ \sin\theta_{S2} \end{pmatrix} V_{S2} \cdot \frac{H_{AC}}{H_{S2}} \tag{8}$$

The heading of the aircraft θ_{AC} is given by the aircraft orientation angles θ_m, $m = 0, ..., 9$ as described in Section 2.2 and [7]. The aircraft heading angle θ_{AC} must be determined independently for the ten multispectral bands θ_m. The four high resolution bands generally provide a consistent and robust average aircraft heading angle θ_{AC}. When contrails are visible (see Figure 3b), the contrail and aircraft directions are the same $\theta_{CT} = \theta_{AC}$, and they provide a more accurate heading.

From the two equations in (8), we obtain the aircraft velocity

$$V_{AC} = \frac{\sin(\theta_V - \theta_{S2})}{\sin(\theta_{AC} - \theta_{S2})} \cdot V \tag{9}$$

and the aircraft altitude

$$H_{AC} = \frac{\sin(\theta_V - \theta_{AC})}{\sin(\theta_{AC} - \theta_{S2})} \cdot \frac{V}{V_{S2}} \cdot H_{S2} \tag{10}$$

These relations can also be obtained from simple sine relations for the triangles in Figures 4 and 5.

When the aircraft and satellite directions are parallel, $\theta_{S2} = \theta_{AC}$, the aircraft velocity and altitude cannot be determined separately. At zero altitude (as will be discussed below for ships) there is no parallax effect so that $\theta_{AC} = \theta_V$ and $V_{AC} = V$.

Note that Equations (9) and (10) are invariant to the orientation of the coordinate system as only relative angles appear. They are also clock- vs. counter clockwise invariant.

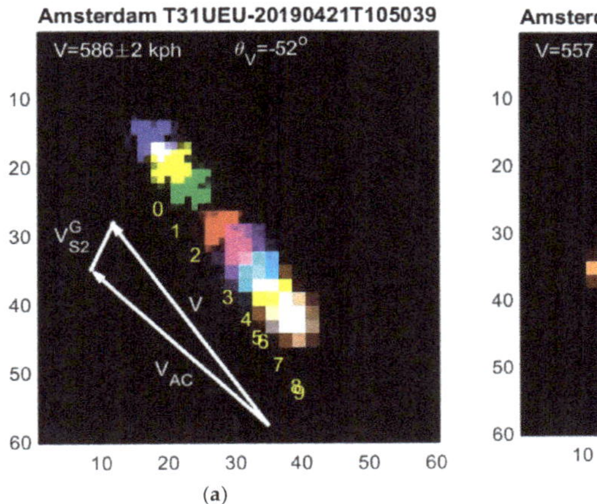

(a)

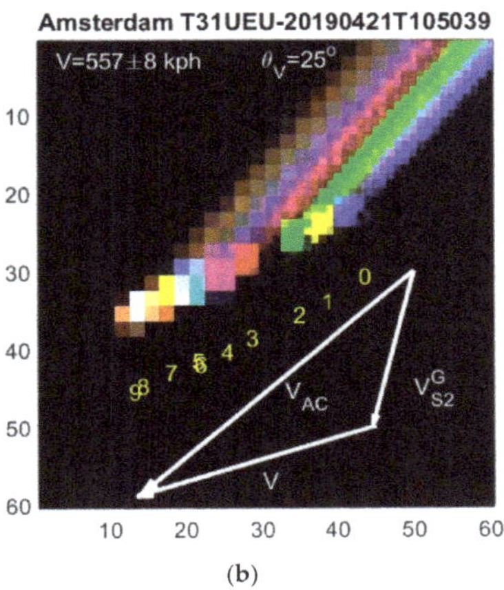

(b)

Figure 4. Planebows. Scale in pixels. (**a**) A slow and low flying aircraft near Amsterdam airport after takeoff recorded in 10 multispectral images with offsets. Red, green and blue are true colors whereas the seven remaining colors are overlayed with false colors, e.g., m = 1 is yellow. Numbers 0, ..., 9 indicate the central aircraft position in each band (but moved 10 pixels down). The triangle shows the apparent ($\vec{V}$), aircraft ($\vec{V}_{AC}$) and satellite ($\vec{V}^G_{S2} = \vec{V}_{S2}H_{AC}/H_{S2}$) velocity vectors. (**b**) Planebow of a fast flying aircraft at high altitude near Amsterdam with strong contrails from each wing motor.

In Figure 4a, a slow and low flying aircraft is shown after take-off from Amsterdam. From linear regression of Equation (3), we find its apparent velocity V = 586 ± 2 kph and direction θ_V = 128°. The aircraft orientation angle θ_{AC} = 133° is determined by the object orientation. From Equations (9) and (10), we find an aircraft speed of V_{AC} = 552 kph, at altitude H_{AC} = 1.764 m. According to live flight tracking, ADS-B, the aircraft at that time and position, has a speed of 507 kph at altitude 1.875 m. Considering delays in the flight tracking updates and neglecting wind speeds, the agreement is fair.

In Figure 4b another aircraft with strong contrails is captured near Amsterdam, where we expect little jet stream. Thus the long contrails show the aircraft heading $\theta_{CT} = \theta_{AC}$ = −134°. The contrails do, however, corrupt the determination of the aircraft central positions, and we therefore remove them in the object images by setting the threshold T sufficiently high above the contrail but below the aircraft reflections. The central object positions in each band can then be used for determining the aircraft positions $\vec{r}_m$ as shown in Figure 4b. The resulting apparent velocity is V = 557 ± 8 kph, and the apparent direction θ_V = −155°. From Equations (9) and (10), we find an aircraft speed of V_{AC} = 855 kph and altitude H_{AC} = 11.231 m. According to the flight tracking system, an aircraft at that time and position has speed V_{AC} = 833 kph and altitude H_{AC} = 11.582 m, both in good agreement with our calculations.

The large apparent velocities V can in turn be exploited for a search for aircrafts in S2 images as they are the only fast moving objects.

3.4. Jet Stream and Contrails

The polar jet stream circulates eastward in a meandering way as illustrated in Figure 5a. It lies between latitudes 50–60° at altitudes 9–12 km, and are a few hundred km wide. The jet typically has a speed of ~100 kph but can exceed 400 kph. Flight information systems show that aircrafts benefitting from the jet stream typically fly a hundred kph faster east- than westwards.

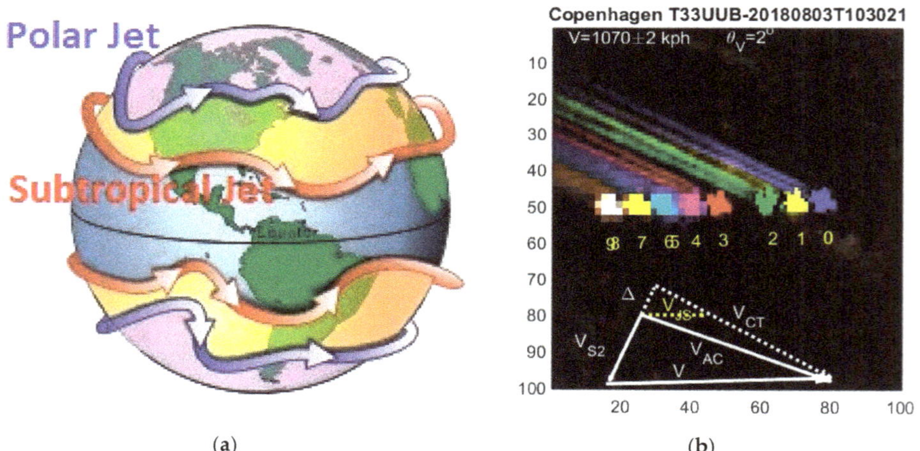

(a)   (b)

Figure 5. (a) Illustration of Earth's meandering polar and subtropical jet streams. (b) An aircraft over Copenhagen where the contrails are affected by the polar jet stream (see text).

The jet stream (and winds in general) with velocity $\vec{V}_{JS}$ will sweep an aircraft and its contrails along. The aircraft orientations θ_m are the same as the contrail direction θ_{CT}. The aircraft orientation in the image no longer matches the true aircraft heading θ_{AC} with respect to ground as shown in Figure 5b. If the contrail angle is used in Equations (9) and (10) for the aircraft angle, it can erroneously lead to supersonic aircrafts flying well above the altitude of most commercial airliners. Such extreme values only apply to the Concorde and a few other aircrafts. Fighter jets can be excluded in Figure 5b as the size of the aircraft is too large.

In order to correct for wind speeds, we introduce two auxillary quantities (see Figure 5b), namely the velocity in the contrail direction

$$V_{CT} = \frac{sin(\theta_V - \theta_{S2})}{sin(\theta_{CT} - \theta_{S2})} \cdot V \tag{11}$$

and the jet stream correction to the parallax effect

$$\Delta = \frac{sin(\theta_{JS} - \theta_{CT})}{sin(\theta_{CT} - \theta_{S2})} \cdot V_{JS} \tag{12}$$

The aircraft altitude is then (see Equation (10))

$$H_{AC} = \frac{H_{S2}}{V_{S2}} \left(\frac{sin(\theta_V - \theta_{CT})}{sin(\theta_{CT} - \theta_{S2})} \cdot V - \Delta \right) \tag{13}$$

The aircraft velocity can be determined from

$$V_{AC}^2 = V_{CT}^2 + \Delta^2 - 2V_{CT} \cdot \Delta \cdot cos(\theta_{CT} - \Theta_{S2}) \tag{14}$$

Finally, the aircraft heading angle can be determined from Equation (9).

In Figure 5b the aircraft engines under each wing create two long contrails with angle $\theta_{CT} = -24°$. The apparent velocity is $V = 1070 \pm 2$ kph with direction $\theta_V = 2°$. From Equation (11) we find $V_{CT} = 1035$ kph. Assuming that the jet stream heads east $\theta_{JS} = 0°$ with speed $V_{JS} = 200$ kph, we obtain $\Delta = 82$ kph, $H_{AC} = 11.514$ m, $V_{AC} = 1.025$ kph and $\theta_{AC} = -19°$. The aircraft speed relative to the jet stream is thus 825 kph. These numbers are compatible with normal aircraft cruising altitude and speed, however, the unknown jet stream speed was fitted.

4. Ship Velocities

Low objects on the surface of the Earth, such as ships, have no parallax. In addition, ocean currents are usually slow and can be neglected. The apparent velocity is then simply the ship velocity

$$V_{Ship} = V \tag{15}$$

Likewise, the vessel heading angle is the apparent direction θ_V and is the same for all bands. By overlaying all multispectral images as shown in Figures 6 and 7, the heading angle θ_V can be determined accurately, especially for large ships and/or when ship wakes are long.

Ships sail much slower than aircrafts, typically ~10 m/s (19.4 knots or 36 kph). Therefore, the ships in Figures 5 and 6 move less than two pixels in the temporal offset interval $t_9 = 2.085$ s, which requires accurate determination of ship positions.

4.1. Short Wakes

When ship wakes are short, they have less effect on the estimated central positions and both the central and the corrected positions of Equation (1) can be used for determining the ship velocity. An example is shown in Figure 6 where $\vec{c}_m$ and $\vec{r}_m$ are plotted with black and red numbers, respectively. Both are temporally ordered correctly and yield the ship velocity $V = 15 \pm 1$ kph. According to AIS, a ship at that time and position is sailing at a speed of 15 kph in the same direction.

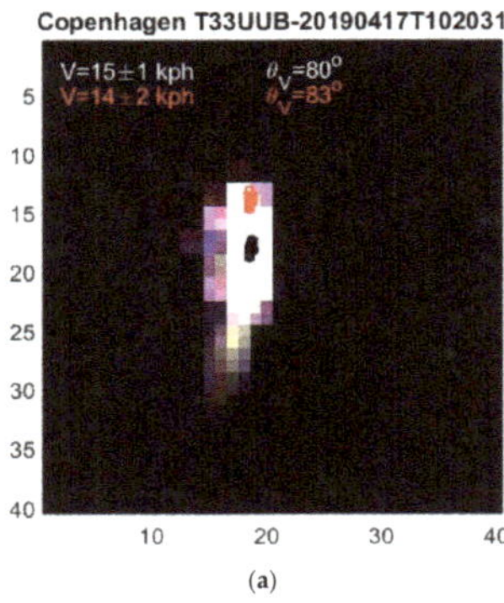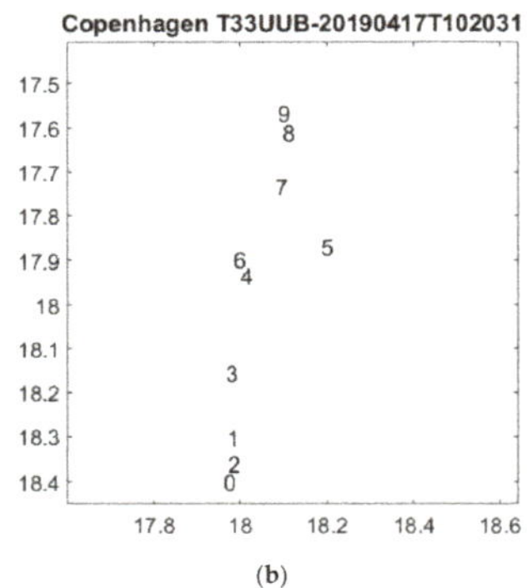

Figure 6. (**a**) Ship with short wake. As all ten multispectral ship images almost overlap, they appear white. The central positions shown with black numbers $m = 0, ..., 9$ are expanded in (**b**). The temporal ordering appears approximately correct.

4.2. Wake Corrections

Unfortunately, ship wakes can be longer than the ship and we find that they can corrupt the position determination considerably. Ship wakes move the apparent central position backwards with respect to vessel direction by an amount that typically is larger for the lower wavelengths. This is observed in Figure 7 where the ordering is not temporal but rather spectral, i.e., according to band wavelength as: $m = 0, 2, 3, 4, 6, 7, 1, 8, 5, 9$ (see Table 1). Wake reflection seems to decrease gradually towards the infrared [9,10,15]. If one simply uses the central positions for determining V, one obtains a ship speed of 69 ± 9 kph, which is much too large. Using the front positions of Equation (1), their ordering is closer to temporal and the ship speed is only 30 ± 4 kph. According to AIS, a ship at that time and position has a speed of 25 kph in the same direction. The length correction to central positions not only improves the accuracy of the velocity determination but also reduces the uncertainty.

The uncertainty in velocity is typically 5–10 kph, which sets the limit on how slow ship speeds can be determined. To improve the position accuracy, the more advanced calculation of $\vec{r}_m$ and study of the spectral dependence of wakes is required.

The two examples in Figures 6 and 7 may give the wrong impression, that faster ships create longer wakes, which is not always the case. Wake length may depend on e.g., ship type and size, surface winds and background. The temporal offsets are therefore useful complementary but are also partly correlated information.

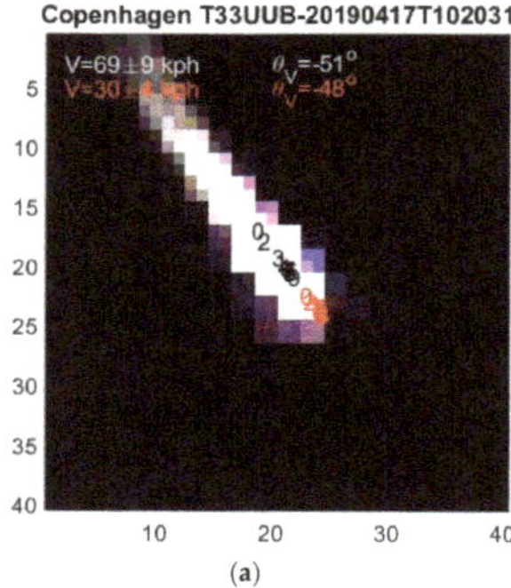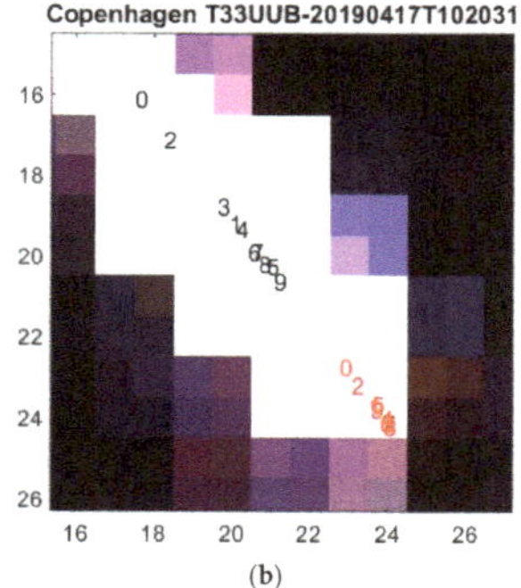

Figure 7. (**a**) As Figure 6 but for a fast ship creating a long wake, which corrupts the temporal ordering. Adding ship lengths L_m to central position as shown with red numbers in (**b**) improves the temporal ordering and velocity determination.

4.3. Kelvin Waves

Kelvin waves from large and fast ships are occasionally observed in S2 images [7]. A sailing ship creates Kelvin waves bounded by cusp-lines separated by an angle of $\pm$arcsin(1/3) = $\pm$19.47° on each side of the ship and its wake. The Kelvin wavelength is related to the ship speed V as [17]

$$\lambda = \frac{2\pi V^2}{g} \tag{16}$$

where g = 9.8 m/s^2 is the gravitational acceleration at the surface of the Earth. The wavelength can be determined by a Fourier analysis of the image, and the ship speed follows from Equation (16).

In Figure 8, 10 Kelvin waves are observed within ca. 50 pixels, i.e., λ = 50 m, and we obtain the ship speed V = 8.8 m/s or 32 kph. The apparent ship speed from temporal delays is V = 29 $\pm$ 3 kph, when corrected for wakes. According to AIS, a ship at that position and time is sailing north with 30 kph in good agreement with both satellite results.

The Kelvin waves are stationary relative to the ship, i.e., travel with ship velocity as seen from the satellite. The time delay of 2 s between the first and last multispectral corresponds to ~20 m in this case or almost half a wavelength. The Kelvin wave in the last image is therefore in anti-phase with the first and tends to interfere destructively if the multispectral images are added or plotted on top of each other. The Fourier transform of the individual band images is therefore optimal. It also improves the transform if the ship is masked, which is straight forward to do as the ship central position, width, breadth and heading are known.

The time offsets also cause a Kelvin wave phase shift in the multispectral bands

$$\phi_m = \frac{2\pi V t_m}{\lambda} = \frac{g t_m}{V} \tag{17}$$

Slow ships result in large phase shifts. The multispectral phases can also be found in the Fourier analysis.

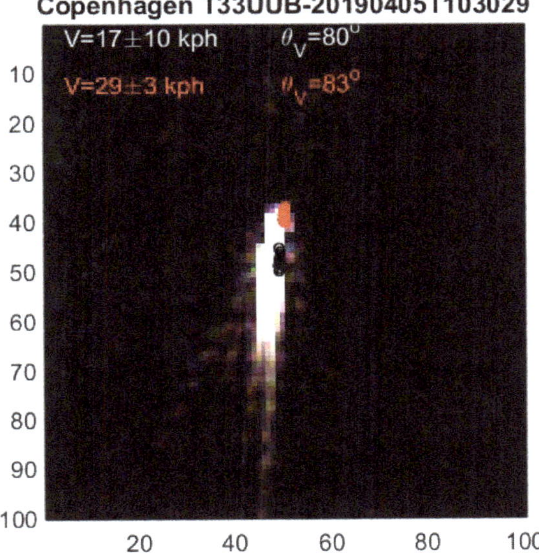

Figure 8. Ship with up to a dozen observable Kelvin waves in its wake.

5. Discussion

The above results can be taken as proof of principle for our novel method utilizing the temporal offsets in the Sentinel-2 multispectral imager for determining velocities and altitudes of moving objects. For aircrafts in particular, the accuracy is excellent; a few kph compared to cruise velocities of the order of 1000 kph, and a few hundred meters uncertainty in altitude compared to the standard 10 km cruise altitude of commercial airliners. Contrails improve heading direction determination but can also debase the positions unless the threshold is set correctly between aircraft and contrail reflections. Jet streams when present were shown to affect velocities and altitudes significantly. Unfortunately, the jet stream velocity cannot be determined from the S2 images but requires separate atmospheric information or alternatively, the jet stream can be estimated by requiring the aircraft to fly with standard cruise speed or altitude of commercial airliners.

The method was also shown to apply to ships with similar uncertainty of a few kph, which is sufficient for fast ships but comparable to slow ships. However, wakes cause a serious systematic error as they corrupt the temporal towards spectral ordering. Utilizing measurements of ship length and adding them can correct for part of this error. Yet, a much better understanding of the correlations between wakes, speed and spectral reflections is required and needs further investigation.

Our method is limited in the sense that it only utilizes the central position and length of objects for each band. The image contains much more spectral information on the object extent and form that is not utilized. Yet, it works surprisingly well even for small and slow objects such as ships with a variety of complex wakes.

Moving objects over land was not considered in this work. The more complex background will require better algorithms for removing the background in each spectral image, which is outside the scope of this work.

6. Summary and Outlook

We have described the basic physics behind moving objects in satellite multispectral images with temporal offsets, parallax effects and influence of jet streams. The basic formulas were derived and as proof of principle, a number of representative examples were shown for aircrafts and ships. The analysis serves as a proof of principle and provides a working model.

From apparent velocities the resulting aircraft speed, heading, and altitudes were calculated accurately and compared to data from the navigation systems ADS-B with good agreement. Jet streams can influence aircraft speeds and altitudes and the jet stream velocity must be determined independently or fitted.

Ship velocities are not affected by parallax but difficult to determine accurately for slow ship speeds or when long wakes are present. We described a simple but effective correction method, which improves the calculation of ship speeds considerably when compared to AIS. The detailed influence of thresholds, backgrounds, object lengths and contrails and wake reflections in the different multispectral bands should be studied in more detail in order to further improve the position determination. In addition, wake lengths may depend on e.g., ship type and size, surface winds and background. The temporal offsets are useful complementary information and the correlation to wake and ship lengths and widths provides additional information. Comparison to wake detection and velocity determination in Synthetic Aperture Radar radar images [18] should also be studied.

For better statistics, a large number of ships and aircrafts is required, where velocities and altitudes of aircrafts and ships are calculated and compared to AIS and ADS-B data with improved trajectory prediction [19]. This would also allow for improving the model by fine-tuning and optimizing parameters such as thresholds, better wake corrections and possibly introduce non-equal weights in the linear regression analysis of Equation (3). The large set of ships and aircrafts would also build an annotated database necessary for training machine learning algorithms [20–23]. Convolutional neural networks could be trained on this database so as to attempt to extract many more parameters and possibly refine the estimation of altitudes and velocities.

Funding: This work received no external funding.

Acknowledgments: We acknowledge the Arctic Command Denmark for support and interest. Images contain modified Copernicus Sentinel data from 2019 [1].

References

1. Copernicus Program, Sentinel Scientific Data Hub. Available online: https://schihub.copernicus.eu https://sentinel.esa.int/documents/247904/690755/Sentinel_Data_Legal_Notice (accessed on 27 June 2019).
2. Gascon, F.; Bouzinac, C.; Thépaut, O.; Jung, M.; Francesconi, B.; Louis, J.; Lonjou, V.; Lafrance, B.; Massera, S.; Gaudel-Vacaresse, A.; et al. Copernicus Sentinel-2A Calibration and Products Validation Status. *Remote Sens.* **2017**, *9*, 584. [CrossRef]
3. Skakun, S.V.; Vermote, E.; Roger, J.-C.; Justice, C. Multispectral Misregistration of Sentinel-2A Images: Analysis and Implications for Potential Applications. *IEEE Geosci. Remote Sens. Lett.* **2017**, *14*, 1–5. [CrossRef] [PubMed]
4. Daniel, B.; Schaum, A.; Allman, E.; Leathers, R.; Downes, T. Automatic ship detection from commercial multispectral satellite imagery. *Proc. SPIE* **2013**, *8743*. [CrossRef]
5. Burgess, D.W. Automatic ship detection in satellite multispectral imagery. *Photogram. Eng. Remote Sens.* **1993**, *59*, 229–237.
6. Corbane, C.; Marre, F.; Petit, M. Using SPOT-5 HRG data in panchromatic mode for operational detection of small ships in tropical area. *Sensors* **2008**, *8*, 2959–2973. [CrossRef] [PubMed]
7. Heiselberg, H. A Direct and Fast Methodology for Ship Recognition in Sentinel-2 Multispectral Imagery. *Remote Sens.* **2016**, *8*, 1033. [CrossRef]
8. Heiselberg, P.; Heiselberg, H. Ship-Iceberg discrimination in Sentinel-2 multispectral imagery. *Remote Sens.* **2017**, *9*, 1156. [CrossRef]
9. Lapierre, F.D.; Borghgraef, A.; Vandewal, M. Statistical real-time model for performance prediction of ship detection from microsatellite electro-optical imagers. *EURASIP J. Adv. Signal Process.* **2009**, *2010*, 1–15. [CrossRef]
10. Bouma, H.; Dekker, R.J.; Schoemaker, R.M.; Mohamoud, A.A. Segmentation and Wake Removal of Seafaring Vessels in Optical Satellite Images. *Proc. SPIE* **2013**, *8897*. [CrossRef]
11. De Michele, M.; Raucoules, D.; Arason, P.; Spinetti, C.; Corradini, S.; Merucci, L. Volcanic Plume Elevation Model Derived from Landsat 8: Examples on Holuhraun (Iceland) and Mount Etna (Italy). In Proceedings of the European Geoscience Union, Vienna, Austria, 17–22 April 2016.
12. Merucci, L.; Zakšek, K.; Carboni, E.; Corradini, S. Stereoscopic Estimation of Volcanic Ash Cloud-Top Height from Two Geostationary Satellites. *Remote Sens.* **2016**, *8*, 206. [CrossRef]
13. Markuse, P. Hurricane Jose as seen by Sentinel-2—More than meets the Eye. Available online: https://pierre-markuse.net/2017/09/11/hurricane-jose-seen-sentinel-2-meets-eye/ (accessed on 11 September 2017).
14. Ericson, T. Planespotting. *Google Earth Engine*. 2017. Available online: https://medium.com/google-earth/planespotting-465ee081c168 (accessed on 27 July 2017).
15. Eismann, M.T. *Hyperspectral Remote Sensing*; SPIE: Bellingham, WA, USA, 2012; Volume PM210, p. 748.
16. Selva, M.; Aiazzi, B.; Butera, F.; Chiarantini, L.; Baronti, S. Hyper-sharpening: A first approach on SIM-GA data. *IEEE J. Sel. Top. Appl. Earth Obs. Remote Sens.* **2015**, *8*, 3008–3024. [CrossRef]
17. Thomson, W. On ship waves. *Proc. Inst. Mech. Eng.* **1887**, *38*, 409–434. [CrossRef]
18. Graziano, M.D.; D'Errico, N.; Rufino, G. Wake Component Detection in X-Band SAR Images for Ship Heading and Velocity Estimation. *Remote Sens.* **2016**, *8*, 498. [CrossRef]
19. Borkowski, P. The ship movement trajectory prediction algorithm using navigational data fusion. *Sensors* **2017**, *17*, 1–12. [CrossRef] [PubMed]
20. Ship-Iceberg Classifier Challenge in Machine Learning. Available online: https://www.kaggle.com/c/statoil-iceberg-classifier-challenge (accessed on 15 February 2018).
21. Tang, J.; Deng, C.; Huang, G.B.; Zhao, B. Compressed-domain ship detection on spaceborne optical image using deep neural network and extreme learning machine. *IEEE Trans.Geosci. Remote Sens.* **2015**, *53*, 1174–1185. [CrossRef]

22. Mogensen, N. Ship-Iceberg Discrimination in Sentinel-1 SAR Imagery using Convoluted Neural Networks and Transfer Learning. Master's Thesis, Technical University of Denmark, Anker, Denmark, 2019.
23. Lin, Z.; Ji, K.; Leng, X.; Kuang, G. Squeeze and Excitation Rank Faster R-CNN for Ship Detection in SAR Images. *IEEE Geosci. Remote Sens. Lett.* **2019**, *16*, 751–755. [CrossRef]

Article

Assessment of the Steering Precision of a Hydrographic Unmanned Surface Vessel (USV) along Sounding Profiles Using a Low-Cost Multi-Global Navigation Satellite System (GNSS) Receiver Supported Autopilot

Mariusz Specht [1,*], Cezary Specht [2], Henryk Lasota [3] and Piotr Cywiński [3,4]

[1] Department of Transport and Logistics, Gdynia Maritime University, Morska 81-87, 81-225 Gdynia, Poland
[2] Department of Geodesy and Oceanography, Gdynia Maritime University, Morska 81-87, 81-225 Gdynia, Poland; c.specht@wn.umg.edu.pl
[3] Department of Marine Electronic Systems, Gdańsk University of Technology, 11/12 Gabriela Narutowicza, 80-233 Gdańsk, Poland; henlasot@pg.edu.pl (H.L.); peter@navinord.com (P.C.)
[4] NAVINORD Piotr Cywiński, Długa 47, 83-260 Kaliska, Poland
* Correspondence: m.specht@wn.umg.edu.pl

Received: 24 July 2019; Accepted: 10 September 2019; Published: 12 September 2019

Abstract: The performance of bathymetric measurements by traditional methods (using manned vessels) in ultra-shallow waters, i.e., lakes, rivers, and sea beaches with a depth of less than 1 m, is often difficult or, in many cases, impossible due to problems related to safe vessel maneuvering. For this reason, the use of shallow draft hydrographic Unmanned Surface Vessels (USV) appears to provide a promising alternative method for performing such bathymetric measurements. This article describes the modernisation of a USV to switch from manual to automatic mode, and presents a preliminary study aimed at assessing the suitability of a popular autopilot commonly used in Unmanned Aerial Vehicles (UAV), and a low-cost multi-Global Navigation Satellite System (GNSS) receiver cooperating with it, for performing bathymetric measurements in automated mode, which involves independent movement along a specified route (hydrographic sounding profiles). The cross track error (XTE) variable, i.e., the distance determined between a USV's position and the sounding profile, measured transversely to the course, was adopted as the measure of automatic control precision. Moreover, the XTE value was statistically assessed in the publication.

Keywords: Unmanned Surface Vessel (USV); multi-Global Navigation Satellite System (GNSS) receiver; bathymetric measurements; cross track error (XTE)

1. Introduction

Hydrography is a branch of applied science which deals with the measurement and description of physical features of the navigable portion of the Earth's surface and adjoining coastal areas, with special reference to their use for the purpose of navigation [1]. The basic activity of national hydrographic institutions is the creation of nautical charts based on measurement data obtained from direct depth measurements performed by hydrographic vessels. The precision of steering an unmanned hydrographic vessel along sounding profiles with the aim of measuring the depth and creating a bathymetric chart of an inland or marine waterbody is the main factor determining the quality of the bathymetric chart under creation. In traditional (manned) solutions, a vessel is controlled manually or in automatic mode using a vessel autopilot. The methodology of performing such measurements has been described comprehensively and in detail in the literature on the subject [2–5] and includes the following issues and processes: the selection of technical equipment for the vessel [6,7], planning of measurements [8], their performance [9,10], and the processing of results [11–14].

Contemporary bathymetric measurements are performed in a traditional (manned) manner using complex bathymetric systems based on MultiBeam EchoSounders (MBES) [15]. This enables the simultaneous measurement of a wide swath of the seabed, and Global Navigation Satellite Systems (GNSS) with various positioning accuracy, starting from the marine Differential Global Positioning System (DGPS) ensuring a positioning accuracy of 1–2 m ($p = 0.95$) [16,17], through to the European Geostationary Navigation Overlay Service (EGNOS), with an accuracy of 2–3 m ($p = 0.95$) [18,19], ending with solutions based on GNSS geodetic networks ensuring an accuracy of 2–5 cm ($p = 0.95$).

The beginning of the 21st century saw the development of autonomous vessels [20,21] and the use of unmanned vessels in a variety of measurement applications [22], including hydrography. Contemporary Autonomous/Unmanned Surface Vessels (ASV/USV) are design solutions which differ from each other mainly in the construction of the hull and the type of propulsion. Single- or multi-hulled vessels with a screw or screwless propulsion are commonly used in these fields [23]. Their particular common feature is a shallow draft, which allows them to enter hard-to-reach waterbodies, including those with a shallow depth [7,24]. For this reason, the use of small unmanned vessels in both inland and marine [25] hydrography is nowadays becoming increasingly widespread (Figure 1).

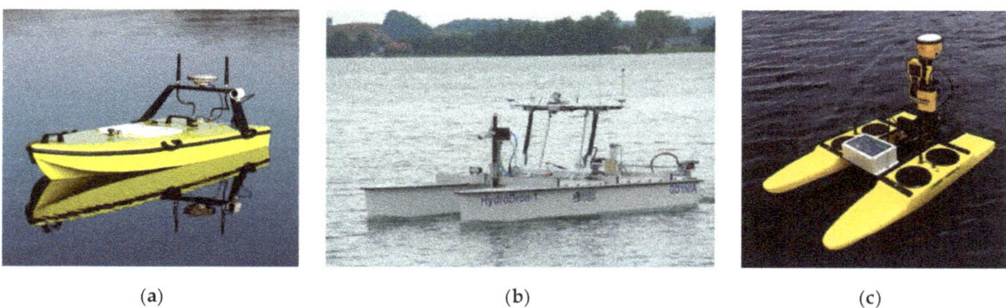

(a)   (b)   (c)

Figure 1. Hydrographic unmanned surface vessels. A single-hulled CEE-USV vessel (**a**), a double-hulled HydroDron (**b**), and a Seafloor Systems vessel (**c**).

Small unmanned vessels, including hydrographic ones, are most often controlled either directly using a radio controller or automatically using an autopilot. For unmanned vessels, measurements are planned with software, which enables the design of sounding profiles (directions and distances between them) based on a specific measurement waterbody—usually on the basis of an orthophotomap from the Google Maps service (Figure 2a). Some of them also have a vessel position real-time monitoring capability. However, one should note that software of this type is used only by highly advanced structures of hydrographic unmanned vessels (i.e., those with an autonomous navigation mode) [26]. On the other hand, when it comes to manned vessels, there are a number of programs used to display maps and to navigate during (inland and maritime) sailing and to plan them. They include free software, such as OpenCPN and Transas iSailor, and proprietary software, such as HYPACK or QINSy (Figure 2b).

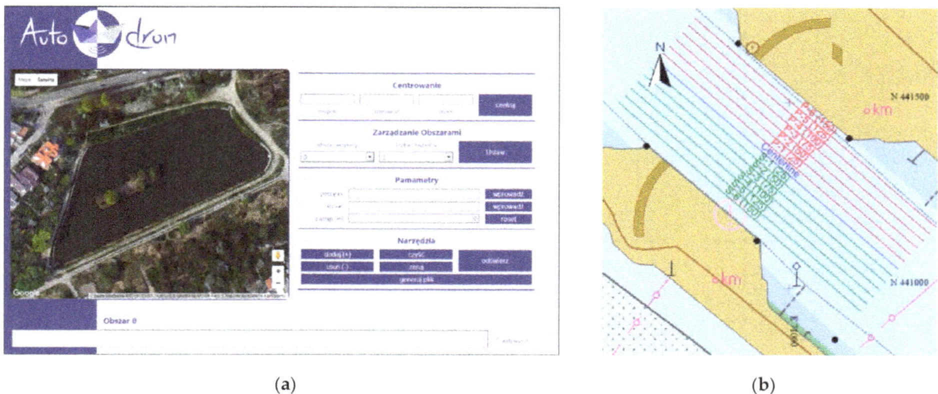

(a)   (b)

Figure 2. Applications used to plan sounding work, such as AutoDron (**a**) [26] and QINSy (**b**).

Contemporary hydrographic unmanned vessels use autopilots with Proportional-Integrative-Derivative (PID) controllers, which, after the tuning process, allow the vessel to be maintained on a specified trajectory (course), referred to as a hydrographic sounding profile. A PID controller is a commonly used feedback controller, which is popular because of its simplicity in design and robustness in performance. In a PID controller, which is sometimes called a three-term controller, three separate constants work together to minimise the error between the desired commands and the actual system response [27]. The distinguishing feature of the PID controller is the ability to use the three control terms of proportional, integral, and derivative influence on the controller output to apply accurate and optimal control of the USV's course between the points located on the sounding profile. The PID controller continuously calculates an error value, e(t), as the difference between the calculated course between the (start and end) points of the sounding profile, r(t), and the current course calculated between the current USV's position from the Global Positioning System (GPS) and the turning point on the profile, y(t). On this basis, the autopilot calculates the control signal, u(t), as a weighted sum of the outputs of three PID control terms (Figure 3) and feeds it to the vessel steering system. The K_p, K_i, and K_d coefficients are values for determining the effect of particular terms on the automatic process control. Their values are determined in the autopilot tuning process.

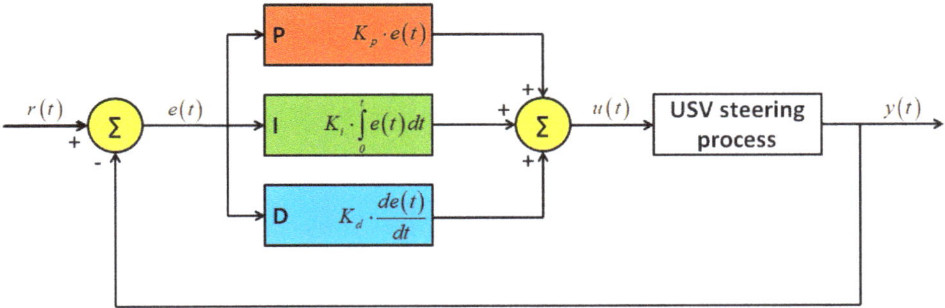

Figure 3. USV autopilot steering process. Own study based on [28].

The values of variables r(t) and y(t) presented in Figure 3 are, in navigation terms, two angles. The former is the sounding profile azimuth (the angle between the line connecting the beginning and the end of the profile and the geographic north), while the latter is the current USV course, i.e., the angle between the instantaneous course line and the direction of geographic north (Figure 4).

In automatic mode, the USV autopilot should control in such a manner so that, based on the course y(t) being continuously determined, the rudder angle could be changed to obtain the end point of the profile along the shortest route. This is only possible provided that the current vessel's position is determined using GNSS systems. Hence, the accuracy of USV positioning using a GNSS system is crucial due to the control accuracy. On the other hand, the referenced study proposed that the measure of steering precision should be the value of the transverse, lateral deviation from the course (cross track error—XTE). It should be minimised, i.e., the vessel ought to be located as close as possible to the sounding profile. For calculating XTE, the discrete location samples representing the vehicle's path must be represented in a local Cartesian (x, y) coordinate system. While some location recording equipment may directly provide this information, others (such as that used in this work) provide output in the form of latitude and longitude coordinates. These were simply transformed using formulas and methods described in International Organization for Standardization (ISO) 12188-2:2012 standard [29]. This provided input to the XTE calculation procedure that consisted of a list of points in Cartesian coordinates for both the outbound and return paths [30]. In Figure 4, the XTE parameter, i.e., the lateral distance (transverse to the course) between the USV's position and the sounding profile, is added.

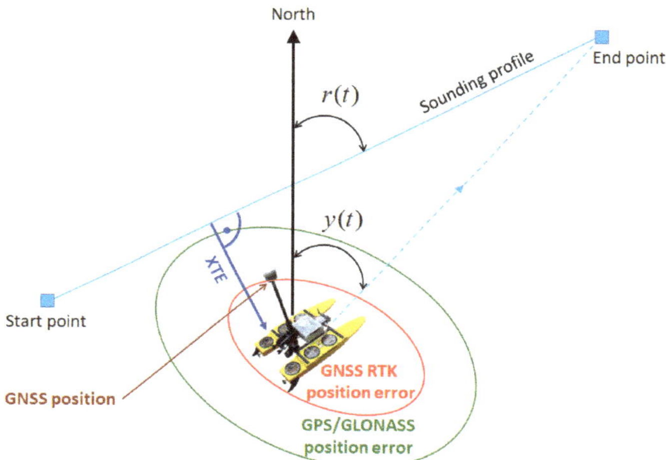

Figure 4. Graphic (navigation) interpretation of the automatic course adjustment system using an autopilot.

In view of the fact that contemporary USVs use various GNSS system solutions (with various positioning accuracies) for the positioning, it is reasonable to ask: what GNSS solutions should be applied to ensure a positioning accuracy that is acceptable from the steering accuracy perspective? This study attempted to assess the accuracy of controlling a small automatic hydrographic vessel using an autopilot with a low-cost multi-GNSS receiver that is typical of recreational and Unmanned Aerial Vehicles (UAV). This solution is an alternative to the commonly applied positioning based on the GNSS geodetic network, which requires a very expensive receiver to ensure the positioning accuracy at a level of 2–5 cm ($p = 0.95$), a control system based on a marine gyrocompass or a satellite gyrocompass, and a professional vessel autopilot.

2. Materials and Methods

2.1. USV Modernisation

The first stage of research work was the technical modernisation of a hydrographic USV. The modernised USV was a small vessel controlled directly using Radio Control (RC) equipment, with no automatic mode (understood as the ability to independently accomplish a planned mission). It had a length of 1.1 m and a width of 0.7 m, with a total weight of approximately 18 kg. The USV was classified as X class, in accordance with the adopted global American nomenclature [31]. The hydrographic vessel was a commercial HyDrone vehicle manufactured by the Seafloor Systems Inc. (United States of America—USA). Its design included two hulls made from High-Density Polyethylene (HDPE) resistant to environmental conditions and mechanical factors. They were connected by an H-shaped frame made from aluminium (Figure 5a). It was normally equipped with a hydrographic system comprising a geodetic GNSS receiver (e.g., Trimble R10) and a Single Beam Echo Sounder (SBES) (e.g., SonarMite BTX). The main aims of the USV modernisation were to:

- Enable the performance of hydrographic measurement campaigns in automatic mode involving independent (with no operator's participation) sailing along the planned sounding profiles;
- Improve the operating parameters and functionality through increasing the operating range, extending the operation time, and enhancing the reliability characteristics of both particular components and the entire system.

(a)   (b)

Figure 5. USV before (**a**) and after (**b**) the modernisation.

The USV modernisation concerned the following components: a PixHawk Cube autopilot was installed; an RC microwave transmission was modernised to increase the operation range by three times; for the autopilot control, a low-cost multi-GNSS receiver (u-blox NEO-M8N) with a built-in Fluxgate compass was used; the drive system was replaced and its power was increased (2×50 N); the previously used Absorbent Glass Mat (AGM) batteries (2×108 Wh) were replaced by LiPo batteries (2×326 Wh) to increase the time of operation by four times while reducing the weight by 28% (Figure 5b). Table 1 lists the basic technical characteristics and operational properties which were subject to modernisation. They were divided into two categories, related to automation (A) and operation (O).

The scope of the conducted modernisation of the hydrographic USV included all its main systems: drive, control, telemetric, and positioning. The only component that was not modified was the hydrographic system, comprising a geodetic GNSS receiver and an SBES, which met the International Hydrographic Organization (IHO) requirements. Figure 6 shows the basic components subjected to modernisation and a schematic diagram of electrical connections.

Table 1. A summary comparison of the USV functionality parameters before and after the modernisation.

A/O	Functionality	Before Modernisation	After Modernisation
A	Control	Direct RC	Direct RC, semi-automatic, automatic
A	RC operation range	300 m (2400 MHz)	1 km (868 MHz)
A	Telemetry monitoring	Additional PC	Integrated with the RC equipment
A	Positioning system	u-blox NEO-7: 56 channels; GPS, GLONASS, Galileo, QZSS, SBAS	u-blox NEO-M8N: 72 channels, GPS, GLONASS, BDS, Galileo, QZSS, SBAS
O	Possibility for hull replacement	No	Yes
O	Type of drive	Engines, main engine shafts, screws	Integrated pushing propellers 2×50 N
O	Engine cooling and ESC	Yes - forced, water	Not required
O	Battery bank	2×9Ah AGM (2×108 Wh)	2×22 Ah LiPo (2×326 Wh)
O	Time of operation until battery replacement time	1.5 h	6 h
O	Vehicle weight	25 kg	18 kg
O	Ingress protection class	IP44	IP56

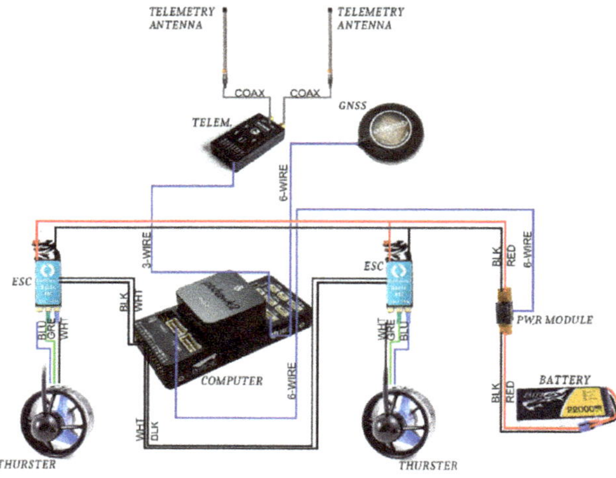

Figure 6. A schematic diagram of the main components subjected to the USV modernisation.

2.2. Measurements

The assessment of the effect of the conducted unmanned vessel modernisation on the operational properties was carried out based on experimental testing. On 7 March 2019, hydrographic measurements were performed to assess the accuracy of steering an ASV/USV in automatic mode along sounding profiles (Figure 7) designed in accordance with the principles included in the IHO S-44 standard [6]. Since the hydrometeorological conditions are an important factor affecting the obtained results, the measurements were performed in windless weather and the sea state was equal to 0 according to the Douglas scale (no wind waves or sea currents). The waterbody selected for the study was a small storage reservoir located in Gdańsk with a constant depth of 1.5 m, an orthophotomap of which was obtained from the Google Earth platform.

The study used an USV modernised to include automatic mode and equipped with a low-cost multi-GNSS receiver, whose task was to sail along the pre-set test routes. The sounding profiles were designed in four variants. For the first two routes (Figure 7a,b), they were designed in such a manner that the sounding lines were parallel to each other. The mutual distance between the profiles was assumed to be 5 and 10 m. The arrangement of the sounding profiles of two other routes (Figure 7c,d) resembled "narrowing squares" (a spiral of) towards the centre of the waterbody being sounded. The distance between the successive polygons was constant and amounted to 5 and 10 m for both variants, similar to the first two routes.

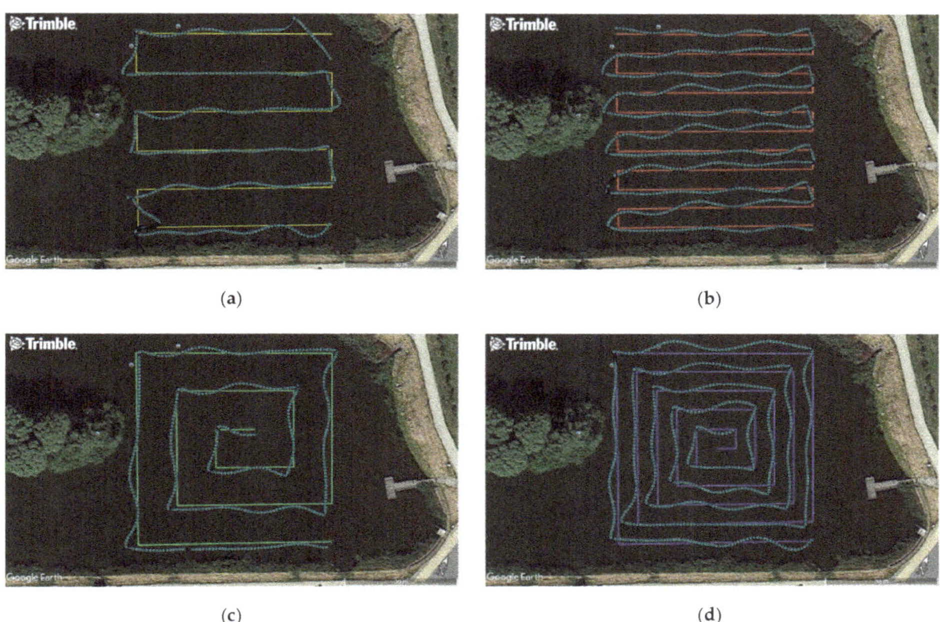

(a)

(b)

(c)

(d)

Figure 7. Sounding profiles designed in two variants: parallel (**a**,**b**) and spiral (**c**,**d**), along with the actual plotted USV route. Measurement points were recorded by a geodetic GNSS receiver (sky-blue colour).

The test routes, i.e., hydrographic sounding profiles along which the USV moved, were planned using a typical geodetic software, Trimble Business Center, and satellite images obtained from the Google Earth platform. The coordinates of the route turning points were then exported to *.kml files. They were recorded as geodetic coordinates, referred to the WGS84 ellipsoid with a precision of nine decimal places. The planned USV routes set in this manner were converted to *.shp files dedicated to the ArduPilot Mission Planner software operating the 3DR PX4 Pixhawk autopilot installed on the vessel, thanks to telemetry (TBS Crossfire TX LITE).

A navigation multi-GNSS receiver NEO-M8N (GPS, GLONASS (GLObal NAvigation Satellite System), BDS (BeiDou Navigation Satellite System), Galileo, QZSS (Quasi-Zenith Satellite System), SBAS (Satellite Based Augmentation System)), which was installed along the vessel's axis, was connected to the autopilot. In order to assess the steering accuracy, a geodetic receiver (Trimble R10) using the GNSS geodetic network was applied. The installation of two different receivers on the USV resulted from the fact that the dedicated accuracy of determining a position's coordinates by the navigation multi-GNSS receiver was 2.5 m (CEP (Circular Error Probable), $p = 0.50$), while the geodetic receiver using the VRSNet.pl network enabled the performance of measurements in real time with an accuracy of 2–5 cm ($p = 0.95$) in both a horizontal and vertical plane. The geodetic GNSS receiver was a reference receiver based on which the XTE value was determined.

The navigation multi-GNSS receiver was installed at the front of the mounting frame connecting both hulls of the vessel, and the geodetic receiver was placed on a 1 metre long pole attached to the rear part of the mounting frame using a special grip. The data recorded at the time of the measurements by the geodetic receiver were recorded on the controller's internal card, while the position of the navigation receiver was sent in real time to a laptop with u-center software installed by u-blox (Figure 8).

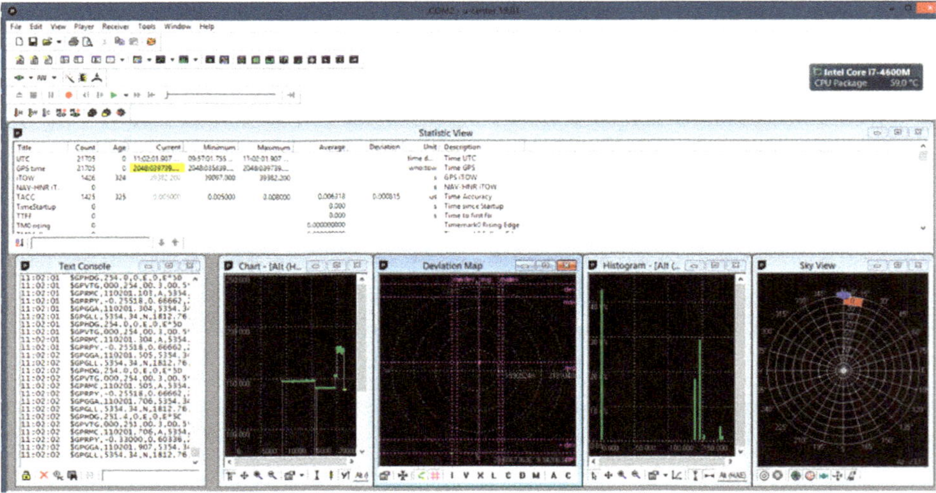

Figure 8. The window of the u-center software used to save messages recorded by the navigation multi-GNSS receiver in the National Marine Electronics Association (NMEA) standard.

3. Results

The position coordinates recorded by the geodetic GNSS receiver, i.e., the latitude and longitude, were provided in an angular (curvilinear) measurement, preventing the direct determination of the XTE value in metres, in relation to the designed sounding profiles. For this reason, the geographic (angular) coordinates were projected from the surface of the World Geodetic System 1984 (WGS84) ellipsoid (with the parameters of a = 6,378,137.00 m, b = 6,356,752.314) [32] onto a flat surface using the Gauss–Krüger transformation commonly applied in geodesy [33]. The calculations yielded two-dimensional coordinates (x, y), where the x value represents the distance (in metres) between the point and the equator, calculated along the meridian arc (on the WGS84 ellipsoid), and the variable y is the distance (in metres) to the arbitrarily set central meridian. The minus sign indicates that the point is located to the west of the meridian, while the plus sign corresponds to a location to the east of the meridian. In order to avoid negative values of the coordinates on the y axis, a constant value of, for example, 500,000 m (each zone in the PL-2000 system spanned 3° of longitude) was frequently added to the result [34,35].

The studies used the PL-2000 system, and the replacement of angular coordinates into Cartesian coordinates was conducted based on the following mathematical relationships [36]:

$$x = m_0 \cdot N \cdot \left[\frac{S(B)}{N} + \frac{(\Delta L)^2}{2} \cdot \sin(B) \cdot \cos(B) + \frac{(\Delta L)^4}{24} \cdot \sin(B) \cdot \cos^3(B) \cdot (5 - t^2 + 9 \cdot \eta^2 + 4 \cdot \eta^4) + \right.$$
$$\left. + \frac{(\Delta L)^6}{720} \cdot \sin(B) \cdot \cos^5(B) \cdot (61 - 58 \cdot t^2 + t^4 + 270 \cdot \eta^2 - 330 \cdot \eta^2 \cdot t^2 + 445 \cdot \eta^4) \right] \tag{1}$$

$$y = m_0 \cdot N \cdot \left[\Delta L \cdot \cos(B) + \frac{(\Delta L)^3}{6} \cdot \cos^3(B) \cdot (1 - t^2 + \eta^2) + \right.$$
$$\left. + \frac{(\Delta L)^5}{120} \cdot \cos^5(B) \cdot \left(5 - 18 \cdot t^2 + t^4 + 14 \cdot \eta^2 - 58 \cdot \eta^2 \cdot t^2 + 13 \cdot \eta\right) \right] \tag{2}$$
$$+ 500000 + \frac{L_0}{3} \cdot 1000000$$

where m_0 is the scale factor (-), N is the ellipsoid normal (radius of curvature perpendicular to the meridian) (m), $S(B)$ is the meridian arc length from the equator to the arbitrary latitude (B) (m), ΔL is the distance between the point and the central meridian (rad), B and L are the ellipsoidal coordinates of the point (°), and L_0 is the longitude of the central meridian (°). The other parameters of projection to the two-dimensional Cartesian coordinates in the PL-2000 system include:

$$t = \tan(B), \tag{3}$$

$$\eta = \sqrt{\frac{e^2 \cdot \cos^2(B)}{1 - e^2}}, \tag{4}$$

where e is the first eccentricity (-).

Then, in order to determine the XTE distances, it was necessary to describe each route profile in a mathematical manner. Since the routes were comprised of several straight sections, each profile was presented as a linear function expressed using the following formula:

$$x_{i,j} = a_{i,j} \cdot y_{i,j} + b_{i,j}, \tag{5}$$

where i is the route number, j is the section number for the i-th route, $x_{i,j}$ and $y_{i,j}$ are the flat rectangular coordinates PL-2000 of the point j recorded by a particular receiver on the i-th route, $a_{i,j}$ is a slope of the straight line j for the i-th route, defined as follows:

$$a_{i,j} = \frac{x_{i,j+1} - x_{i,j}}{y_{i,j+1} - y_{i,j}}, \tag{6}$$

$b_{i,j}$ is a y-intercept of the straight line j for the i-th route, defined as follows:

$$b_{i,j} = x_{i,j} - a_i \cdot y_{i,j}. \tag{7}$$

The distances were then calculated between the recorded coordinates of the geodetic GNSS receiver (converted into two-dimensional x, y coordinates) and the designed route sections. During the calculations, two variants needed to be considered. The first variant assumed that for the measured points, straight lines perpendicular to the designed hydrographic profiles could be drawn (Figure 9a), while in the second variant, it was not possible to determine the perpendicular line (Figure 9b). In the first case, the XTE value was calculated using the following formula:

$$XTE_{i,k} = \frac{\left| -a_{i,j} \cdot y_{p_{i,k}} + x_{p_{i,k}} - b_{i,j} \right|}{\sqrt{\left(-a_{i,j} \right)^2 + 1}}, \tag{8}$$

where k is the point number recorded by a particular receiver, $x_{pi,k}$ and $y_{pi,k}$ are the flat rectangular coordinates PL-2000 of the point k recorded by a particular receiver on the i-th route.

On the other hand, in the second variant the distance to the nearest route turning point was determined using the following formula:

$$XTE_{i,k} = \sqrt{\left(x_{pi,k} - x_{i,j} \right)^2 + \left(y_{pi,k} - y_{i,j} \right)^2}. \tag{9}$$

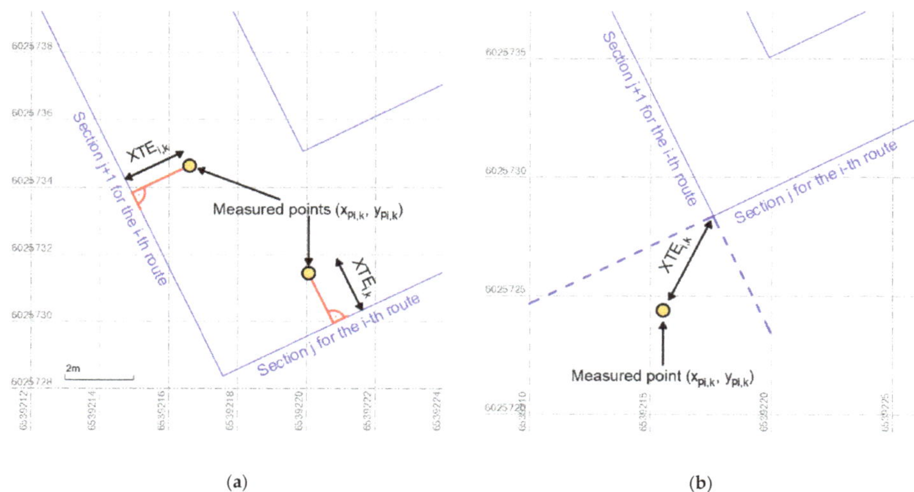

(a)   (b)

Figure 9. A graphical method for determining the distance of the measured point to the designed route, if the straight lines perpendicular to the hydrographic profiles could be drawn (**a**) and it is not possible to determine the perpendicular line (**b**).

After calculating the XTE value, this variable was statistically analysed. Two accuracy measures were adopted as the assessment criteria: XTE ($p = 0.68$) and XTE ($p = 0.95$) which corresponded to the probabilities of 68% and 95%. Table 2 shows the values of these measures.

Table 2. Statistical accuracy measures (XTE) of steering an USV for particular routes.

Accuracy Measure	Route a (10 m)	Route b (5 m)	Route c (10 m)	Route d (5 m)
Route type				
Number of measurements	458	848	506	869
XTE ($p = 0.68$)	0.92 m	1.15 m	1.40 m	1.27 m
XTE ($p = 0.95$)	2.01 m	2.38 m	2.20 m	2.39 m

Based on the results presented in Table 2, it can be concluded that the statistics for the accuracy of steering an USV in the automatic mode, measured by the XTE variable for four representative routes, were very similar to each other and amounted to 0.92–1.4 m for $p = 0.68$ and 2.01–2.39 m for $p = 0.95$. The measurements demonstrated that the accuracy of maintaining the vessel along the profile was not affected by the route shape (parallel sounding lines and the "narrowing squares") and the mutual distance between the profiles (5 and 10 m) (Figure 7).

As part of the study, an analysis of the XTE random variable distribution was conducted as well. To this end, Easy Fit software, which verifies the conformity of the measurements with typical statistical distributions using conformity tests (Kolmogorov–Smirnov, Anderson–Darling, and chi-square) was applied. The conducted analyses show that the empirical statistical distribution of the XTE values was most similar to four-parameter generalised gamma distribution with the probability density function [37]:

$$f(XTE; \chi, \varepsilon, \phi) = \frac{\varepsilon}{\chi \cdot \Gamma(\phi)} \cdot \left(\frac{XTE}{\chi}\right)^{\varepsilon \cdot \phi - 1} \cdot e^{-\left(\frac{XTE}{\chi}\right)^{\varepsilon}}, \tag{10}$$

where χ is the scale parameter ($\chi > 0$), ε is the shape parameter ($\varepsilon > 0$), and ϕ is the shape parameter ($\phi > 0$), where $\Gamma(\phi)$ is the Euler's Gamma function with the following form:

$$\Gamma(\phi) = \int_0^\infty u^{\phi-1} \cdot e^{-u} du, \tag{11}$$

and the cumulative distribution function with the following form:

$$F(XTE; \chi, \varepsilon, \phi) = \frac{\Gamma_n\left[\phi, \left(\frac{XTE}{\chi}\right)^\varepsilon\right]}{\Gamma(\phi)}, \tag{12}$$

where the introduced auxiliary variable amounts to:

$$\gamma = \left(\frac{XTE}{\chi}\right)^\varepsilon, \tag{13}$$

and the incomplete gamma function is presented by the relationship:

$$\Gamma_n(\phi, \gamma) = \int_0^\gamma u^{\phi-1} \cdot e^{-u} du. \tag{14}$$

Figure 10 shows empirical and theoretical functions of the XTE variables: probability density function and cumulative distribution function.

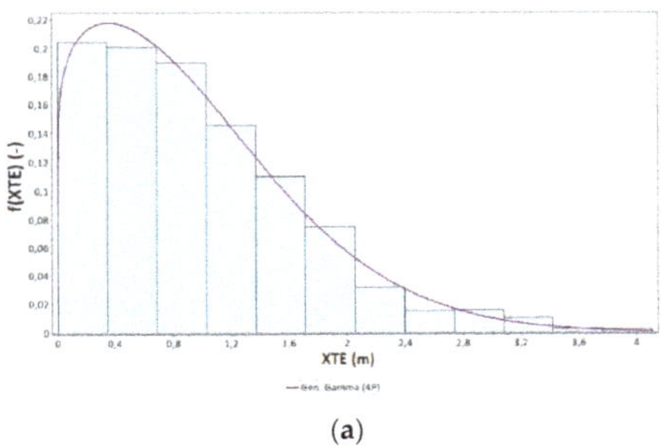
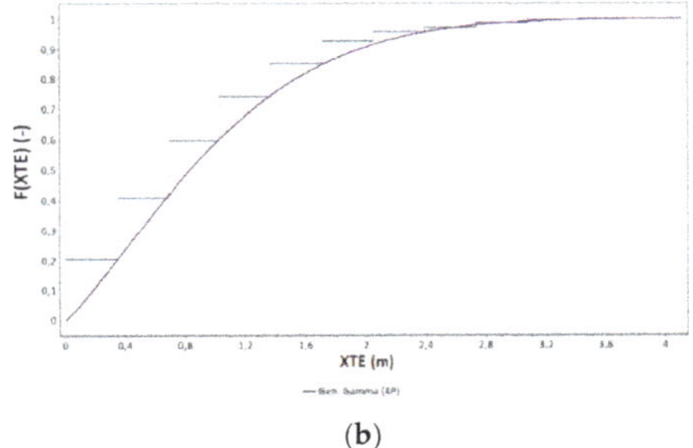

(a) (b)

Figure 10. Probability density function (**a**) and cumulative distribution function (**b**) of the generalised gamma distribution of the XTE variable.

Based on Figure 10, it can be concluded that the quantile of an order of 0.68 (percentile) in the analysed population amounted to 1.21 m, while the quantile of an order of 0.95 amounted to 2.34 m. This should be interpreted as meaning that 68% of the distance between the route recorded by geodetic GNSS receiver and the designed route was no greater than 1.21 m and, analogously, 95% of elements of this population did not exceed the value of 2.34 m. Next, the values of empirical and theoretical cumulative distribution function of the generalised gamma distribution of the XTE variable were compared using the following formula:

$$\Delta p(XTE) = F_n(XTE) - F(XTE), \tag{15}$$

where $F_n(XTE)$ is the empirical cumulative distribution function of the XTE variable, and $F(XTE)$ is the theoretical cumulative distribution function of the XTE variable.

A graph of the $\Delta p(XTE)$ values is shown in Figure 11. At the same time, it was compared with the theoretical normal distribution.

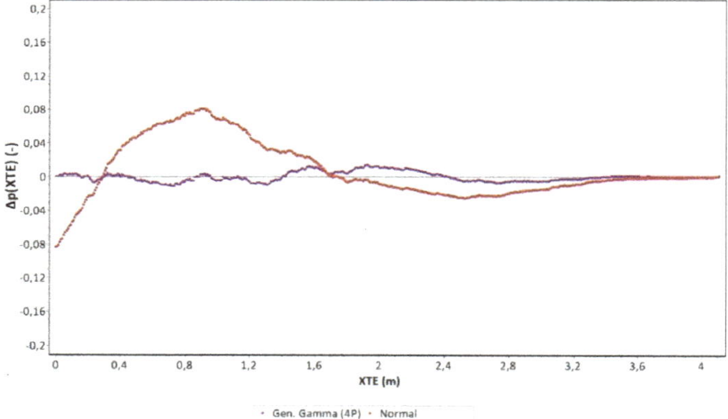

Figure 11. The difference between the probabilities of the occurrence of variable XTE for the empirical and theoretical generalised gamma distribution.

Based on Figure 11, it can be observed that the empirical distribution was almost identically fitted to the theoretical generalised gamma distribution. The maximum difference between the probabilities of the occurrence of variable XTE for the empirical and theoretical generalised gamma distribution was less than ±0.01–0.015. On the other hand, for the normal (Gaussian) distribution, which was the most important probability distribution applied in practice, the analysed measure was several times greater and amounted to approximately ±0.08.

4. Discussion

The conducted study proved that the use of a low-cost multi-GNSS receiver as a position source for the autopilot popular in unmanned systems makes it possible to obtain a very high precision of steering for a hydrographic USV. Such a solution allows a hydrographic vessel to be maintained along the sounding profile with a transverse error (XTE) not exceeding 2 m. In traditional (manned) hydrography, sailing along sounding profiles is typically performed using systems based on a professional autopilot cooperating with a gyrocompass, satellite compass, or, less frequently, a magnetic compass. The typical accuracy of the course maintenance by gyrocompass devices is 1–1.5° [38], with that of satellite compasses being similar. However, under unfavourable conditions, it can even temporarily reach a value of 10° [39]. Moreover, expensive geodetic GNSS receivers are most frequently used in professional hydrography as a positioning system [9].

The proposed solution in this paper is an alternative to the professional measuring systems used in hydrography. The main factor affecting the accuracy of steering a hydrographic USV along a sounding profile using an autopilot is the operational characteristics of the multi-GNSS receiver. Thanks to the continuous development and extension of the GNSS satellite system constellation (GPS, GLONASS, BDS, Galileo), multi-GNSS receivers continue to improve their operational characteristics, including positioning accuracy. Many studies have attempted to verify the positioning accuracy, availability, continuity, and reliability with multi-GNSS receivers from urban [40–42] and maritime [43] to pedestrian positioning applications [44] using different methods [45].

It should be assumed that multi-GNSS receivers used in mobile devices will soon significantly increase their positioning accuracy thanks to the use of dual-frequency receivers [46] and other measurement techniques [40]. On 21 September 2017, Broadcom announced the world's first mass-market, dual-frequency GNSS receiver device, the BCM47755. It is a very strong innovation destined to bring a revolution in the field of survey and geolocalisation. With these kinds of sensors, an accuracy of a few centimetres could be obtainable even with mobile devices [45]. The widespread introduction of multi-frequency GNSS receivers with an accuracy of a few centimetres is undoubtedly a major factor contributing to the improvement of unmanned vessel automatic control precision.

5. Conclusions

The preliminary study demonstrated that the accuracy of maintaining an USV carrying out hydrographic surveys in automatic mode using a popular Pixhawk autopilot and a multi-GNSS receiver supported by a Fluxgate magnetic compass enables the efficient performance of bathymetric measurements. The analysed solution allows a surface vessel to be maintained on sounding profiles with an accuracy of the transverse deviation from a course of 0.92–1.4 m ($p = 0.68$) and 2.01–2.39 m ($p = 0.95$). This means that it meets the IHO category 2 requirement [6] in terms of hydrographic system positioning accuracy. On this basis, it can be concluded that for the performance of selected hydrographic surveys, it is possible to use low-cost and widely available measuring equipment to successfully replace traditional solutions based on a professional autopilot. Moreover, the shallow draft of the vessel (20 cm) enables it to take bathymetric measurements of ultra-shallow waters in an automatic manner, which significantly increases their efficiency.

This publication proposed a number of modernisation solutions for a typical USV, previously controlled directly in a manual manner. It included all its basic components, such as the drive system, GNSS positioning, course control automatics, and the data transmission system. As a result, the vessel became able to take measurements in automatic mode and significantly increased its operational properties.

Research results have shown that low-cost multi-GNSS receivers can be successfully used in USVs for applications related to, among others, hydrographic surveys [47], in supporting the navigation process [21], in underwater photogrammetry [48], or in geological works [49].

Author Contributions: Conceptualisation, M.S., C.S. and H.L.; Data curation, M.S. and P.C.; Formal analysis, M.S. and C.S.; Investigation, M.S., H.L. and P.C.; Methodology, C.S. and H.L.; Supervision, C.S. and H.L.; Visualisation, M.S. and P.C.; Writing—original draft, M.S. and C.S.; Writing—review & editing, H.L. and P.C.

Funding: This research has been financed from the science budget for 2016-20, as a research project within the *"Diamentowy Grant"* Programme No. DI2015 008545.

Conflicts of Interest: The Authors declare no conflict of interest.

References

1. IHO. *Hydrographic Dictionary*, 5th ed.; Vol. I, Special Publication No. 32; IHO: Monaco, Monaco, 1994.
2. CHS. *CHS Hydrographic Survey Standards*, 2nd ed.; CHS: Ottawa, ON, Canada, 2013.
3. IHO. *Manual on Hydrography*, 1st ed.; Publication C-13; IHO: Monaco, Monaco, 2005.
4. NOAA. *NOS Hydrographic Surveys Specifications and Deliverables*; NOAA: Silver Spring, MD, USA, 2017.
5. Umbach, M.J. *Hydrographic Manual*, 4th ed.; NOAA: Silver Spring, MD, USA, 1976.
6. IHO. *IHO Standards for Hydrographic Surveys*, 5th ed.; Special Publication No. 44; IHO: Monaco, Monaco, 2008.
7. Stateczny, A.; Grońska, D.; Motyl, W. Hydrodron—New Step for Professional Hydrography for Restricted Waters. In Proceedings of the 2018 Baltic Geodetic Congress, Gdańsk, Poland, 21–23 June 2018.
8. Bouwmeester, E.C.; Heemink, A.W. Optimal Line Spacing in Hydrographic Survey. *Int. Hydrogr. Rev.* **1993**, *70*, 37–48.
9. Makar, A. Determination of Inland Areas Coastlines. In Proceedings of the 18th International Multidisciplinary Scientific GeoConference SGEM 2018, Albena, Bulgaria, 2–8 July 2018.
10. Specht, C.; Weintrit, A.; Specht, M. Determination of the Territorial Sea Baseline—Aspect of Using Unmanned Hydrographic Vessels. *TransNav-Int. J. Mar. Navig. Saf. Sea Transp.* **2016**, *10*, 649–654. [CrossRef]
11. Elema, I.A.; Kwanten, M.C. Introduction of Vertical Reference Level Lowest Astronomical Tide (LAT) in the Products of the Netherlands Hydrographic Service. In Proceedings of the 15th International Congress of the International Federation of Hydrographic Societies, Antwerp, Belgium, 6–9 November 2006.

12. El-Hattab, A.I. Investigating the Effects of Hydrographic Survey Uncertainty on Dredge Quantity Estimation. *Mar. Geod.* **2014**, *37*, 389–403. [CrossRef]

13. Peters, R.; Ledoux, H.; Meijers, M. A Voronoi-Based Approach to Generating Depth-Contours for Hydrographic Charts. *Mar. Geod.* **2014**, *37*, 145–166. [CrossRef]

14. Russom, D.; Halliwell, H.R.W. Some Basic Principles in the Compilation of Nautical Charts. *Int. Hydrogr. Rev.* **1978**, *55*, 11–19.

15. Stateczny, A.; Włodarczyk-Sielicka, M.; Grońska, D.; Motyl, W. Multibeam Echosounder and LiDAR in Process of 360-Degree Numerical Map Production for Restricted Waters with HydroDron. In Proceedings of the 2018 Baltic Geodetic Congress, Gdańsk, Poland, 21–23 June 2018.

16. Moore, T.; Hill, C.; Monteiro, L. Is DGPS Still a Good Option for Mariners? *J. Navig.* **2001**, *54*, 437–446. [CrossRef]

17. Dziewicki, M.; Specht, C. Position Accuracy Evaluation of the Modernized Polish DGPS. *Pol. Marit. Res.* **2009**, *16*, 57–61. [CrossRef]

18. GSA. *EGNOS Open Service (OS) Service Definition Document*; Version 2.3; GSA: Prague, Czech Republic, 2017.

19. Specht, C.; Pawelski, J.; Smolarek, L.; Specht, M.; Dąbrowski, P. Assessment of the Positioning Accuracy of DGPS and EGNOS Systems in the Bay of Gdansk Using Maritime Dynamic Measurements. *J. Navig.* **2019**, *72*, 575–587. [CrossRef]

20. Wróbel, K.; Montewka, J.; Kujala, P. System-Theoretic Approach to Safety of Remotely-Controlled Merchant Vessel. *Ocean Eng.* **2018**, *152*, 334–345. [CrossRef]

21. Stateczny, A.; Kazimierski, W.; Burdziakowski, P.; Motyl, W.; Wisniewska, M. Shore Construction Detection by Automotive Radar for the Needs of Autonomous Surface Vehicle Navigation. *Int. J. Geo-Inf.* **2019**, *8*, 80. [CrossRef]

22. Stateczny, A.; Burdziakowski, P. Universal Autonomous Control and Management System for Multipurpose Unmanned Surface Vessel. *Pol. Marit. Res.* **2019**, *26*, 30–39. [CrossRef]

23. Li, C.; Jiang, J.; Duan, F.; Liu, W.; Wang, X.; Bu, L.; Sun, Z.; Yang, G. Modeling and Experimental Testing of an Unmanned Surface Vehicle with Rudderless Double Thrusters. *Sensors* **2019**, *19*, 2051. [CrossRef] [PubMed]

24. Romano, A.; Duranti, P. Autonomous Unmanned Surface Vessels for Hydrographic Measurement and Environmental Monitoring. In Proceedings of the FIG Working Week, Rome, Italy, 6–10 May 2012.

25. Specht, C.; Specht, M.; Cywiński, P.; Skóra, M.; Marchel, Ł.; Szychowski, P. A New Method for Determining the Territorial Sea Baseline Using an Unmanned, Hydrographic Surface Vessel. *J. Coast. Res.* **2019**, *35*, 925–936. [CrossRef]

26. Specht, C.; Świtalski, E.; Specht, M. Application of an Autonomous/Unmanned Survey Vessel (ASV/USV) in Bathymetric Measurements. *Pol. Marit. Res.* **2017**, *24*, 36–44. [CrossRef]

27. Beirami, M.; Lee, H.Y.; Yu, Y.H. Implementation of an Auto-Steering System for Recreational Marine Crafts Using Android Platform and NMEA Network. *J. Korean Soc. Marit. Eng.* **2015**, *39*, 577–585. [CrossRef]

28. Rajinikanth, V.; Latha, K. I-PD Controller Tuning for Unstable System Using Bacterial Foraging Algorithm: A Study Based on Various Error Criterion. *Appl. Comput. Intell. Soft Comput.* **2012**, *2012*, 1–10. [CrossRef]

29. ISO. *ISO 12188-2:2012—Tractors and Machinery for Agriculture and Forestry—Test Procedures for Positioning and Guidance Systems in Agriculture—Part 2: Testing of Satellite-Based Auto-Guidance Systems during Straight and Level Travel*; ISO: Geneva, Switzerland, 2010.

30. Rounsaville, J.; Dvorak, J.; Stombaugh, T. Methods for Calculating Relative Cross-Track Error for ASABE/ISO Standard 12188-2 from Discrete Measurements. *Trans. ASABE* **2016**, *59*, 1609–1616.

31. USN. The Navy Unmanned Surface Vehicle (USV) Master Plan. Available online: https://www.navy.mil/navydata/technology/usvmppr.pdf (accessed on 24 July 2019).

32. NGA. *Department of Defense World Geodetic System 1984, Its Definition and Relationships with Local Geodetic Systems*, 3rd ed.; NGA: Springfield, VA, USA, 2004.

33. Deakin, R.E.; Hunter, M.N.; Karney, C.F.F. The Gauss-Krüger Projection. In Proceedings of the 23rd Victorian Regional Survey Conference, Warrnambool, Australia, 10–12 September 2010.

34. Gajderowicz, I. *Map Projections: Basics*; Publishing House of the University of Warmia and Mazury in Olsztyn: Olsztyn, Poland, 2009.

35. Kadaj, R.J. Polish Coordinate Systems. Transformation Formulas, Algorithms and Softwares. Available online: http://www.geonet.net.pl/images/2002_12_uklady_wspolrz.pdf (accessed on 24 July 2019).

36. Hofmann-Wellenhof, B.; Lichtenegger, H.; Collins, J. *Global Positioning System: Theory and Practice*; Springer: New York, NY, USA, 1994.

37. Stacy, E.W. A Generalization of the Gamma Distribution. *Ann. Math. Stat.* **1962**, *33*, 1187–1192. [CrossRef]

38. Jaskólski, K.; Felski, A.; Piskur, P. The Compass Error Comparison of an Onboard Standard Gyrocompass, Fiber-Optic Gyrocompass (FOG) and Satellite Compass. *Sensors* **2019**, *19*, 1942. [CrossRef]

39. Felski, A. Exploitative Properties of Different Types of Satellite Compasses. *Annu. Navig.* **2010**, *16*, 33–40.

40. Liu, W.; Shi, X.; Zhu, F.; Tao, X.; Wang, F. Quality Analysis of Multi-GNSS Raw Observations and a Velocity-Aided Positioning Approach Based on Smartphones. *Adv. Space Res.* **2019**, *63*, 2358–2377. [CrossRef]

41. Szot, T.; Specht, C.; Specht, M.; Dabrowski, P.S. Comparative Analysis of Positioning Accuracy of Samsung Galaxy Smartphones in Stationary Measurements. *PLoS ONE* **2019**, *14*, e0215562. [CrossRef]

42. Wang, L.; Li, Z.; Zhao, J.; Zhou, K.; Wang, Z.; Yuan, H. Smart Device-Supported BDS/GNSS Real-Time Kinematic Positioning for Sub-Meter-Level Accuracy in Urban Location-Based Services. *Sensors* **2016**, *16*, 2201. [CrossRef]

43. Specht, C.; Dąbrowski, P.S.; Pawelski, J.; Specht, M.; Szot, T. Comparative Analysis of Positioning Accuracy of GNSS Receivers of Samsung Galaxy Smartphones in Marine Dynamic Measurements. *Adv. Space Res.* **2019**, *63*, 3018–3028. [CrossRef]

44. Fissore, F.; Masiero, A.; Piragnolo, M.; Pirotti, F.; Guarnieri, A.; Vettore, A. Towards Surveying with a Smartphone. In *New Advanced GNSS and 3D Spatial Techniques*; Springer: Cham, Switzerland, 2010; pp. 167–176.

45. Dabove, P.; Di Pietra, V. Towards High Accuracy GNSS Real-Time Positioning with Smartphones. *Adv. Space Res.* **2019**, *63*, 94–102. [CrossRef]

46. Chen, B.; Gao, C.; Liu, Y.; Sun, P. Real-Time Precise Point Positioning with a Xiaomi MI 8 Android Smartphone. *Sensors* **2019**, *19*, 2835. [CrossRef]

47. Specht, M. Method of Evaluating the Positioning System Capability for Complying with the Minimum Accuracy Requirements for the International Hydrographic Organization Orders. *Sensors* **2019**, *19*, 3860. [CrossRef]

48. Giordano, F.; Mattei, G.; Parente, C.; Peluso, F.; Santamaria, R. MicroVEGA (Micro Vessel for Geodetics Application): A Marine Drone for the Acquisition of Bathymetric Data for GIS Applications. *Int. Arch. Photogramm. Remote Sens. Spat. Inf. Sci.* **2015**, *40*, 123–130. [CrossRef]

49. Giordano, F.; Mattei, G.; Parente, C.; Peluso, F.; Santamaria, R. Integrating Sensors into a Marine Drone for Bathymetric 3D Surveys in Shallow Waters. *Sensors* **2016**, *16*, 41. [CrossRef]

Article

Accuracy of Trajectory Tracking Based on Nonlinear Guidance Logic for Hydrographic Unmanned Surface Vessels

Andrzej Stateczny [1], Pawel Burdziakowski [1,*], Klaudia Najdecka [2] and Beata Domagalska-Stateczna [2]

[1] Department of Geodesy, Faculty of Civil and Environmental Engineering, Gdansk University of Technology, Narutowicza 11-12, 80-233 Gdansk, Poland; andrzej.stateczny@pg.edu.pl

[2] Marine Technology Ltd., Roszczynialskiego 4/6, 81-521 Gdynia, Poland; k.najdecka@marinetechnology.pl (K.N.); b.domagalska@marinetechnology.pl (B.D.-S.)

[*] Correspondence: pawel.burdziakowski@pg.edu.pl

Received: 6 January 2020; Accepted: 2 February 2020; Published: 4 February 2020

Abstract: A new trend in recent years for hydrographic measurement in water bodies is the use of unmanned surface vehicles (USVs). In the process of navigation by USVs, it is particularly important to control position precisely on the measuring profile. Precise navigation with respect to the measuring profile avoids registration of redundant data and thus saves time and survey costs. This article addresses the issue of precise navigation of the hydrographic unit on the measuring profile with the use of a nonlinear adaptive autopilot. The results of experiments concerning hydrographic measurements performed in real conditions using an USV are discussed.

Keywords: trajectory tracking; unmanned surface vehicle; navigation; bathymetry; hydrographic survey

---

1. Introduction

The importance of hydrographic measurement in recent years has been growing constantly due to the increasing use of water transport, including movements in restricted areas. A particular challenge is shallow-water measurements near land or navigational obstacles where the use of larger hydrographic units is impracticable or not justified economically. An important aspect in this respect is the desire to shift loads from roads and motorways to waterways, which, by definition, are safer and transport is more ecologically friendly and economical.

The process of conducting hydrographic measurements in land-restricted waterways requires precise acquisition of the measurement profile, and a limit to the acquisition of redundant data on the measurement strip tabs in the case of multi-beam echosounder (MBES) measurements and excessive gaps between measurement profiles in the case of single-beam sonar measurements. The International Hydrographic Organization (IHO) does not define the accuracy of maintaining the measuring unit on the profile, but only determines the percentage of searching the bottom of the reservoir and the accuracy of determining the position of the measuring unit. For instance, in the case of the most restrictive special category, 100% coverage and an accuracy of 2 m with respect to determination of position with a 95% confidence level is required [1].

Traditionally, for manned hydrographic units, the helmsman follows the position of the unit on the profile on the monitor screen of the measurement system, adjusting the parameters of the unit's movement to the current weather conditions. Many measurement systems used throughout the world provide the helmsman with an indicator which shows the current distance from the planned measurement profile. Continuous tracking of the position of the measuring unit on the profile requires a high degree of concentration and an appropriate response from the helmsman; this is a tedious and challenging task. It requires a relatively frequent change of the watch in the case of the helmsman controlling the hydrographic unit.

A disadvantage of manual control is also the quite low accuracy of maintaining the measuring unit on the profile and a moment of inattention or distraction may result in deviation from the measuring profile thus requiring repetition of registration on the incorrectly registered profile. For this reason, significant amounts of redundant data are recorded, which hamper the processing of the measurement data. The issue of achieving precise control of the measurement profile on a hydrographic unit can be solved by using an unmanned surface vehicle (USV) with an adaptive autopilot which realizes precise control of both course and speed.

Generally, unmanned mobile platforms, including surface, aerial, and ground vehicles have been increasingly employed for numerous operations. Currently, the unmanned platforms are widely used around the world, giving rise to many new research opportunities. In this context, various control schemes have been also developed to perform predefined tasks. Unmanned surface vessels suffer from many uncertainties and unknown external disturbances like winds, waves, currents, etc., which inevitably lead to a very challenging task [2] like accurate trajectory tracking within this complex navigation environment [3]. Due to this, various advanced control techniques and schemes for surface vehicles have been developed, and the following groups can be highlighted: backstepping technique [4–7], chaos control approach [8–10], fuzzy control [2,3,11,12], neural control [13–18], and finite-time control [19–21].

This paper considers the problems of precise control of the measurement profile on a hydrographic unit using a nonlinear adaptive autopilot originally designed for unmanned aerial vehicles. The results of experiments carried out under real conditions are presented.

2. Trajectory Tracking

Precise tracking of the trajectory during maneuvering is very important, especially for an autonomous multi-purpose water platform [22,23] engaged in hydrographic survey missions of restricted access areas like ports, embankments, anchorages, bays and lakes, and rivers. Generally, in such areas, maneuvering requires execution of precise, previously planned track information. Additionally, in any difficult navigational situation such as caused by recreational boats and other traffic, precise execution of the planned or preplanned track becomes crucial.

The evaluated track following method, termed guidance logic, was designed originally for aerial applications [24] and unmanned aerial vehicles (UAVs) and was successfully implemented in many UAV applications [24–26]. This approach is also used by small unmanned surface vessels [27–29]; however, the platforms applied in that research are significantly smaller than the platform used in the present work. Studies on autonomous navigation algorithms and navigation strategies have been reported [30–32] while interesting studies for an adaptive system for steering strategy are available [33–35]. Aspects on the safety of vehicle navigation have been discussed [36–38] while actual problems concerning control of trajectory tracking for marine vehicles have also been reported [2–21,39,40].

An outline of the steps and the computer programing methods used for trajectory tracking is given in Figure 1. Implementation follows that of a previous study [41] and where the mission control module is similar to a system described elsewhere [22].

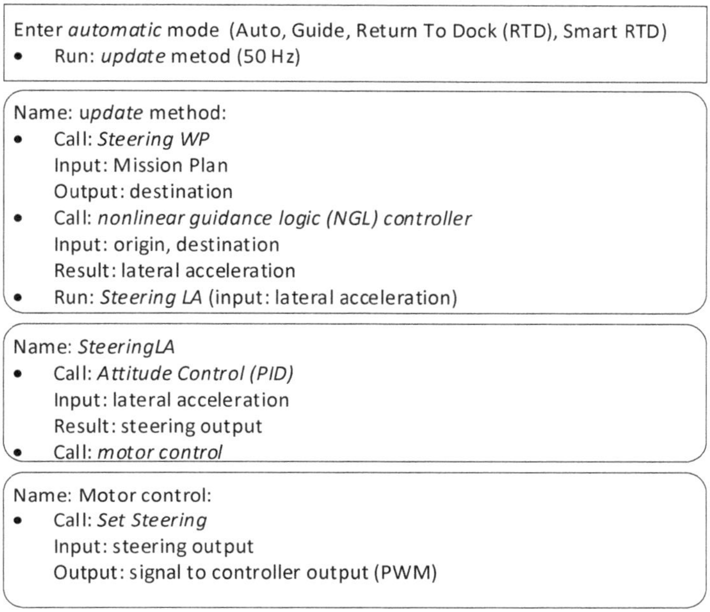

Enter *automatic* mode (Auto, Guide, Return To Dock (RTD), Smart RTD)
- Run: *update* metod (50 Hz)

Name: *update* method:
- Call: *Steering WP*
  Input: Mission Plan
  Output: destination
- Call: *nonlinear guidance logic (NGL) controller*
  Input: origin, destination
  Result: lateral acceleration
- Run: *Steering LA* (input: lateral acceleration)

Name: *SteeringLA*
- Call: *Attitude Control (PID)*
  Input: lateral acceleration
  Result: steering output
- Call: *motor control*

Name: Motor control:
- Call: *Set Steering*
  Input: steering output
  Output: signal to controller output (PWM)

Figure 1. Trajectory tracking algorithm scheme.

This algorithm requires a frequency of 50 Hz (method name: update), meaning that the update method is used 50 times per second. The method is activated only in automatic mode, meaning that only the automatic modes Auto, Guided, Return to Dock (RTD), and Smart RTD use this method. The automatic modes make use of all navigation sensors, and valid sensors readings are compulsory for automatic mode activation. Any missing sensors' data are unable to activate this function. Moreover, the Auto mode requires a programmed route (route plan), the RTD mode requires a dock (home) position and a saved travelled route to perform smart RTD, and the Guided mode requires a desired point location. All data are provided in the form of geographical waypoints (WPs).

The update method uses, in order, the following methods:

- Steering WP—the basic route plan and vehicle destination. The method provides the location that the vehicle should achieve (destination).
- NLG controller—the method where the navigation process is implemented. The method, based on destination and vehicle origin, calculates the lateral acceleration.
- Steering LA—the method, based on the PID (proportional–integral–derivative) controller, calculates the steering output.
- Set Steering—the method converts the steering output to the appropriate navigation controller output signal. The output signal (PWM: Pulse-Width Modulation) is the electrical signal interpreted by the electric motor's controller and the rudder's linear actuator controller.

All the methods enumerated above participate in trajectory tracking and have been tuned separately during the platform development and validation process.

The guidance logic [24] selects the desired point on the planned trajectory and generates a lateral acceleration (a_s) using this point, in accordance with following formula:

$$a_s = 2\frac{V^2}{L_1}\sin\eta \tag{1}$$

where the desired point is on the planned track at distance L_1 from the actual origin of the vehicle (Figure 2). Knowing that:

$$L_1 = 2R \sin \eta \qquad (2)$$

where the lateral acceleration calculated by Equation (1) equals the centripetal acceleration required to follow instantaneously a circular segment and a_s is the property used to track a circle of any radius R. This characteristic predisposes the guidance logic to work faithfully with a curved path. Moreover, as shown elsewhere [24], the method shows better capabilities than the PID controller when used on the UAV in the presence of wind. In this research, the method will be examined on the unmanned surface vessel (USV) using state-of-the-art geodetic grade measurements. The reference position of the USV was measured using independently acquired GPS navigation data (GPS RTK (Real Time Kinematic) geodetic receiver).

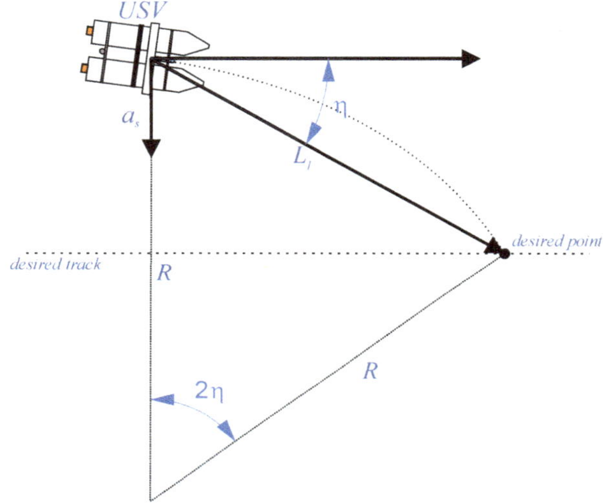

Figure 2. Guidance logic trajectory tracking.

The original Equations (1) and (2) were modified [41] for USV navigation purposes. The implementation required the addition of two parameters (L_1 Damping Factor and L_1 Period). Therefore, for the new L_1:

$$L_{1t} = \frac{1}{\pi} \zeta T v \qquad (3)$$

ζ is the L_1 Damping Factor, T is the L_1 Period (s), and v is the speed of the unit. Finally, substituting Equation (3) into Equation (1) can be written as:

$$a_{st} = \frac{k\,v^2}{L_{1t}} \sin(\eta_1 - \eta_2) \qquad (4)$$

where k is the L_1 Control Gain, defined as:

$$k = 4\zeta^2 \qquad (5)$$

The cross-track error (XTE), corresponding to $l_\perp$ on Figure 3 and defined as a distance between actual USV position perpendicular to the intended (desired) track, can be approximated as a second order differential equation [42]:

$$\ddot{d} + 2\zeta\omega_n \dot{d} + \omega_n^2 d = 0 \qquad (6)$$

where the natural frequency ω_n (natural frequency (eigenfrequency) is the frequency at which a structure or system have the tendency to oscillate in the absence of any driving or damping forces), is related to T:

$$\omega_n = \frac{2\pi}{T} \tag{7}$$

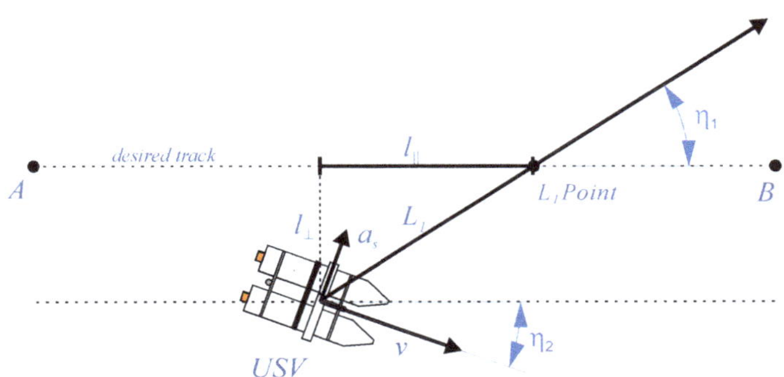

Figure 3. Guidance logic trajectory tracking for an unmanned surface vehicle (USV).

The dynamics of Equation (6) indicate that changing the values of ζ and ω_n adjusts the control response of XTE. The ζ and ω_n values are specified by the user and, for this research, were adjusted experimentally after long and extensive trials.

The parameter tuning requires some user experience regarding the USV's parameters and the response observations during the trials. As a general rule for tuning, the following instructions are given. The L_1 period T is given in seconds with a range from 1 to 60 (increments of 1) for the L_1 tracking loop and is the primary control parameter for aggressive turns in auto mode (Figure 1). This parameter should be larger for less responsive USV platforms. For smaller and more maneuverable USVs a lesser value can be set. The starting value was adjusted experimentally as 20 s. The L_1 control damping ratio ζ with range from 0.6 to 1 (increments of 0.05) should be increased if the USV overshoots the track being followed.

Significant changes made to the original Equation (1) enable the length L_1 to be calculated dynamically by the navigation loop depending on the USV ground speed changes and enable the user to specify a constant period for the tracking loop.

The L1 Control Gain was changed from a fixed value of 2 (Equation (1)) to be calculated based on the ζ value set by the user. This enables additional damping to be specified to compensate for delays in the velocity measurement and for the USV frame to respond.

Figure 4 presents all possible tracking movements. The current tracking mode depends on the area where the USV is located in relation to the route plan. Figure 4 presents a simple survey plan based on 6 WPs. When the vehicle is located in USV_1, the first WP is acknowledged to be tracked and vehicle i proceeds to point A. The point is considered to be reached when the unit is within the WP_{Radius} parameter. The WP_{Radius} is the distance in meters from a WP when the algorithm considers the WP has been reached and determines when the unit will proceed to the next WP. After reaching the first WP, the unit starts the L1 tracking mode (positions USV_2, USV_4). In the L1 tracking mode a point L is tracked and this point is dynamically located on the line between the last and next WP. The L point is the intersection point between the track line and the circle with the radius equal to the distance L_1 defined by the user. When point B is reached the next WP (C) is acknowledged to be tracked (Next WP Tracking Area). The procedure is repeated until the unit reaches the last WP when it stops and waits for the next command, either from the operator or the autonomous system.

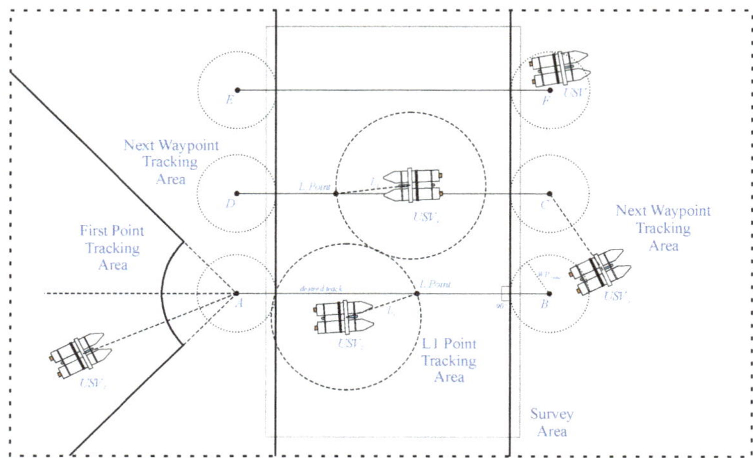

Figure 4. Tracking areas for USV hydrographic survey using L1 controller.

The L1 method calculates the lateral acceleration to be executed by the unit. The L_1 acceleration is translated to the motors and the steer command using the Steering LA and Motor Control methods (Figure 1). The Steering LA method is based on a PID controller. The automatic (mission control) algorithms, according to research [25] were demonstrated to control speed and course of the surface vessel. The PID controller continuously attempts to minimize the error $e(t)$ over time by adjustment of a control variable $u(t)$. The error value function $e(t)$ is the difference between a desired set point $r(t)$ and the measured process variable $y(t)$ ($e(t) = r(t) - y(t)$). Process variable is represented by the value that is being controlled (e.g., actual speed or actual course). The PID controller can be expressed as:

$$u(t) = K_p e(t) + \frac{1}{T_i} \int_0^t e(t)dt + T_d \frac{de(t)}{dt} \tag{8}$$

where K_p, K_i, and K_d are non-negative and denote the coefficients for the proportional, integral, and derivative terms, respectively. The parameters were set experimentally during test trials. The PID desired and the PID achieved were monitored and displayed and the coefficients can be adjusted to achieve the appropriate object response and parameter (course and speed) stabilization.

3. System Specification

The steering system specification for the unit (Table 1) is based on a combination of skid steering and a traditional rudder (Figure 5).

Table 1. USV technical specification.

Specification	Data
Dimensions (L × W × H)	4230 × 2080 × 1390 mm
Draft	500 mm
Weight	360 kg
Power supply	48 V 200 Ah lithium iron phosphate battery (LiFePO$_4$) (16 cells) for propulsion, 24 V lead-acid battery for electronics
Endurance	12 h (cruise speed)
Motors	2 × Torqeedo Cruise 4.0
Remote control range	40 km
Telemetry data range	50 km
Payload data range	6 km
Max speed	14 knots

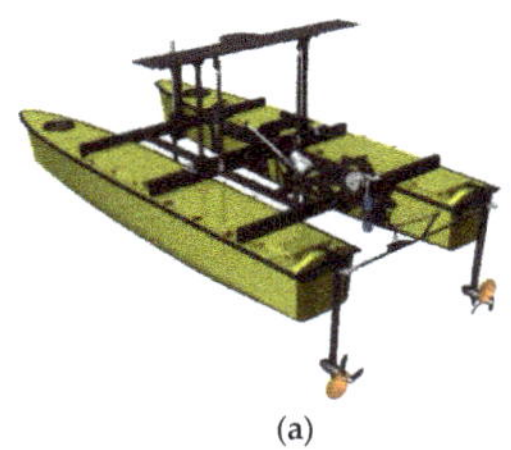
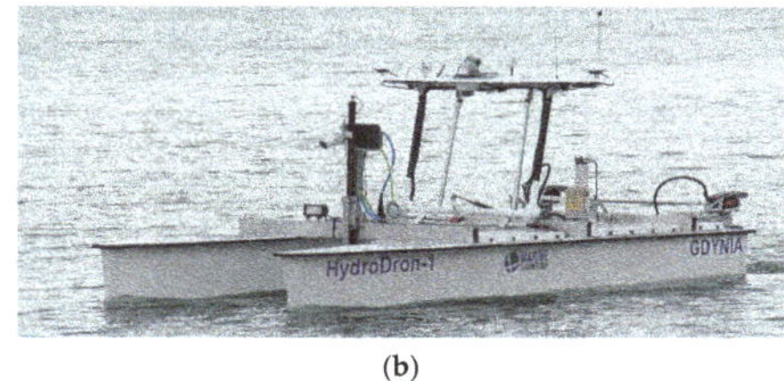

(a)   (b)

Figure 5. Combined skid and traditional steering system—motors and rudder design plan (**a**), photo of USV HydroDron during experiments (**b**).

Skid-Steering is a type of vehicle steering where rotation (yaw) is obtained by a difference in the speeds of the left and right propellers (wheel) and is typical for vehicles with non-orientable propellers (wheels). Electric motors installed on the platform and situated in the right hull are turned using a linear actuator, therefore, maneuverability is enhanced, and the unit is more responsive to steering, when compared to steering using only pure skid steering. Additionally, this USV uses pivot turns when the angle of turn is greater than a specific angle (can be set in the Pivot Turn Angle parameter). The pivot turn angle was selected as a result of extensive trials and was equal to 45°.

The USV is equipped with an autopilot and the navigation position is calculated based on the autopilot's internal sensors: three magnetic compasses (nine MEMS magnetometers), an inertial navigation system (INS) based on simple MEMS (microelectromechanical system) sensors consisting of nine gyroscopes and nine accelerometers (compasses and INS embedded in autopilot) and an external GNSS receiver based on an UBlox M8N module. In that configuration, the USV position is calculated using an EKF (extended Kalman filter). The EKF is a 24-state extended Kalman filter and the autopilot's filter estimates the following states: altitude, velocity, position, gyro bias offsets, gyro scale factors, Z accel. bias, Earth's magnetic field, platform body magnetic field, and wind velocity. For calculations in this study, the navigation autopilot used the EKF output position, which means that this was not a pure GPS reading, but filtered and estimated based on other internal sensors. Only one GNSS module based on the UBlox M8N module was used (Table 2).

Table 2. Navigation GPS specification.

Parameter Name	Specification
Channels	72
Signal tracking:	GPS: L1C/A SBAS: L1C/A QZSS: L1C/A GLONASS: L1OF BeiDou: B1 Galileo: E1B/C2
Horizontal position accuracy [1]	GPS and GLONAS: 2.5 m SBAS: 2.0 m
Velocity accuracy [2]	0.05 m/s
True heading accuracy [2]	0.3°
Operating limits	Altitude: 50,000 m Velocity: 500 m/s Acceleration: 4 g
Time to first fix	Cold start: <26 s Warm start: <1 s
Max output frequency	5 Hz

[1] CEP, 50%, 24 h static, −130 dBm, >6 SVs; [2] 50% at 30 m/s.

The GPS RTK module readings are not used for navigation; the module is part of the hydrographic equipment and additionally, the acquired RTK readings were used as a reference and as an independent USV position registration. The GPS RTK receiver specifications are given in Table 3. The GPS RTK used for position registration was a state-of-the-art survey grade receiver, embedded in the SPLITBOX-STD-T hydrographic equipment and based on the Trimble GNSS receiver.

Table 3. GPS RTK specification.

Parameter Name	Specification
Channels	220
Signal tracking	GPS: L1 C/A, L2E, L2C, L5 GLONASS: L1 C/A, L2 C/A, L2 P, L3 CDMA Galileo: 1 BOC, E5A, E5B, E5AltBOC Beidou B1, B2 SBAS, QZSS L-Band OmniSTAR VBS, HP, XP
Horizontal position accuracy (1 sigma)	SBAS/DGPS: 0.5 m/0.25 m PPP: 10 cm RTK: 0.8 cm + 1 ppm
Velocity accuracy	0.7 cm/s RMS
True heading accuracy	0.09 ° at 2 m baseline 0.05 ° at 1 0m baseline
Operating limits	Altitude: 18,000 m Velocity: 515 m/s Acceleration: 11 g
Time to first fix	Cold start: <45 s Warm start: <30 s
Signal reacquisition	L1/L2/L5: <2.0 s
Max output frequency	50 Hz

4. Experiments

The experiments were divided into three phases. Phase one was data acquisition. In this phase three different patterns were planned and executed. The patterns were planned to represent typical hydrographic surveys based on the present unit (Pattern 1) (Figure 6) and the bottom object investigation plan (Patterns 2 and 3) (Figure 6). All the profiles were numbered in accordance with their execution order. Profiles 1 to 10 belong to pattern no. 1, profiles from 11 to 14 belong to pattern no. 2, and profiles from 15 to 19 belong to pattern no. 3. All data were recorded within their typical hardware configuration, meaning that no additional technical rearrangements of the unit were conducted. The unit in this configuration was prepared to undertake surveys based on best knowledge and practice including use of state-of-the-art hydrographic equipment, therefore, all equipment was calibrated, and all GNSS and INS equipment offsets were measured, and data entered into all hardware units.

Phase two of the experiment concerned logged data filtration and preparation and evaluation of navigation GPS accuracy in the dynamic measurements. To prepare for this evaluation, studies described elsewhere [1] were used.

The approach used the PL-2000 system (ETRS89/Poland CS2000 zone 6) which afforded the replacement of angular coordinates recorded by the GPS and RTK by Cartesian coordinates (in meters). This conversion (from angular GPS coordinates to Cartesian) allowed the calculations to be simplified and the results to be presented in meters. The PL-2000 coordinate system is Cartesian 2D coordinate system with northing (x) and easting (y) axes with orientations fixed to north, east, and units of measurement in meters.

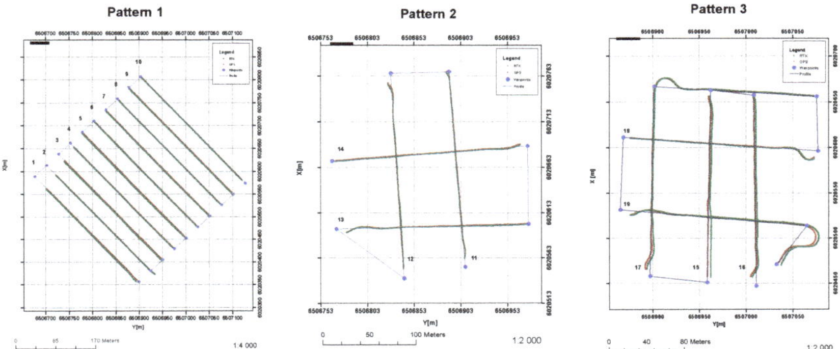

Figure 6. Survey pattern plans.

4.1. Data Synchronization

Navigation GPS maximal output frequency was declared as 5 Hz, which means that the GPS position (p_{GPS}) was reported a maximum of five times per second (Figure 7a). To compare both the registered tracks, i.e., RTK and GPS, the data rate for both RTK and GPS should be the same (Figure 7b). However, the RTK system reported position (p_{RTK}) with a maximum of 50 points every second and in fact both systems registered tracks at different rates.

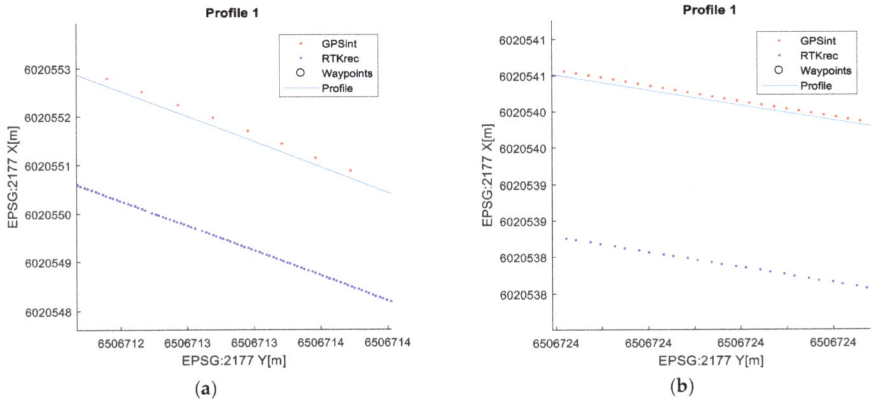

Figure 7. Recorded GPS and RTK tracks (**a**) different frequency (**b**) interpolated GPS positions.

To align the track rates a linear interpolation was applied to the lower rate track, i.e., the GPS track. All GPS positions (p_{GPS}) were interpolated with the maximum number of points equal to that for the RTK (p_{RTK}).

Assuming a data set consisting of independent data values x_i and dependent data values y_i, where $x = 1, \ldots, n$, we can find an interpolation function $\hat{y}(x)$ such that $\hat{y}(x_i) = y_i$ for every point in our data set. This means that the interpolation function goes through the given data points. Given a new x∗, we can interpolate its function value using $\hat{y}(x*)$ [43]. In the present study, a linear interpolation was used.

The estimated positions are assumed to lie on the line joining the nearest registered positions of the estimated track with n_{GPS} registered positions. It is assumed, without loss of generality, that the

GPS coordinates transformed to PL-2000 (X_G and Y_G) are in ascending order, then new interpolated position coordinates $\hat{X}_{G_i}$ and $\hat{Y}_{G_i}$ are calculated according to:

$$\hat{X}_{G_i} = X_{G_i} + \frac{\left(X_{G_{i+1}} - X_{G_i}\right)\left(n_R - n_G\right)}{\left(n_{G+1} - n_G\right)} \tag{9}$$

$$\hat{Y}_{G_i} = Y_{G_i} + \frac{\left(Y_{G_{i+1}} - Y_{G_i}\right)\left(n_R - n_G\right)}{\left(n_{G+1} - n_G\right)} \tag{10}$$

where: n_R represents the RTK measurement number, that is, $n_R = 1, \ldots, n_{RTK}$, n_G represents the GPS measurement number, that is, $n_G = 1, 1 + r, \ldots, n_{GPS}$, where r is a RTK to GPS measurement ratio calculated according to the following equation:

$$r = \frac{n_{RTK}}{n_{GPS}} \tag{11}$$

An example of a result for the GPS position interpolation is presented in Figure 6. In every case the number of GPS measurements was increased and equals the number of RTK measurements.

4.2. System Offsets

The position of the unit for navigation purposes is taken from the navigation GPS placed on the top of the mast located in the geometric center of the USV. The GPS antenna location is determined by experience and typical recommendations for unmanned units; that is, the best place for navigations system is the geometric center of the unit and at the highest possible place to diminish interference with board electronics and ensure best satellite visibility. These actions were taken in the present study.

For the hydrographic equipment, this consisted of a multi beam echosounder (MBES), a precise SGB IMU, and an RTK receiver. The best practice for that equipment localization is that the IMU sensors and the MBES antenna should be as close as possible, if not, all offsets should be entered into the system. In that case, the offsets represent coordinates in the local unit coordinate system of all hydrographic equipment including the GPS, the RTK antennas, and the IMU unit, where the IMU sensor is the coordinate system origin with the Y axis parallel to the unit long axis of symmetry (Figure 8). The hydrographic equipment is calibrated, and offsets are entered into the hydrographic system in accordance with values presented in Table 4. This means that the GPS RTK position was already reported with offsets by the hydrographic system at the origin.

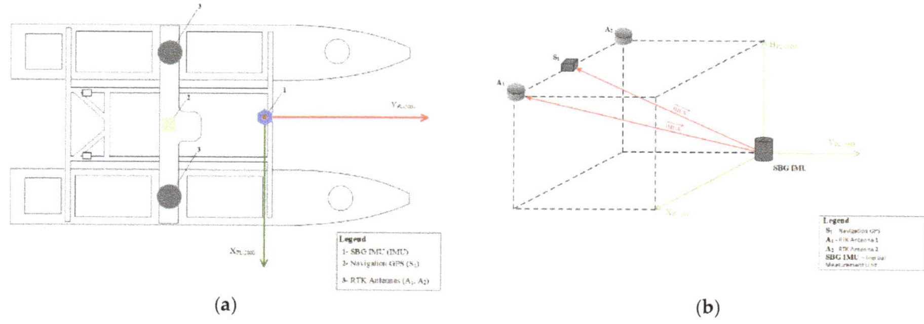
(a)   (b)

Figure 8. Measurement equipment location on the unit (**a**) and local unit coordinate system (**b**).

Table 4. Navigation equipment offsets.

	dy	**dx**	**d$_H$**	**|COGA$_1$|**	**|COGS$_1$|**	**Remarks**
GPS	−1.22	0.16	0.89	16.402		For Equations (12) and (13)
Origin	0	0	0		15.186	
Antenna RTK 1	−1.22	−0.64	0.89			In hydrographic system
Antenna RTK 2	−1.22	0.96	0.89			In hydrographic system

Given that the system navigation GPS is working out of the hydrographic system and for proper evaluation a vector between the RTK and GPS readings has to be included. The corrected GPS position was calculated from already interpolated GPS positions according to the following formula:

$$\hat{X}^*_{G_i} = \hat{X}_{G_i} - \left(d_{xG}\ \cos(HDG_i) - d_{yG}\sin(HDG_i)\right) \tag{12}$$

$$\hat{Y}^*_{G_i} = \hat{Y}_{G_i} - \left(d_{xG}\sin(HDG_i) + d_{yG}\cos(HDG_i)\right) \tag{13}$$

where $\hat{X}^*_{G_i}$, $\hat{Y}^*_{G_i}$ are the antenna coordinates of the GPS receiver in the national plane rectangular coordinate system, d_{xG}, d_{yG} are the antenna offset values of the GPS antenna unit on the vehicle coordinate system, defined at the center of IMU(RTK) unit with the y axis being parallel to the unit symmetry axis, and the x axis being perpendicular to the y axis, and HDG is the vessel's actual course reported by the hydrographic system (Figure 9).

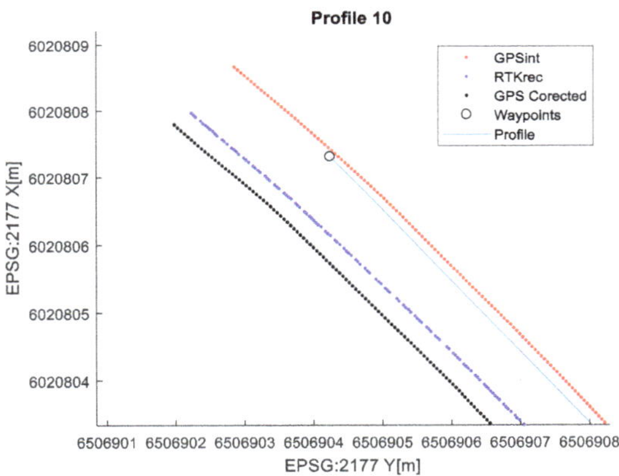

Figure 9. Example of corrected position for GPS at profile no. 10 (black dots).

4.3. GPS Evaluation

Having obtained all data at an equal rate, with the same coordinate system (PL-2000) and corrected the coordinates using antenna offsets, the navigation GPS accuracy may be calculated. Initially a Euclidean distance between the navigation GPS interpolated position ($\hat{p}_{GPS}$) and the RTK referenced position was calculated, in accordance with following equation:

$$d\left(P_{GPS}, P_{RTK}\right) = \sum_{n_R=1}^{n_R\ =n_{RTK}} \sqrt{\left(\hat{P}_{GPS_{n_R}} - p_{RTK_{n_R}}\right)^2} \tag{14}$$

Consequently, the dynamic navigation GPS accuracy was calculated in accordance with formula in [44].

4.4. Cross Track Error

The cross track error (XTE) is defined as the distance between the planned sounding profile (planned USV track) and the actual unit position. The sounding profiles are represented by a line connecting two defined WPs. Figure 3 presents the XTE, which equals $l_\perp$. Assuming that the sounding profile is defined by two waypoints $WP_i\left(X_{W_i}, Y_{W_i}\right)$ and the next waypoint $WP_{i+1}\left(X_{W_{i+1}}, Y_{W_{i+1}}\right)$, and the actual unit position is an interpolated GPS position with coordinates $\hat{p}_{GPS_{n_R}}\left(\hat{X}_{G_i}, \hat{Y}_{G_i}\right)$ at the measurement number n_R, then the actual XTE can be calculated in accordance with the following formula:

$$XTE_{n_R} = \frac{\left|\left(Y_{W_{i+1}} - Y_{W_i}\right)\hat{X}_{G_i} - \left(X_{W_{i+1}} - X_{W_i}\right)\hat{Y}_{G_i} + X_{W_{i+1}}Y_{W_i} - Y_{W_{i+1}}X_{W_i}\right|}{\sqrt{\left(Y_{W_{i+1}} - Y_{W_i}\right)^2 + \left(X_{W_{i+1}} - X_{W_i}\right)^2}} \tag{15}$$

5. Results

The final results represent the calculation process outlined above. Each profile was calculated separately. Table 5 is a graphical example of the results for three representative profiles, np. 2, 6, and 10. The coordinates difference graph represents differences for the X and Y coordinates between the RTK and GPS registered positions along the selected profile. The Euclidean distance represents the results of Equation (14) along selected profiles in meters. The XTE represents the cross track error along selected profiles as a result of applying Equation (15).

Table 5. Graphical results for example profiles no. 2, 6, and 10.

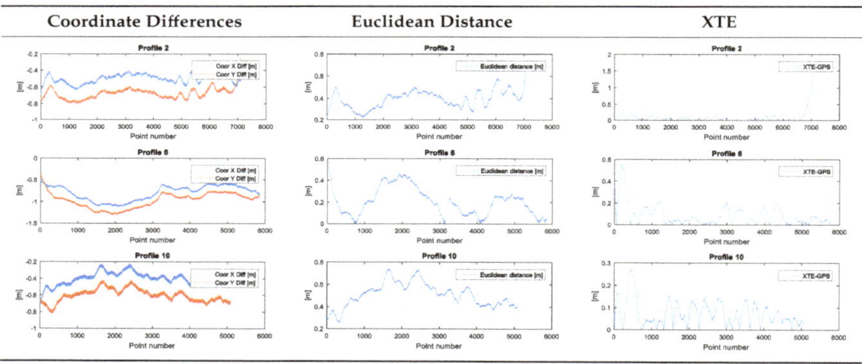

Based on the above calculation, the accuracy of dynamic navigation GPS was calculated in accordance with equations described elsewhere [44,45] for each profile separately. Table 6 presents the results of the calculations for each profile. Additionally, in Table 6 the mean XTE was added to evaluate the accuracy of the trajectory tracking for each profile.

The mean values for the research study are presented in Table 7. The mean accuracy for trajectory tracking (mean unit XTE) of the USV is very satisfactory, particularly if we compare it with traditional manual profile tracking or trajectory tracking realized by a typical autopilot used for ship navigation.

Table 6. Results for all profiles.

Profile No.	σ_x	σ_y	DRMS	2DRMS	CEP	R95	Mean XTE
1	0.4490	0.4788	0.6564	1.3129	0.5483	1.1405	0.1389
2	0.0599	0.0698	0.0920	0.1840	0.0768	0.1598	0.1080
3	0.1650	0.2108	0.2677	0.5354	0.2231	0.4640	0.0934
4	0.1029	0.0977	0.1419	0.2839	0.1182	0.2459	0.0578
5	0.1756	0.1975	0.2643	0.5286	0.2208	0.4592	0.0582
6	0.1698	0.1770	0.2453	0.4906	0.2048	0.4260	0.0881
7	0.1013	0.1447	0.1766	0.3533	0.1464	0.3046	0.0875
8	0.0857	0.0853	0.1209	0.2418	0.1009	0.2098	0.0911
9	0.0946	0.1044	0.1409	0.2817	0.1177	0.2448	0.0783
10	0.0783	0.0773	0.1101	0.2201	0.0918	0.1909	0.0750
11	0.2164	0.1547	0.2660	0.5320	0.2171	0.4516	0.2546
12	0.3149	0.0575	0.3201	0.6402	0.2120	0.4410	0.2433
13	0.1403	0.2244	0.2646	0.5293	0.2177	0.4528	0.3474
14	0.0591	0.1475	0.1589	0.3177	0.1245	0.2590	0.2315
15	0.2188	0.4595	0.5089	1.0179	0.4074	0.8475	0.3826
16	0.7284	0.2069	0.7572	1.5145	0.5362	1.1153	0.3218
17	0.5927	0.3553	0.6910	1.3821	0.5522	1.1486	1.5335
18	0.1795	0.3014	0.3508	0.7016	0.2874	0.5977	1.1249
19	0.1899	0.5650	0.5961	1.1922	0.4567	0.9499	0.4945

Table 7. Mean accuracy results for USV configuration.

	σ_x	σ_y	DRMS	2DRMS	CEP	R95	Mean Unit XTE
Mean	0.2170	0.2166	0.3226	0.6452	0.2558	0.5321	0.3058

6. Discussion

Nonlinear guidance, originally designed for UAV trajectory tracking, has been adopted and used for USV trajectory tracking. This method allows us to keep a low XTE during all sounding profiles' tracking; more importantly, the USV course changes that keep to the track are very gentle, and do not cause significant disturbances in hydrographic measurements. The trajectory tracking and the USV response for the calculated course inputs depend on the correct PID tuning. As stated above, PID tuning was carried out during extensive field tests. The configuration employed did not show any oscillations, the course changes although robust were gentle, allowing us to keep track within specified limits for the unit to weather conditions and wind direction. The correct PID tuning is very important for course and track keeping. As mentioned elsewhere [28], the USV used in this research showed a regular oscillation for track following, caused by not having ideal PID tuning, and this can significantly affect the unit's endurance and the quality of hydrographic measurement.

The position registered by the navigation GPS is not used directly by the USV for navigation. The pure GPS position readings are filtered using EKF and the final navigation position is calculated using all data available to the autopilot internal sensors. This technique is used widely within the robotics community to estimate a robot's position. Consequently, the position used for navigation, as the results show, has good accuracy. The profile planned by the hydrographers, was followed based on GPS position. If we take into account the antenna position (in the center of the unit) and the offsets between the GPS antenna and the multibeam sonar (main hydrographic sensor) this causes quite significant differences between the real unit track and the sonar antenna track. The displacement differs and depends on the wind speed and direction. Practically, the profile is registered by the RTK system with offsets, therefore, from the hydrographic point of view, all data are registered correctly, however this difference can be diminished.

7. Conclusions

As the results show, the accuracy of trajectory tracking based on nonlinear guidance logic is suitable for hydrographic USV profile tracking, and the presented unit configuration permits very precise track following with a mean XTE of around 30 cm. This is a very good result, if we compare this performance with a traditional manned hydrographic vessel. In the authors' opinion, trajectory tracking based on nonlinear guidance logic for hydrographic measurements can be implemented on a wide variety of USVs providing correct nonlinear guidance logic and PID parameters.

The difference between the track realized by the unit and the real sonar track can be significant and depends on wind speed and direction. This difference can be diminished using either software offsets or antenna physical displacement. The other approach is to use the RTK signal directly for navigation during the hydrographic survey.

Author Contributions: Conceptualization, A.S. and P.B.; methodology, A.S. and P.B.; bibliography review, A.S., P.B. and B.D.-S.; acquisition, analysis, and interpretation of data, A.S., P.B. and K.N., writing—original draft preparation, A.S. and P.B.; writing—review and editing, A.S., P.B., K.N. and B.D.-S. All authors have read and agreed to the published version of the manuscript.

Funding: This research was funded by the European Regional Development Fund under the 2014–2020 Operational Programme Smart Growth as part of the project "Developing of autonomous/remote operated surface platform dedicated hydrographic measurements on restricted reservoirs" implemented as part of the National Centre for Research and Development competition, INNOSBZ.

Conflicts of Interest: The authors declare no conflict of interest. The funders had no role in the design of the study; in the collection, analyses, or interpretation of data; in the writing of the manuscript, or in the decision to publish the results.

References

1. Specht, M. Method of Evaluating the Positioning System Capability for Complying with the Minimum Accuracy Requirements for the International Hydrographic Organization Orders. *Sensors* **2019**, *19*, 3860. [CrossRef] [PubMed]
2. Wang, N.; Er, M.J. Direct Adaptive Fuzzy Tracking Control of Marine Vehicles with Fully Unknown Parametric Dynamics and Uncertainties. *IEEE Trans. Contr. Syst. Technol.* **2016**, *24*, 1845–1852. [CrossRef]
3. Wang, N.; Er, M.J.; Sun, J.; Liu, Y. Adaptive Robust Online Constructive Fuzzy Control of a Complex Surface Vehicle System. *IEEE Trans. Cybern.* **2016**, *46*, 1511–1523. [CrossRef] [PubMed]
4. Liu, T.; Dong, Z.; Du, H.; Song, L.; Mao, Y. Path following control of the underactuated USV based on the improved line-of-sight guidance algorithm. *Pol. Marit. Res.* **2017**, *24*, 3–11. [CrossRef]
5. Liao, Y.; Wan, L.; Zhuang, J. Back stepping dynamical sliding mode control method for the path following of the underactuated surface vessel. *Procedia Eng.* **2011**, *15*, 256–263. [CrossRef]
6. Dong, Z.; Wan, L.; Li, Y.; Liu, T.; Zhang, G. Trajectory tracking control of underactuated USV based on modified backstepping approach. *Int. J. Nav. Archit. Ocean Eng.* **2015**, *7*, 817–832. [CrossRef]
7. Fan, Y.; Mu, D.; Zhang, X.; Wang, G.; Guo, C. Course keeping control based on integrated nonlinear feedback for a USV with pod-like propulsion. *J. Navig.* **2018**, *71*, 878–898. [CrossRef]
8. Huang, Q.; Li, T.; Li, Z.; Hang, Y.; Yang, S. Research on PID control technique for chaotic ship steering based on dynamic chaos particle swarm optimization algorithm. In Proceedings of the 10th World Congress on Intelligent Control and Automation, Beijing, China, 6–8 July 2012; pp. 1639–1643.
9. Li, Y.; Yang, S.; Yu, Y.; Liu, M. Study on optimization and simulation of hydrofoil USV propulsion intelligent control based on chaos algorithm. In Proceedings of the 2017 2nd International Conference on Materials Science, Machinery and Energy Engineering (MSMEE 2017), Dalian, China, 13–14 May 2017. [CrossRef]
10. Huang, Q.; Liu, X.; Li, T.; Wang, K.; Wang, S. On impulsive parametric perturbation control techniques for chaotic ship steering. In Proceedings of the Proceedings of 2014 IEEE Chinese Guidance, Navigation and Control Conference, Yantai, China, 8–10 August 2014; pp. 428–433.
11. Wang, N.; Sun, J.; Er, M.J. Tracking-Error-Based Universal Adaptive Fuzzy Control for Output Tracking of Nonlinear Systems with Completely Unknown Dynamics. *IEEE Trans. Fuzzy Syst.* **2018**, *26*, 869–883. [CrossRef]

12. Wang, N.; Su, S.; Yin, J.; Zheng, Z.; Er, M.J. Global Asymptotic Model-Free Trajectory-Independent Tracking Control of an Uncertain Marine Vehicle: An Adaptive Universe-Based Fuzzy Control Approach. *IEEE Trans. Fuzzy Syst.* **2018**, *26*, 1613–1625. [CrossRef]

13. Patino, H.D.; Liu, D. Neural network-based model reference adaptive control system. *IEEE Trans. Syst. Man Cybern. Part B (Cybern.)* **2000**, *30*, 198–204. [CrossRef]

14. Dai, S.-L.; Wang, C.; Luo, F. Identification and learning control of ocean surface ship using neural networks. *IEEE Trans. Ind. Inform.* **2012**, *8*, 801–810. [CrossRef]

15. Zhang, Y.; Hearn, G.E.; Sen, P. A neural network approach to ship track-keeping control. *IEEE J. Ocean. Eng.* **1996**, *21*, 513–527. [CrossRef]

16. Brown, M.; Harris, C.J. *Neurofuzzy Adaptive Modelling and Control*; Prentice Hall: Upper Saddle River, NJ, USA, 1994.

17. Wang, N.; Joo Er, M. Self-Constructing Adaptive Robust Fuzzy Neural Tracking Control of Surface Vehicles with Uncertainties and Unknown Disturbances. *IEEE Trans. Control Syst. Technol.* **2015**, *23*, 991–1002. [CrossRef]

18. Wang, N.; Sun, J.; Er, M.J.; Liu, Y. A Novel Extreme Learning Control Framework of Unmanned Surface Vehicles. *IEEE Trans. Cybern.* **2016**, *46*, 1106–1117. [CrossRef] [PubMed]

19. Wang, N.; Karimi, H.R.; Li, H.; Su, S. Accurate Trajectory Tracking of Disturbed Surface Vehicles: A Finite-Time Control Approach. *IEEE/ASME Trans. Mechatron.* **2019**, *24*, 1064–1074. [CrossRef]

20. Wang, N.; Qian, C.; Sun, J.; Liu, Y. Adaptive Robust Finite-Time Trajectory Tracking Control of Fully Actuated Marine Surface Vehicles. *IEEE Trans. Control Syst. Technol.* **2016**, *24*, 1454–1462. [CrossRef]

21. Wang, N.; Pan, X. Path Following of Autonomous Underactuated Ships: A Translation–Rotation Cascade Control Approach. *IEEE/ASME Trans. Mechatron.* **2019**, *24*, 2583–2593. [CrossRef]

22. Stateczny, A.; Burdziakowski, P. Universal autonomous control and management system for multipurpose unmanned surface vessel. *Polish Marit. Res.* **2019**, *1*, 30–39. [CrossRef]

23. Stateczny, A.; Kazimierski, W.; Gronska-Sledz, D.; Motyl, W. The Empirical Application of Automotive 3D Radar Sensor for Target Detection for an Autonomous Surface Vehicle's Navigation. *Remote Sens.* **2019**, *11*, 1156. [CrossRef]

24. Park, S.; Deyst, J.; How, J. A new nonlinear guidance logic for trajectory tracking. *AIAA Guid. Navig. Control Conf. Exhib.* **2004**. [CrossRef]

25. Guo, W.; Wang, S.; Dun, W. The Design of a Control System for an Unmanned Surface Vehicle. *Open Autom. Control Syst. J.* **2015**, *7*, 50–156. [CrossRef]

26. Moreno, D.; Chaos, D.; Aranda, J.; Muñoz, R.; Díaz, J.M.; Dormido-Canto, S. Application of an aeronautic control for ship path following. *J. Marit. Res.* **2009**, *6*, 71–82.

27. Specht, C.; Specht, M.; Cywinski, P.; Skóra, M.; Marchel, Ł.; Szychowski, P. A New Method for Determining the Territorial Sea Baseline Using an Unmanned Hydrographic Surface Vessel. *J. Coast. Res.* **2019**, *35*, 925–936. [CrossRef]

28. Specht, M.; Specht, C.; Lasota, H.; Cywinski, P. Assessment of the Steering Precision of a Hydrographic Unmanned Surface Vessel (USV) along Sounding Profiles Using a Low-Cost Multi-Global Navigation Satellite System (GNSS) Receiver Supported Autopilot. *Sensors* **2019**, *19*, 3939. [CrossRef]

29. Seto, M.L.; Crawford, A. Autonomous shallow water bathymetric measurements for environmental assessment and safe navigation using USVs. In Proceedings of the OCEANS 2015-MTS/IEEE Washington, Washington, DC, USA, 19–22 October 2015.

30. Alessandri, A.; Donnarumma, S.; Martelli, M.; Vignolo, S. Motion Control for Autonomous Navigation in Blue and Narrow Waters Using Switched Controllers. *J. Mar. Sci. Eng.* **2019**, *7*, 196. [CrossRef]

31. Munoz-Banon, M.; del Pino, I.; Candelas, F.; Torres, F. Framework for Fast Experimental Testing of Autonomous Navigation Algorithms. *Appl. Sci. Basel* **2019**, *9*, 1997. [CrossRef]

32. Kunicka, M.; Litwin, W. Energy Demand of Short-Range Inland Ferry with Series Hybrid Propulsion Depending on the Navigation Strategy. *Energies* **2019**, *12*, 3499. [CrossRef]

33. Borkowski, P. Adaptive System for Steering a Ship along the Desired Route. *Mathematics* **2018**, *6*, 196. [CrossRef]

34. Borkowski, P. Inference Engine in an Intelligent Ship Course-Keeping System. *Comput. Intell. Neurosci.* **2017**. [CrossRef]

35. Li, C.; Jiang, J.; Duan, F.; Liu, W.; Wang, X.; Bu, L.; Sun, Z.; Yang, G. Modeling and Experimental Testing of an Unmanned Surface Vehicle with Rudderless Double Thrusters. *Sensors* **2019**, *19*, 2051. [CrossRef]

36. Zhan, W.; Xiao, C.; Wen, Y.; Zhou, C.; Yuan, H.; Xiu, S.; Zhang, Y.; Zou, X.; Liu, X.; Li, Q. Autonomous Visual Perception for Unmanned Surface Vehicle Navigation in an Unknown Environment. *Sensors* **2019**, *19*, 2216. [CrossRef] [PubMed]

37. Lisowski, J. The sensitivity of state differential game vessel traffic model. *Polish Marit. Res.* **2016**, *23*, 14–18. [CrossRef]

38. Dudojc, B.; Mindykowski, J. New Approach to Analysis of Selected Measurement and Monitoring Systems Solutions in Ship Technology. *Sensors* **2019**, *19*, 1775. [CrossRef] [PubMed]

39. Li, J.; Du, J.; Sun, Y.; Lewis, F.L. Robust adaptive trajectory tracking control of underactuated autonomous underwater vehicles with prescribed performance. *Int. J. Robust Nonlinear Control* **2019**, *29*, 4629–4643. [CrossRef]

40. Paliotta, C.; Lefeber, E.; Pettersen, K.; Pinto, J.; Costa, M.I.C.; de Sousa, J.T.D.B. Trajectory Tracking and Path Following for Underactuated Marine Vehicles. *IEEE Trans. Control Syst. Technol.* **2019**, *27*, 1423–1437. [CrossRef]

41. GITHUB. Available online: https://github.com/ArduPilot/ardupilot/commit/a3c2851120f3572893bdf29ddc0e e24dac67cbe1 (accessed on 12 December 2019).

42. Jang, T.; Han, S. Analysis for VTOL Flight Software of PX4. In Proceedings of the 2018 18th International Conference on Control, Automation and Systems (ICCAS), Daegwallyeong, South Korea, 17–20 October 2018; pp. 872–875.

43. Siauw, T.; Bayen, A. *An Introduction to MATLAB® Programming and Numerical Methods for Engineers*; Academic Press: Cambridge, MA, USA, 2015. [CrossRef]

44. Specht, C.; Dabrowski, P.S.; Pawelski, J.; Specht, M.; Szot, T. Comparative analysis of positioning accuracy of GNSS receivers of Samsung Galaxy smartphones in marine dynamic measurements. *Adv. Space Res.* **2019**, *63*, 3018–3028. [CrossRef]

45. NovAtel Positioning Leadership. GPS Position Accuracy Measures. APN-029 Revision 1. 2003. Available online: https://www.novatel.com/assets/Documents/Bulletins/apn029.pdf (accessed on 12 December 2019).

MDPI

St. Alban-Anlage 66

4052 Basel

Switzerland

Tel. +41 61 683 77 34

Fax +41 61 302 89 18

www.mdpi.com

Sensors Editorial Office

E-mail: sensors@mdpi.com

www.mdpi.com/journal/sensors